Teacher's
Resource Book

Maria Karyda

C1

Business
Partner

FT Publishing

FINANCIAL TIMES

GSE

Global Scale of English

Contents

Overview

Business Partner is a flexible course designed for a variety of learners. It is suitable for students with mixed abilities, requirements and interests and for varied class sizes where the common requirement is to learn professional English language and develop key skills for the workplace.

When talking to learners, their reasons for studying business English almost always relate to their employability. Many tertiary students want to maximise their chances of finding a job in an international environment, while in-work professionals want to communicate more effectively in their workplace and improve their future career prospects. Other learners may simply need to study and pass a business English exam in order to complete their overall degree.

In all three cases, teachers need to be able to engage and motivate by providing learning materials which:

- are interesting and relevant to their life experiences.
- match their learning needs and priorities.
- are appropriate for the amount of study time available.

Business Partner has been designed to enable teachers to meet these needs without spending many hours researching their own materials. The content and structure of the course is based on three key concepts: **employability**, **flexibility** and **learner engagement**.

Course aims and key concepts

Employability

Balance between language and business skills training

In order to achieve their employability goals, learners need to improve their knowledge of English language as it is used in the workplace and also develop key skills for the international workplace. *Business Partner* provides this balance.

In addition to building their vocabulary and grammar and developing their writing skills, *Business Partner* trains students in Communication and Business skills. Language being only one aspect of successful communication, students also require an understanding of different business situations and an awareness of different communication styles, especially when working across cultures.

- 'Communication skills' (Lesson 3) provides the soft skills needed in order to work effectively with people whose personality and culture may be different from your own. This includes pitching your ideas, handling performance reviews and networking skills.
- 'Business skills' (Lesson 4) provides the practical skills needed in different business situations, such as taking part in meetings, presentations and negotiations.

Flexibility

The modular approach means that *Business Partner* can be adapted to suit a variety of teaching requirements, from extensive lessons to intensive short courses. In addition to the Coursebook, a wide variety of additional optional activities and resources are provided which can be used to focus on, and extend, material which is most useful to learners' needs.

Extra activities and extra grammar points

You can extend your lessons or focus in more depth on certain areas by using the large bank of extra activities in MyEnglishLab (clearly signposted for you throughout the Coursebook). These include extra vocabulary and grammar practice exercises for use in class, as well as activities which draw attention to useful language in reading texts.

 Teacher's resources: extra activities

These are PDFs in MyEnglishLab that you can download and print or display on-screen.

 Teacher's resources: alternative video and activities

Alternative videos with worksheets are available for some units and are clearly signposted. You can use these in the classroom as an alternative approach to the topic in Lesson 1, depending on your students' needs.

 The summary contains examples of how to order information in sentences. Go to MyEnglishLab for optional grammar work.

Business Partner offers a flexible approach to grammar depending on whether you want to devote a significant amount of time to a grammar topic, or focus on consolidation only when you need to. There is one main grammar point in each unit, presented and practised in Lesson 2.

In addition, the Writing section (Lesson 5) includes a link to an optional second grammar point in MyEnglishLab, where students can watch short video presentations of the grammar points and do interactive activities.

 page 112 Spoken English   There is one Spoken English section per unit at the back of the book. For details see page 5.

Teacher's Resource Bank: Photocopiables, Writing bank, Reading bank and Useful language bank

You can use these resources as and when needed with your classes:

- the Photocopiables further activate and practise vocabulary from Lesson 1 and grammar from Lesson 2.
- the Reading bank for each unit gives students more reading practice and can be also used for self-study. The activity types reflect those found in a range of business English exams.
- the Writing bank provides supplementary models of professional communication.
- the Useful language bank extends useful phrases for a range of business situations.

Learner engagement

Video content: We all use video more and more to communicate and to find out about the world and we have put video at the heart of *Business Partner*. There are two videos in every unit with comprehension and language activities:

- an authentic video package in Lesson 1, based on real-life video clips and interviews suitable for your learners' level of English.
- a dramatised communication skills training video in Lesson 3 which follows characters in an international team as they deal with different professional challenges.

Authentic content: Working with authentic content really helps to engage learners, and teachers can spend many hours searching for suitable material online. *Business Partner* has therefore been built around authentic videos and articles from leading media organisations such as the *Financial Times* and news channels. These offer a wealth of international business information as well as real examples of British, U.S. and non-native-speaker English.

Relevance for learners without work experience: Using business English teaching materials with learners who have little or no work experience can be particularly challenging. *Business Partner* has been carefully designed to work with these students as well as with in-work professionals. In the case of collaborative speaking tasks and roleplays, the situation used will either be:

- one that we can all relate to as customers and consumers; OR
- a choice of situations will be offered including a mix of professional and everyday situations.

Both will allow learners to practise the skill and language presented in the lesson, but in a context that is most relevant to them.

Business workshops: Learners have the opportunity to consolidate and activate the language and skills from the units in eight business workshops at the end of the book. These provide interesting and engaging scenarios where students simulate real-life professional situations such as roleplaying meetings, negotiations or presentations.

Approach to language and skills

Business Partner offers fully integrated skills, including the essential critical-thinking and higher-order thinking skills, which are built into the activities.

Vocabulary and video The main topic vocabulary set is presented and practised in Lesson 1 of each unit, building on vocabulary from the authentic video. Teachers are given lots of opportunities to use the vocabulary in discussions and group tasks, and to tailor the tasks to their classroom situations.

Useful language (such as techniques for engaging an audience, expressions for trust-building, phrases for workplace mediation) supports learners' capability to operate in real workplace situations in English. Two useful language sets are presented and practised in every unit, in Lessons 4 and 5. You will be able to teach the language in group speaking and writing tasks. There is a Useful language bank at the back of this Teacher's Resource Book which students can also find in MyEnglishLab so that they can quickly refer to useful language support when preparing for a business situation, such as a meeting, presentation or interview.

Listening and video The course offers a wide variety of listening activities (based on both video and audio recordings) to help students develop their comprehension skills and to hear target language in context. All of the video and audio material is available in MyEnglishLab and includes a range of British, U.S. and non-native-speaker English. Lessons 1 and 3 are based on video (as described above). In four of the eight units, Lesson 2 is based on audio. In all units, you also work with a significant number of audio recordings in Lesson 4 and the Business workshop.

Grammar The approach to grammar is flexible depending on whether you want to devote a significant amount of time to grammar or to focus on the consolidation of grammar only when you need to. There is one main grammar point in each unit, presented and practised in Lesson 2. There is a link from Lesson 5 to an optional second grammar point in MyEnglishLab – with short video presentations and interactive practice. Both grammar points are supported by the Grammar reference section at the back of the Coursebook (p.118). This provides a summary of meaning and form, with notes on usage or exceptions, and business English examples.

Reading *Business Partner* offers a wealth of authentic texts and articles from a variety of sources, particularly the *Financial Times*. Every unit has a main reading text with comprehension tasks. This appears either in Lesson 2 or in the Business workshop. There is a Reading bank at the back of this Teacher's Resource Book which students can also find in MyEnglishLab and which has a longer reading text for every unit with comprehension activities.

Speaking Collaborative speaking tasks appear in Lessons 1, 3, 4 and the Business workshop in every unit. These tasks encourage students to use the target language and, where relevant, the target skill of the lesson. There are lots of opportunities to personalise these tasks to suit your own classroom situation.

Writing *Business Partner* offers multiple opportunities to practise writing. Lesson 5 in every unit provides a model text and practice in a business writing skill. The course covers a wide range of genres such as reports, minutes, advertising copy, self-assessment, all for many different purposes. There are also short writing tasks in Lesson 2 which provide controlled practice of the target grammar. There is a Writing bank at the back of this Teacher's Resource Book which students can also find in MyEnglishLab and which provides models of different types of business writing and useful phrases appropriate to their level of English.

Spoken English The Spoken English section at the back of the Coursebook offers additional comprehension practice of spontaneous conversation including a variety of accents. Exercises also focus on specific aspects of spoken English, e.g. attentive listening, vague language and discourse markers.

Approach to Communication skills

A key aspect of *Business Partner* is the innovative video-based communication skills training programme.

The aims of the Communications skills lessons are to introduce students to the skills needed to interact successfully in international teams with people who may have different communication styles from them due to culture or personality. Those skills include pitching ideas, handling performance reviews and networking.

These lessons are based on videos that provide realistic examples of work situations. This is particularly important for pre-work learners who may not have direct experience of the particular situations they are about to see. In each lesson, students are given a work situation with a potential problem to analyse and then engage in a meeting to deal with the situation. They then reflect on their meeting and compare it to a possible solution to the problem in the video. This gives students the opportunity to engage in critical viewing of each video and gain awareness of the impact of different communication styles.

Approach to testing and assessment

Business Partner provides a balance of formative and summative assessment. Both types of assessment are important for teachers and learners and have different objectives. Regular review and on-going assessment allow students to evaluate their own progress and encourage them to persevere in their studies. Formal testing offers a more precise value on the progress made in their knowledge and proficiency.

Formative assessment: Each Coursebook lesson is framed by a clear lesson outcome which summarises the learning deliverable. The lesson ends with a self-assessment section which encourages students to reflect on their progress in relation to the lesson outcome and to think about future learning needs. More detailed self-assessment tasks and suggestions for further practice are available in MyEnglishLab. (See also the section on the Global Scale of English and the Learning Objectives for Professional English on page 6.)

The Coursebook also contains one review page per unit at the back of the book to recycle and revise the key vocabulary, grammar and functional language presented in the unit; these are structured to reflect the modularity of the course.

Summative assessment: Unit tests are provided and activities are clearly labelled to show which section of the unit they are testing to reflect the modular structure of the course. The tests are available in PDF and Word formats so that you can adapt them to suit your purposes. They are also available as interactive tests that you can allocate to your students if you wish to do so.

These Unit tests are based on task types from the major business English exams. There is also an additional LCCI writing task for professional English for every unit. This approach familiarises learners with the format of the exams and gives them practice in the skills needed to pass the exams.

MyEnglishLab also contains extra professional English practice activities. The content and level of the tasks match the Coursebook so they can also be used as additional revision material.

The Global Scale of English

The Global Scale of English (GSE) is a standardised, granular scale from 10 to 90 which measures English language proficiency. The GSE Learning Objectives for Professional English are aligned with the Common European Framework of Reference (CEFR). Unlike the CEFR, which describes proficiency in terms of broad levels, the Global Scale of English identifies what a learner can do at each point on a more granular scale – and within a CEFR level. The scale is designed to motivate learners by demonstrating incremental progress in their language ability. The Global Scale of English forms the backbone for Pearson English course material and assessment.

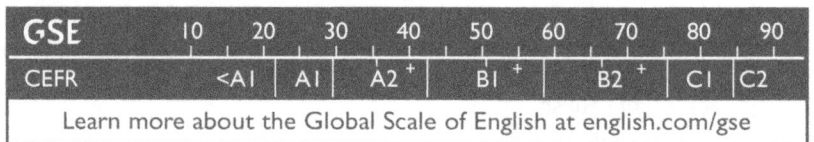

GSE		10	20	30	40	50	60	70	80	90
CEFR			<A1	A1	A2 +	B1 +	B2 +	C1	C2

Learn more about the Global Scale of English at english.com/gse

Business Partner has been written based on these Learning Objectives, which ensure appropriate scaffolding and measurable progress. Each Lesson outcome in each lesson in the Coursebook encapsulates a number of specific Learning Objectives which are listed in this Teacher's Resource Book in the Teacher's notes. These Learning Objectives are also listed in the self-assessment sheets available to students in MyEnglishLab. (See also Formative assessment above in Approach to testing and assessment.) The GSE Learning Objectives for the whole coursebook are listed in the GSE Mapping Booklets, which are available for download from https://www.pearson.com/english/catalogue/business-english/businesspartner/levels.html.

Course structure

Business Partner is an eight-level course based on the Global Scale of English (GSE) and representing the CEFR levels: A1, A2, A2+, B1, B1+, B2, B2+, C1.

	For the teacher	For the student
print		Coursebook with Digital Resources Workbook
blended	Teacher's Resource Book with MyEnglishLab	Coursebook with MyEnglishLab (=interactive workbook practice)
digital	Pearson English Portal	Coursebook ebook

Business Partner is a fully hybrid course with two digital dimensions that students and teachers can choose from. MyEnglishLab is the digital component that is integrated with the book content.

Access to MyEnglishLab is given through a code printed on the inside front cover of this book. As a teacher, you have access to both versions of MyEnglishLab and to additional content in the Teacher's Resource folder.

Depending on the version that students are using, they will have access to one of the following:

Digital Resources includes downloadable coursebook resources, all video clips, all audio files, Lesson 3 additional interactive video activities, Lesson 5 interactive grammar presentation and practice, Reading bank, Useful language bank, Writing bank, and My Self-assessment.

MyEnglishLab includes all of the **Digital Resources** plus the full functionality and content of the self-study interactive workbook with automatic gradebook. Teachers can also create a group or class in their own MyEnglishLab and assign workbook activities as homework.

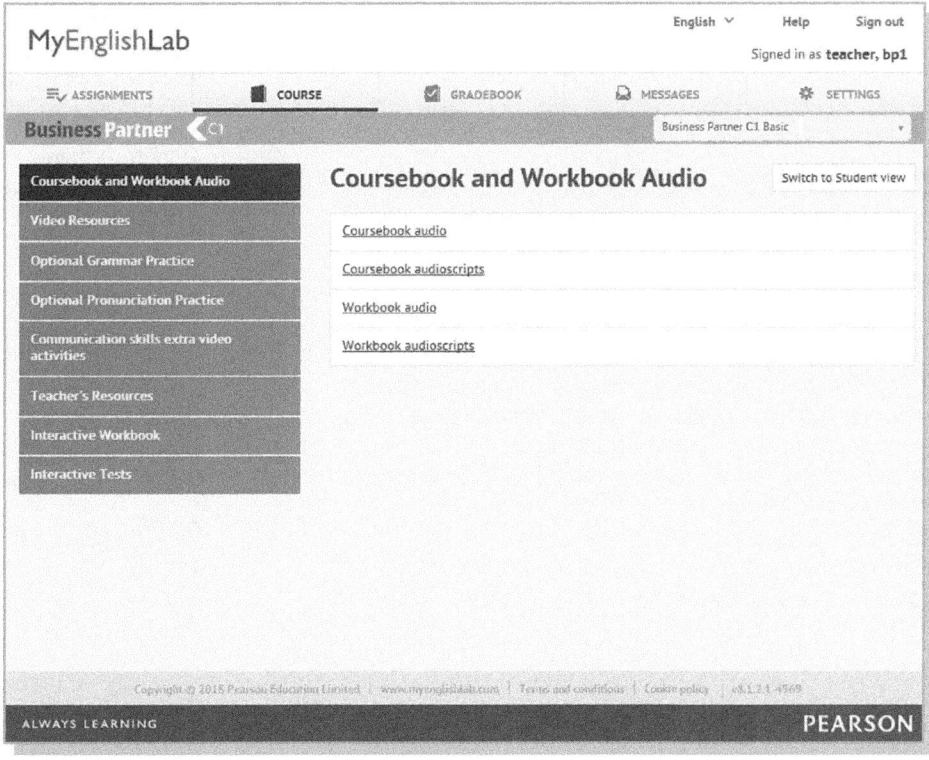

Coursebook
(with access code for MyEnglishLab)

- Eight units, each containing five lessons (see page 10 for unit overview)
- Eight Business workshop lessons relating to each of the eight units
- A one-page Review per unit to revise key language and grammar
- A Spoken English section focusing on colloquial spontaneous speech
- A Grammar reference with detailed explanations and examples
- Videoscripts and audioscripts
- A glossary of key business vocabulary from the book

Coursebook video and audio material is available on MyEnglishLab for all students, regardless of which edition of the Coursebook they are using.

MyEnglishLab digital component

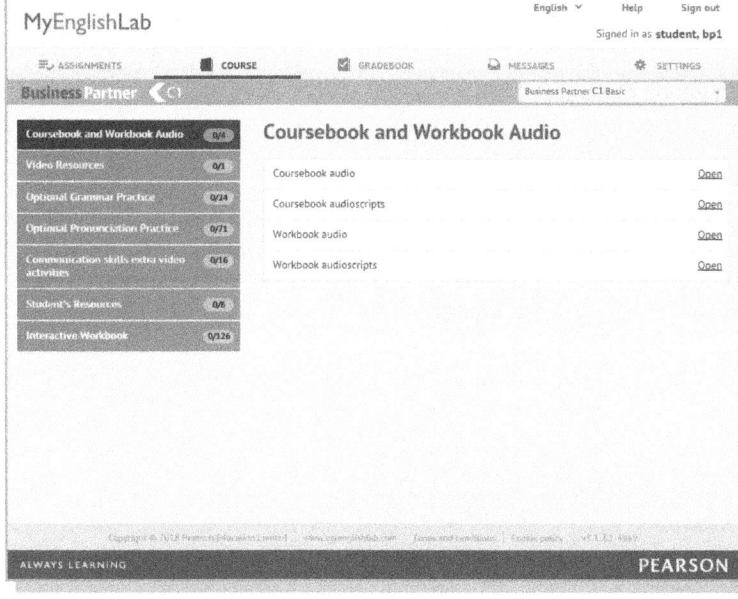

Accessed using the code printed on the inside cover of the Coursebook. Depending on the version of the course that you are using, learners will have access to one of the following options:

Digital resources powered by MyEnglishLab
- Video clips and scripts
- Audio files and scripts
- Extra Coursebook activities (PDFs)
- Lesson 3 extra interactive video activities
- Lesson 5 interactive grammar presentation and practice
- Reading bank
- Writing bank
- Useful language bank
- Extra professional English practice
- My Self-assessment
- Workbook audio files and scripts

Full content of MyEnglishLab
- All of the above
- Interactive self-study Workbook with automatic feedback and gradebook

Workbook

- Additional self-study practice activities, reflecting the structure of the Coursebook. Activities cover vocabulary, grammar, useful language, reading, listening and writing.
- Additional self-study practice activities for points presented in the Spoken English section
- Answer key
- Audioscripts

Workbook audio material is available on MyEnglishLab.

Teacher's Resource Book (with access code for MyEnglishLab)

- Teaching notes for every lesson including warm-ups, background /culture notes and answer keys
- Business brief for every unit with background information on the unit topic and explanations of key terminology; it gives teachers an insight into contemporary business practices even if they have not worked in these particular environments

- Photocopiable activities – two per unit with teaching notes and answer keys
- Reading bank – an extended reading text for every unit with comprehension activities (+ answer keys)
- Writing bank – models of different types of business writing with useful phrases
- Useful language bank – useful phrases for different business situations, e.g. presentations, meetings and negotiations

MyEnglishLab digital component

Accessed using the code printed on the inside cover of the Teacher's Resource Book.

Coursebook resources
- Video clips and scripts
- Audio files and scripts
- Extra Coursebook activities (PDFs)
- Lesson 3 extra interactive video activities for self-study
- Lesson 5 interactive grammar presentation and practice for self-study
- Extra professional English practice
- My Self-assessment: a document that students can use to record their progress and keep in their portfolio

Workbook resources
- Self-study interactive version of the Workbook with automatic feedback and gradebook
- Teachers can assign Workbook activities as homework
- Workbook audio files and audioscripts

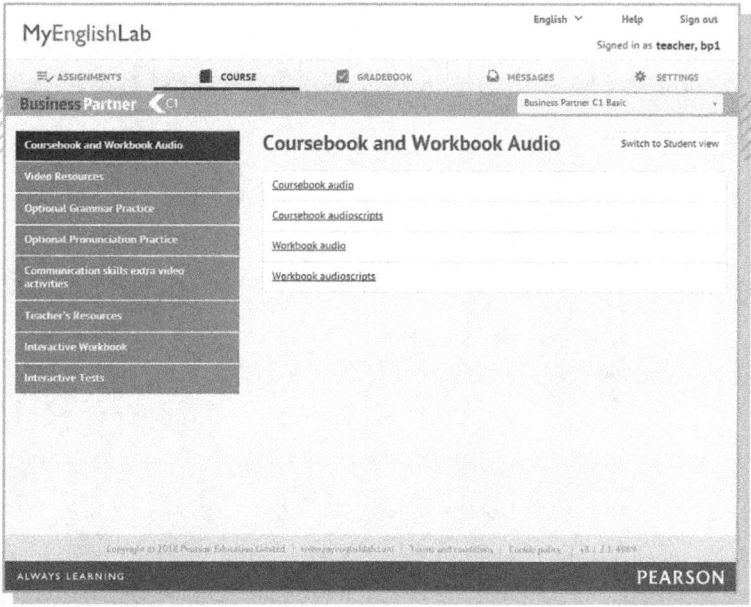

Teacher's Book resources
- Alternative video (Units 1, 2, 3 and 6) and extra activities
- Photocopiable activities + teaching notes and answer keys
- Reading bank + answer keys
- Writing bank
- Functional language bank

Tests
- Unit tests (PDFs and Word), including exam task types
- Interactive Unit tests, with automatic gradebook
- Tests audio files
- Tests answer keys

Pearson English Portal

- Digital version of the Teacher's Resource Book
- Digital version of the Coursebook with classroom tools for use on an interactive whiteboard
- All resources (see page 17).

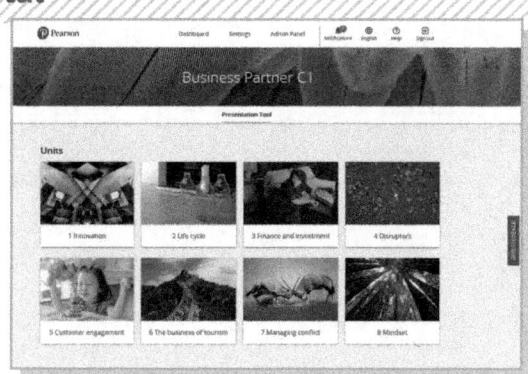

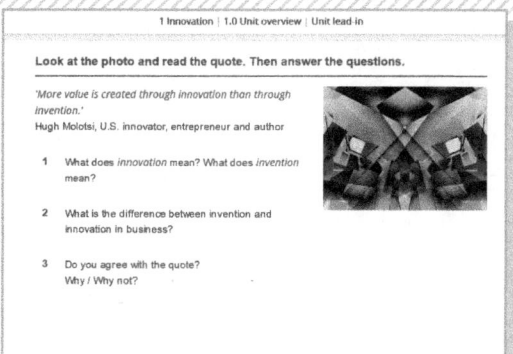

A unit of the Coursebook

Unit overview page ≫

1 ≫ A well-known or provocative quote related to the unit topic is provided as a talking point.
There are suggestions for how to use the quote in the Teacher's Resource Book notes for each unit.

2 ≫ The unit overview summarises the contents of each lesson as well as the lesson outcomes.

3 ≫ There are also references to content at the back of the book, which supplements the main unit.

Innovation 1≫

1 ≫ 'More value is created through innovation than through invention.'
Hugh Molotsi, U.S. innovator, entrepreneur and author

2 ≫ **Unit overview**

1.1	**An innovative approach** **Lesson outcome:** Learners can use vocabulary related to innovation.	**Video:** Innovation Director, Haiyan Zhang **Vocabulary:** Innovation **Project:** The impact of technology
1.2	**How innovators think** **Lesson outcome:** Learners can correctly use or omit articles where necessary.	**Reading:** Understanding what makes inventors tick **Grammar:** Articles: *a/an, the*, no article **Writing:** An intranet post on creativity
1.3	**Communication skills:** Pitching your ideas **Lesson outcome:** Learners can use persuasive techniques to effectively pitch their ideas.	**Preparation:** Pitching an idea for a magazine section **Roleplay:** A pitch **Video:** Pitching your ideas
1.4	**Business skills:** Engaging presentations **Lesson outcome:** Learners can use a range of strategies and expressions for making high-impact openings and conclusions to presentations.	**Listening:** Presentation by a senior sales director **Useful language:** Phrases for presenting **Task:** Opening and closing a presentation
1.5	**Writing:** Investment research **Lesson outcome:** Learners can write a research report that analyses a business opportunity, and can use topic sentences and cohesion to structure paragraphs.	**Model text:** Research report **Useful language:** Topic sentences and cohesion **Grammar:** Substitution of nouns and noun phrases **Task:** Write a research report based on a SWOT analysis

3 ≫ Business workshop 1: p.88 | Review 1: p.104 | Spoken English 1.2: p.112 | Grammar reference: p.116

≫ 7 ≪

Lesson 1 ▶

The aims of this lesson are to:

- engage students with the unit topic through a video.
- present and practise topic business vocabulary, drawing from the video.
- activate the language students have learnt in a group project at the end of the lesson.

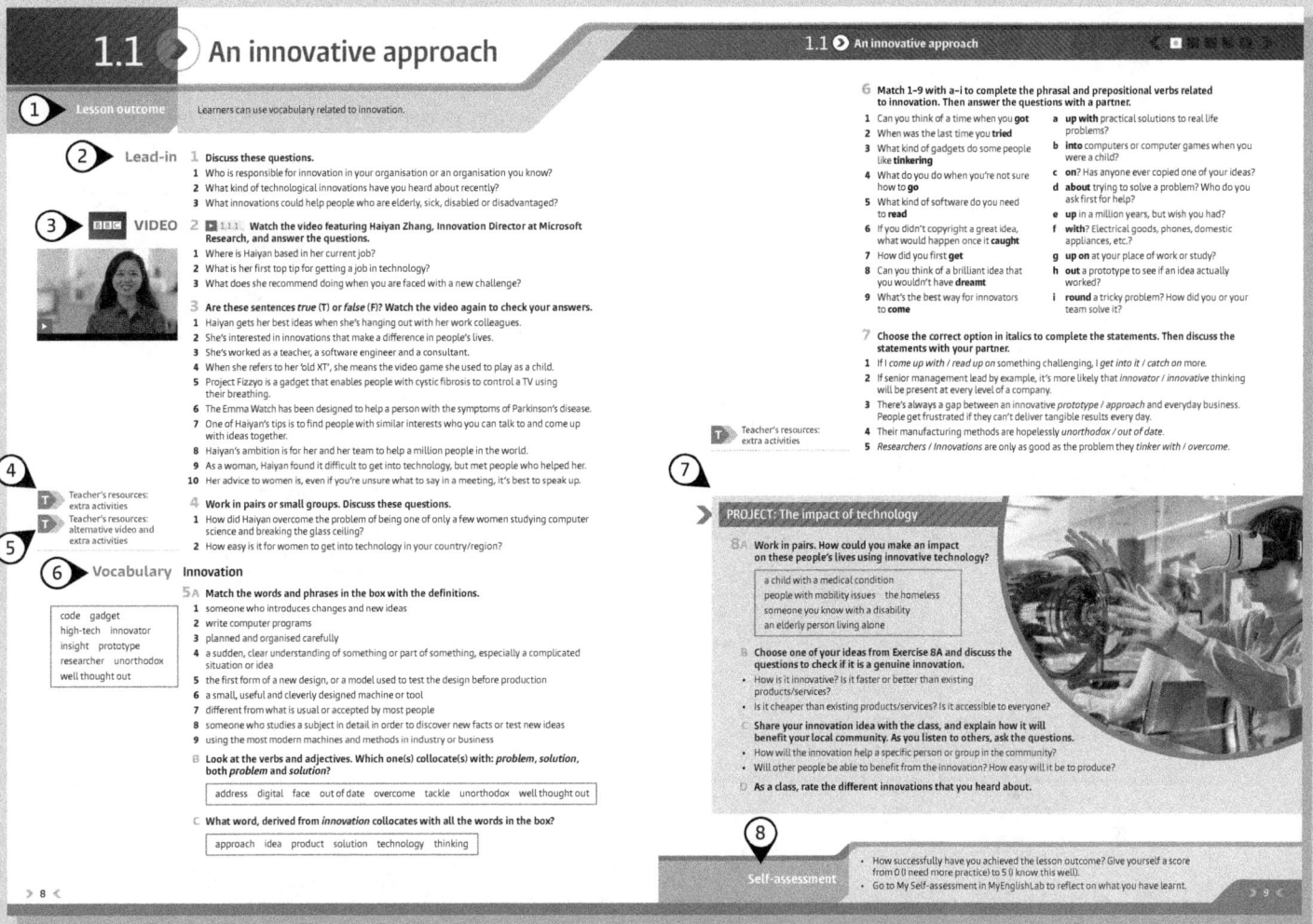

① The lesson outcome defines a clear learning outcome for every lesson. Each lesson outcome encapsulates a number of specific Learning Objectives for Professional English which are listed in this Teacher's Resource Book in the Teaching notes.

② Every lesson begins with a short Lead-in activity to engage learners with the lesson topic.

③ Lesson 1 is based on an authentic video of about 4 minutes with comprehension activities.

④ **T** Teacher's resources: extra activities Extra activities are clearly signposted. These are PDFs available in MyEnglishLab and Pearson English Portal to display on-screen or print. They can be used to extend a lesson or to focus in more depth on a particular section.

⑤ **T** Teacher's resources: alternative video and extra activities In Units 1, 2, 3 and 6, there are Alternative videos and worksheets, which you can use as an alternative to the ones in the unit or in addition to them. The alternative videos and alternative video worksheets are available for download in MyEnglishLab and Pearson English Portal.

⑥ The main unit vocabulary set is presented and practised in Lesson 1, building on vocabulary in the video. Extra activities for the video are available in MyEnglishLab and Pearson English Portal.

⑦ The Project at the end of Lesson 1 is a collaborative group task with a strong emphasis on communication and fluency building. It can be done in class or in more depth in and out of class.

⑧ Every lesson ends with a short Self-assessment section which encourages learners to think about the progress they have made in relation to the lesson outcome.

Lesson 2 **>** Reading or Listening

The aims of this lesson are to:

- provide students with meaningful reading or listening skills practice based on engaging, relevant and up-to-date content.
- present and practise the unit grammar point, drawing on examples from the reading text or audio recording.
- encourage students to activate the grammar point they have practised through communicative speaking or writing activities.

The Coursebook page showing Unit 1.2 'How innovators think'.

(1)> The lesson outcome defines a clear learning outcome for every lesson. Each lesson outcome encapsulates a number of specific Learning Objectives for Professional English which are listed in this Teacher's Resource Book in the Teaching notes.

(2)> Every lesson begins with a short Lead-in activity to engage learners with the lesson topic.

(3)> Questions in the lesson provide an opportunity for personalisation.

(4)> The reading text is an article from the *Financial Times*. The text focuses on a particular aspect of the unit topic which has an interesting angle and contains examples of the grammar point presented. It is followed by comprehension activities and grammar practice.

(5)> **T** Teacher's resources: extra activities   Extra activities are clearly signposted. These are PDFs available in MyEnglishLab and Pearson English Portal to display on-screen or print. They can be used to extend a lesson or to focus in more depth on a particular section.

(6)> The Spoken English section is signposted in the relevant lessons. There is one Spoken English section per unit at the back of the Coursebook, with more practice in the Workbook. This section focuses on spontaneous colloquial spoken English.

(7)> There is one grammar point in each unit, presented in Lesson 2. In general a guided discovery (inductive) approach has been taken to the presentation of grammar. The grammar is presented with reference to the examples in the reading (or listening) text, followed by controlled practice.

(8)> The Grammar reference section at the back of the book offers detailed explanation and examples of the grammar point covered in the unit.

(9)> Discussion questions and communicative practice of vocabulary and grammar is provided in the final Speaking or Writing section of this lesson.

(10)> Every lesson ends with a short Self-assessment section which encourages learners to think about the progress they have made in relation to the lesson outcome.

Lesson 3 ⟩ Communication skills

The aims of this lesson are to:

- introduce students to the skills needed to interact successfully in international teams.
- encourage students to critically assess their communication style by comparing themselves to others and against a model in the video.
- give students the opportunity to practise longer and more complex speaking in realistic work scenarios.

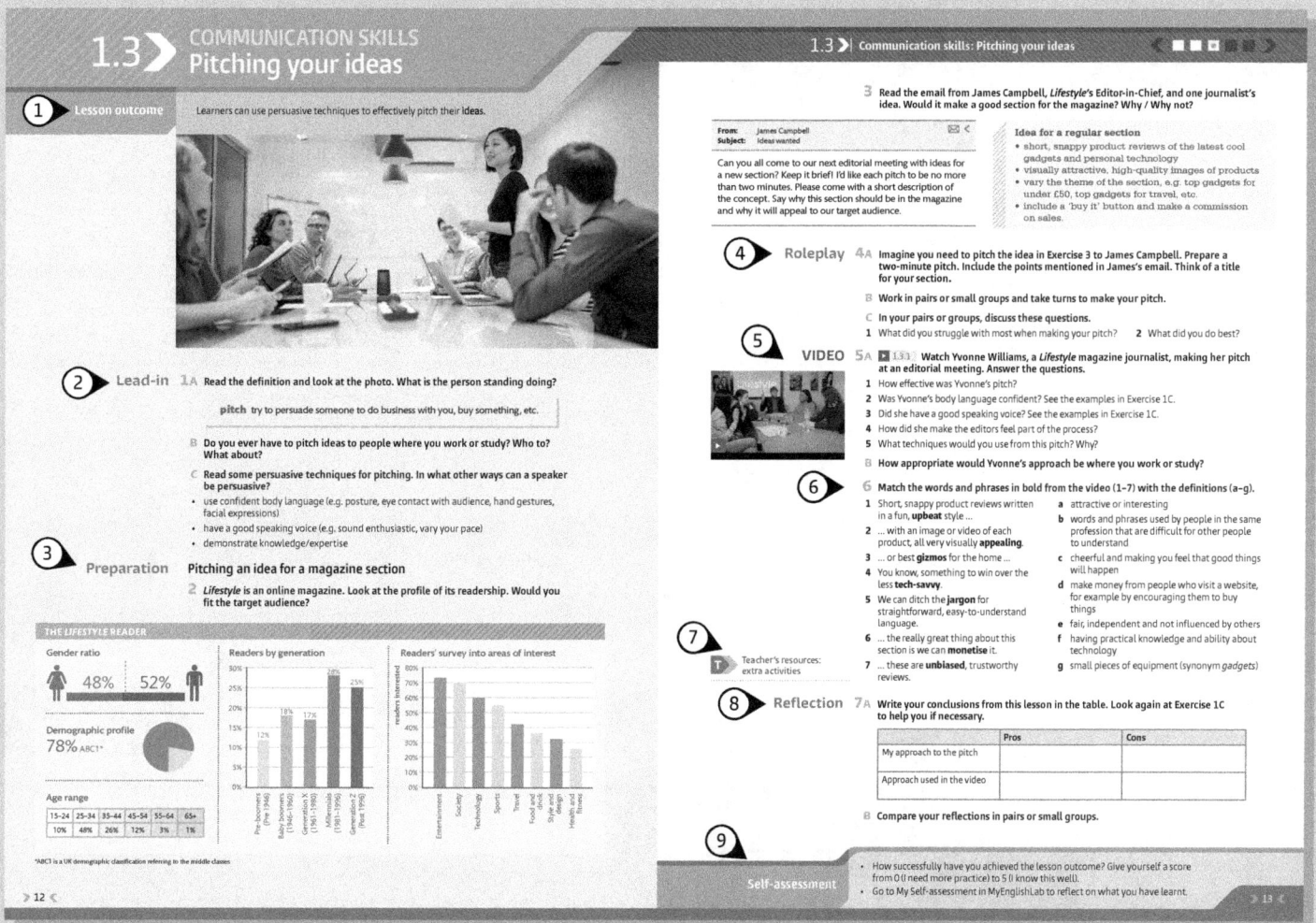

① The lesson outcome defines a clear learning outcome for every lesson. Each lesson outcome encapsulates a number of specific Learning Objectives for Professional English which are listed in this Teacher's Resource Book in the Teaching notes.

② Every lesson begins with a Lead-in activity to engage learners with the lesson topic.

③ In the Preparation section, students analyse information and prepare for a roleplay of a scenario where they have to use their soft skills.

④ In the Roleplay, students act out the scenario and reflect on their performance, before watching a video showing one possible way of communicating in a given situation.

⑤ Students watch the video and analyse how people behaved and communicated in the situation.

⑥ Each lesson includes a vocabulary activity, focusing on the words from the video.

⑦ **T Teacher's resources: extra activities** Extra activities are clearly signposted. These are PDFs available in MyEnglishLab and Pearson English Portal to display on-screen or print. They can be used to extend a lesson or to focus in more depth on a particular section.

⑧ In the final Reflection section, students compare their approach to the problem in the Roleplay with the approach of the people in the video.

⑨ Every lesson ends with a short Self-assessment section which encourages learners to think about the progress they have made in relation to the lesson outcome.

Lesson 4 ⟩ Business skills

The aims of this lesson are to:

- give students exposure to a functional business skill or sub-skill based on a listening activity, encouraging them to notice successful techniques.
- present and practise relevant useful language drawing on examples from the listening.
- encourage students to activate the skill and language they have practised by collaborating on a group task.

① The lesson outcome defines a clear learning outcome for every lesson. Each lesson outcome encapsulates a number of specific Learning Objectives for Professional English which are listed in this Teacher's Resource Book in the Teaching notes.

② Every lesson begins with a short Lead-in activity to engage learners with the lesson topic.

③ An original listening introduces a business skill, related key techniques and key useful language.

④ The Useful language section focuses on the functional language from the listening.

⑤ **T** Teacher's resources: extra activities   Extra activities are clearly signposted. These are PDFs available in MyEnglishLab and Pearson English Portal to display on-screen or print. They can be used to extend a lesson or to focus in more depth on a particular section.

⑥ The final activity is a significant collaborative group task to practise the target business skill and provide an opportunity to use the functional language presented. A scenario or several scenario options are provided to help with mixed classes and often include an opportunity for personalisation.

⑦ Every lesson ends with a short Self-assessment section which encourages learners to think about the progress they have made in relation to the lesson outcome.

Lesson 5 ➤ Writing

The aims of this lesson are to present and practise:

- a specific genre of business writing.
- relevant useful language.

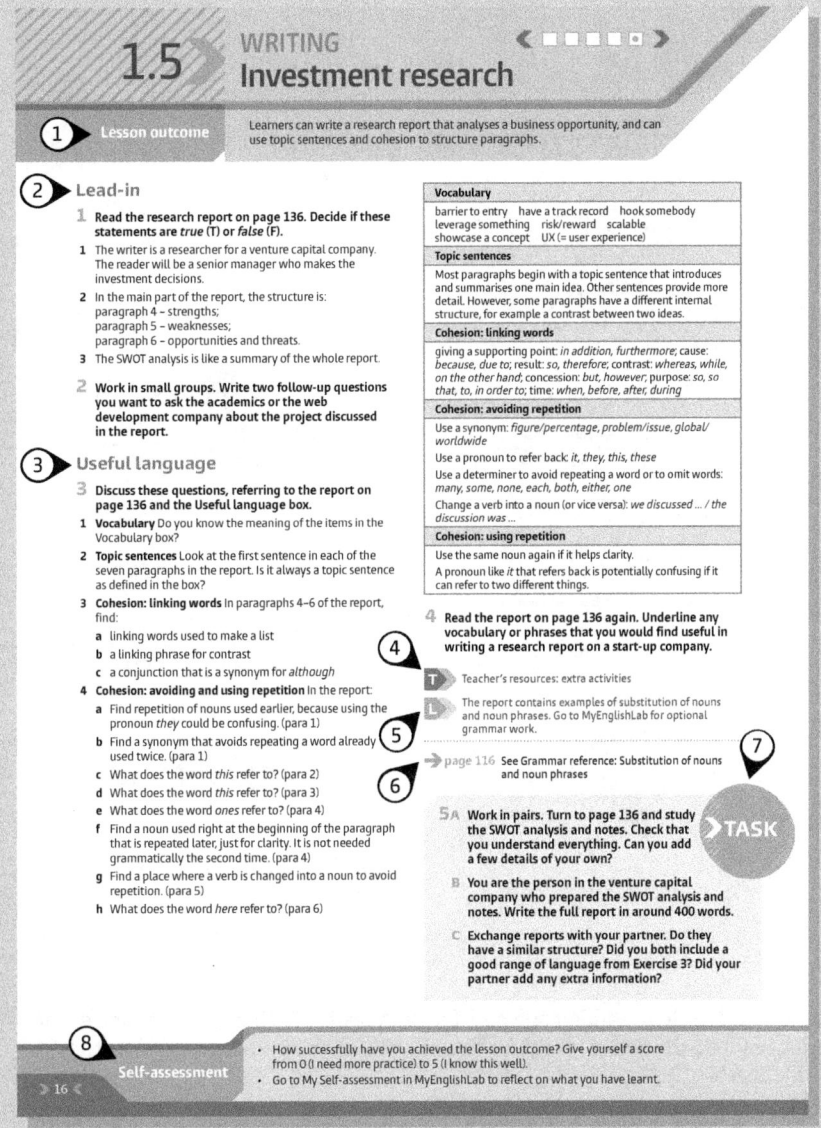

① The lesson outcome defines a clear learning outcome for every lesson. Each lesson outcome encapsulates a number of specific Learning Objectives for Professional English which are listed in this Teacher's Resource Book in the Teaching notes.

② Every lesson begins with a Lead-in in which students analyse a model text in the Additional material for writing at the back of the book.

③ The Useful language section focuses on the vocabulary and linguistic devices typically used in the genre.

④ **T** **Teacher's resources: extra activities** Extra activities are clearly signposted. These are PDFs available in MyEnglishLab and Pearson English Portal to display on-screen or print. They can be used to extend a lesson or to focus in more depth on a particular section.

⑤ **L** **The blog post contains examples of the Past Perfect Continuous. Go to MyEnglishLab for optional grammar work.**

There is an optional second grammar point signposted in every Lesson 5. Examples of the target language are included in the writing model and students can watch a video and do additional practice activities on MyEnglishLab.

⑥ The Grammar reference section includes explanations and further examples of the structures presented in the writing model.

⑦ The final activity is a Writing task, which elicits the useful language presented in the lesson.

⑧ Every lesson ends with a short Self-assessment section which encourages learners to think about the progress they have made in relation to the lesson outcome.

Business workshops ≫

The aims of this lesson are to:

* stimulate a real-life professional situation or challenge which is related to the theme of the unit.
* provide multiple opportunities for free, communicative practice of the language presented in the unit.

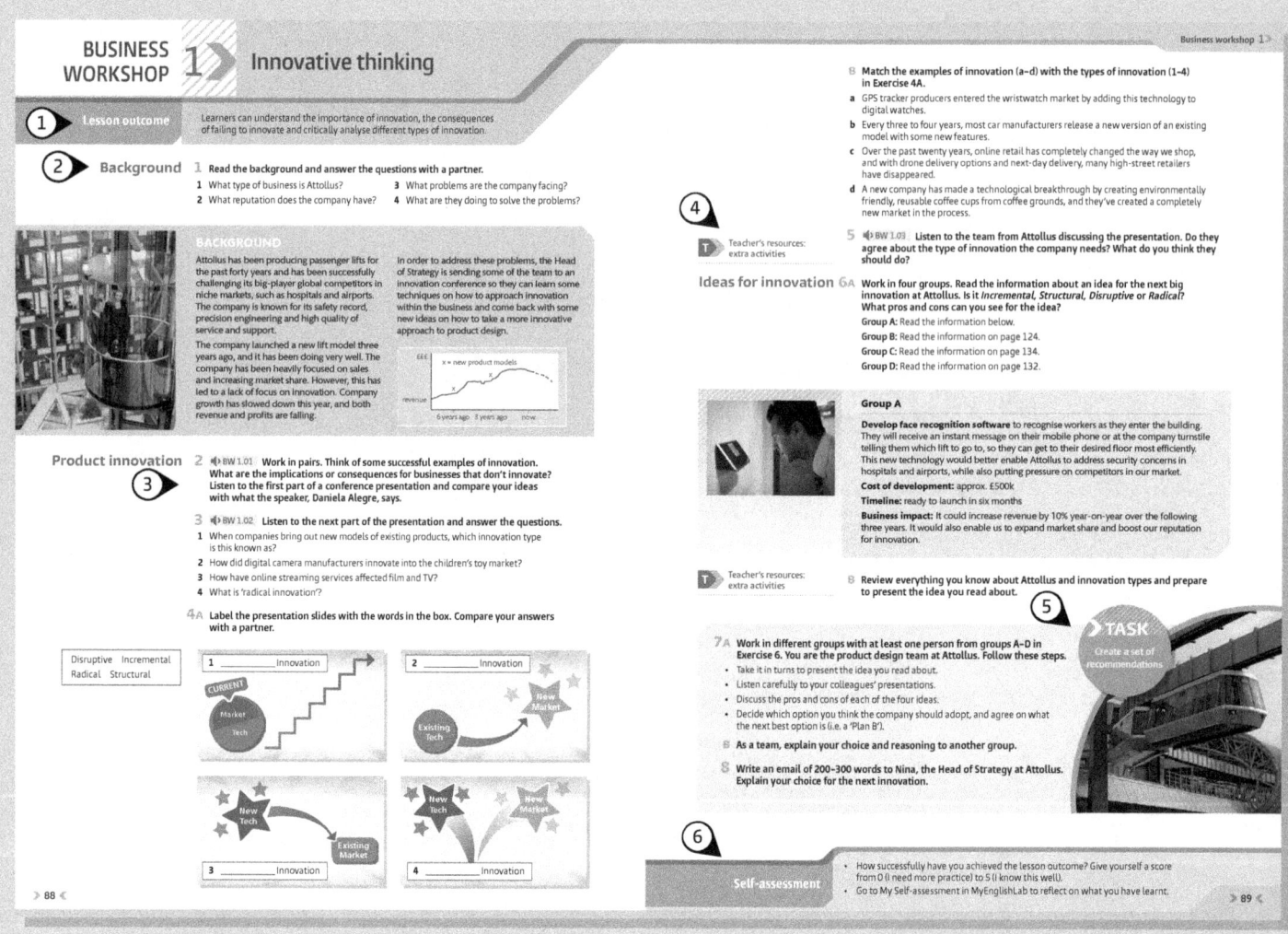

① The lesson outcome defines a clear learning outcome for every lesson. Each lesson outcome encapsulates a number of specific Learning Objectives for Professional English which are listed in this Teacher's Resource Book in the Teaching notes.

② The workshop begins by providing some background information on the company and the situation or challenge the scenario focuses on.

③ In units where Lesson 2 contains a reading text, the Business workshop contains a significant listening section, as in Business workshop 1 here. Where Lesson 2 contains a listening, the Business workshop contains a reading text.

④ **T** Teacher's resources: extra activities   Extra activities are clearly signposted. These are PDFs available in MyEnglishLab and Pearson English Portal to display on-screen or print. They can be used to extend a lesson or to focus in more depth on a particular section.

⑤ The final activity is a practical, collaborative task which addresses the challenge set out in the background section. It focuses on speaking, but usually also includes an element of writing.

⑥ Every lesson ends with a short Self-assessment section which encourages learners to think about the progress they have made in relation to the lesson outcome.

Extra material ▶

Content	For the teacher Available on MyEnglishLab and Pearson English Portal	For the learner Available on MyEnglishLab	Notes
Extra Coursebook activities with answer key	✓	✗	
Photocopiables with teaching notes and answer keys	✓	✗	The Photocopiables further activate and practise vocabulary from Lesson 1 and grammar from Lesson 2.
Reading bank	✓	✓	The Reading bank for each unit gives students more reading practice and can be also used for self-study.
Writing bank	✓	✓	The Writing bank provides supplementary models of professional communication.
Useful language bank	✓	✓	The Useful language bank extends useful phrases for a range of business situations.
Lesson 3 interactive activities	✓	✓	Self-study interactive activities practising the functional language from Lesson 3 videos.
Lesson 5 optional grammar work	✓	✓	Self-study interactive activities practising the grammar points from Lesson 5.
Alternative Lesson 1 lessons	✓	✗	Available for Units 1, 2, 3 and 6.
Extra reading lessons	✓	✗	Available for Units 4, 5, 7 and 8.
Tests – in PDF format	✓	✗	Eight Unit tests consisting of a Language section (testing grammar, vocabulary, and functional language) and a Skills section (testing reading, listening, and writing) and eight LCCI for writing tests.
– in Word	✓	✗	
– interactive tasks	✓	✓	Tests tasks are only visible to students if assigned by the teacher.
Self-assessment	✓	✓	

1 ▶ Innovation

Unit overview

	CLASSWORK	FURTHER WORK
1.1 ❯ An innovative approach	**Lead-in** Students discuss innovation in business. **Video** Students watch a video of an Innovation Director talking about her job. **Vocabulary** Students look at vocabulary related to innovation. **Project** Students discuss the impact of innovative technology on a community.	**MyEnglishLab:** Teacher's resources: extra activities; Reading bank **Teacher's book:** Resource bank Photocopiable 1.1 p.142 **Workbook:** p.4
1.2 ❯ How innovators think	**Lead-in** Students talk about creativity. **Reading** Students read an article about creative thinking. **Grammar** Students study and practise the definite, indefinite and zero article. **Writing** Students practise using articles by writing an intranet post.	**MyEnglishLab:** Teacher's resources: extra activities **Grammar reference:** p.116 Articles: *a/an, the*, no article **Spoken English:** p.112 **Teacher's book:** Resource bank Photocopiable 1.2 p.143 **Workbook:** pp.5–7
1.3 ❯ Communication skills: Pitching your ideas	**Lead-in** Students are introduced to the concept of, and some persuasive techniques for, pitching. **Roleplay** Students make a business pitch. **Video** Students watch a video of a business pitch. **Reflection** Students reflect on the conclusions from the video and their own approach to the pitch.	**MyEnglishLab:** Teacher's resources: extra activities; Interactive video activities
1.4 ❯ Business skills: Engaging presentations	**Lead-in** Students talk about engaging presentations and read an article on the psychology of learning. **Listening** Students listen to the opening and close of a presentation. **Useful language** Students look at techniques and useful phrases for high-impact openings and conclusions to presentations. **Task** Students plan and deliver the opening and close of a presentation.	**MyEnglishLab:** Teacher's resources: extra activities; Useful language bank
1.5 ❯ Writing: Investment research	**Lead-in** Students read and discuss the main content and structure of a research report. **Useful language** Students look at useful vocabulary for research reports and how to use topic sentences and cohesion to structure paragraphs in a research report. **Task** Students write a research report.	**MyEnglishLab:** Teacher's resources: extra activities; Writing bank; Interactive grammar practice **Grammar reference:** p.116 Substitution of nouns and noun phrases **Workbook:** p.8
Business workshop 1 ❯ Innovative thinking	**Listening** Students listen to a presentation on product innovation. **Reading** Students read about and discuss different ideas for innovation within an organisation. **Task** Students discuss and decide on the best innovation idea for an organisation and write an email explaining their choice.	**MyEnglishLab:** Teacher's resources: extra activities

Business brief

The main aim of this unit is to introduce students to the subject of **innovation** and **creative thinking** and the importance of finding ways to foster creativity in business in order to improve and grow. Innovation is key to the development of more effective processes, products and ideas through identifying solutions that meet new needs, unarticulated needs, or perhaps meet existing market needs in different ways. Innovative employees help a company become more **responsive**, assisting it to **diversify** and increase productivity by creating and executing new processes which in turn may increase competitive advantage and provide meaningful differentiation. Creative thinkers act as a catalyst that can make a business grow and help it adapt successfully to the marketplace.

Companies need to innovate constantly to keep up with the competition. Organisations which proactively identify and act on opportunities for change in the volatile business environment will not only survive but also flourish, even in the toughest economic conditions. Using innovation as a technological and strategic tool enables companies to develop more effective ways to achieve their key outcomes. The companies that take the biggest risks and most frequently identify new opportunities are seen as true innovators and leaders by their customers and peers. Responding to external challenges, by developing human as well as technological resources to do things differently, helps businesses respond appropriately to industry disruptors to **transform** productivity and increase market share and profitability.

To facilitate creative thinking among their employees and encourage innovation, companies need to be able to tolerate some degree of uncertainty. The internal structure of the organisation is key to this process: excessively hierarchical companies generally don't have the flexibility to really foster innovation. The most innovative organisations have more fluid set-ups, allowing staff to interact and form and reform teams according to business needs. They reward their employees for thinking experimentally, even if proposed projects don't turn out as planned. They are open to risk and tolerant of ambiguity and know that failing can actually help avoid catastrophic errors. However, it's also important to remember that in business, creativity is never fostered for its own sake, but is always contextualised within a structure that monitors and manages the development of ideas to ensure they're firmly anchored in market requirements and address specific objectives.

Creative people seek novel solutions and new associations. Many of their ideas will never come to fruition, so creative thinkers need to become hardened to disappointment and failure. Understanding their unique ways of thinking is essential to getting the best out of them and giving them space for **blue-sky thinking** – essentially playing with new ideas – is key to successful innovation. The two main qualities that define creativity are **divergent thinking** – thinking outside normal boundaries – and **cognitive flexibility**, the capacity to restructure ideas and see connections between things. Allowing employees time to dream, hypothesise and tinker with ideas frees them up to **think outside the box**, and look at problems and situations from new angles. Similarly, allowing them to work flexibly, forming new teams or working across departments, lets them find like-minded people they can talk to and share interests with and encourages them to brainstorm ideas to create even better ones. Creating prototypes, whether or not these are likely to develop into finished products, also encourages them to try out ideas quickly and identify their pros and cons in the early stages to verify whether they're worth pursuing.

Creative thinking and your students

All students will probably have participated in some sort of brainstorming sessions to encourage blue-sky thinking during their studies. In-work students may also have participated in other simple team-based activities to generate new ideas, such as contributing to a think tank or making a mind map. Some may even have helped develop prototypes or test cases within their organisations. All students should be able to think of examples from their own personal experience of environments which have helped or hindered them in thinking more creatively. They'll also have some knowledge of people from the wider world of business who have been great innovators and creative thinkers in their fields, such as Steve Jobs or James Dyson.

Unit lead-in

Refer students to the unit title and check they understand the meaning of *innovation*. Then look at the quote with the class. Check understanding of *invention* and before discussing the quote, elicit or explain the main difference between *invention* and *innovation* in business: invention is the 'creation' of a new product, service or process for the first time. *Innovation* happens when someone improves, further develops or adds to something that has already been invented in order to make it more profitable, better-suited to a market, etc. In short, invention creates something new and original; innovation turns an existing invention into something which 'sells'. If there is time, let students discuss the quote in pairs or small groups first, then elicit a few ideas around the class.

1.1 ❯ An innovative approach

GSE learning objectives

- Can extract specific details from a TV programme on a work-related topic.
- Can follow a work-related discussion between fluent speakers.
- Can compare the advantages and disadvantages of possible approaches and solutions to an issue or problem.
- Can compare and evaluate different ideas using a range of linguistic devices.
- Can contribute to a group discussion using linguistically complex language.

Warm-up

Refer students to the lesson title, *An innovative approach* and elicit or give a brief definition of it: an approach (way of doing something) which is new, different and better than those that existed before. Ask the class what they would consider *innovative* in their place of work/study. Elicit a few ideas around the class, then move on to the Lead-in questions.

Lead-in

Students discuss innovation in business.

1 Discuss the questions as a class, helping students with any vocabulary they may need. For question 2, you may need to prompt students with ideas and/or teach some relevant vocabulary, e.g. *artificial intelligence*, *augmented reality*, *blockchain* (a system in which a record of transactions made in Bitcoin or another cryptocurrency is maintained across several computers that are linked in a peer-to-peer network), *computing*, *foldable screen*, *hologram*, *incubator* (an organisation which helps new businesses to develop by giving them office space, services and equipment and providing them with business and technical advice), *virtual reality (VR)*, *headset*.

Video

Students watch a video of an Innovation Director talking about her job.

2 ▶ 1.1.1 Explain the activity and before students watch, elicit or explain what an Innovation Director does (their responsibilities focus mainly on the development of new products and services; they oversee the identification, development and production of new products and services for an organisation). Give students time to read the questions, then play the video and check answers with the class.

> **1** Cambridge, UK
> **2** play with technology (get into coding, tinker with electronics, get your hands dirty)
> **3** really understand the problems ('gain insight') and try out ideas early on ('create prototypes')

3 ▶ 1.1.1 Before students watch again, you may wish to provide definitions for the medical conditions mentioned in the video: *sight-impaired* (used to describe someone who cannot see well); *cystic fibrosis* (a serious medical condition, especially affecting children, in which breathing and digesting food is very difficult); *Parkinson('s) (disease)* (a serious illness in which your muscles become very weak and your arms and legs shake). Give students time to read the statements before they watch, then play the video. Alternatively, if you think students may remember some of the information from the first viewing, you could ask them to answer as many of the questions as they can before watching again, then play the video again for them to check/complete their answers. In stronger classes, you could also ask students to correct the false statements.

> **1** T
> **2** T
> **3** F (She was a programmer/software engineer and then became an innovation consultant.)
> **4** F (She is referring to the PC/computer she had.)
> **5** F (Project Fizzyo is a gadget that enables people with cystic fibrosis to control a video game.)
> **6** T
> **7** T
> **8** F (Haiyan says it doesn't matter whether she's helping one person or a million people.)
> **9** F (She found it challenging being one of a few women studying computer science but it was being passionate and demonstrating that passion that helped her, not people.)
> **10** T

4 Put students in pairs or small groups, explain the activity and check understanding of *(breaking the) glass ceiling*. Give them time to discuss the questions in their pairs/groups, then get feedback from the class.

Extra activities 1.1

A This activity provides students with extra listening practice. Explain the task and give students time to look at the gapped sentences and options before they watch. Play the video, twice if necessary, then check answers with the class.

> **1** b **2** a **3** b **4** a **5** c **6** c **7** a **8** b **9** a **10** c

B Explain the activity and give students time to read the extract before they watch the video again. Point out that they need to use between two and four words in each gap. Play the video, then check answers with the class writing (or inviting students to write) the answers on the board, so they can check their spelling.

1 three top tips   **2** with electronics   **3** hands dirty
4 can talk with   **5** brainstorm together
6 bigger, better ideas   **7** solve in your community
8 make a difference

Alternative video worksheet: Open innovation

1 Put students in pairs and give them 2–3 minutes to discuss the questions, then elicit ideas around the class. Alternatively, if time is short, discuss the questions as a class.

1–2 Students' own answers
3 Suggested answer: NASA is well-known for scientific discoveries and space exploration.

2 ▶ ALT1.1.1 Tell students that they are going to watch a video about innovations and go through the instructions with them. Give them time to read the statements, then play the first part of the video (0:00–1:38). Check answers with the class.

1 F (*Postmates is a U.S. company that delivers food using robots.*)
2 T
3 F (*The delivery robots use sensors and cameras to move along pavements and avoid obstacles.*)
4 T

3 ▶ ALT1.1.1 Explain the activity and give students time to read the sentences before they watch. Answer any vocabulary queries they may have, then play the rest of the video (1:39–5:08). If necessary, let students watch a second time to check/complete their answers before class feedback.

1 MG   **2** JL   **3** NH   **4** SR   **5** MP

4–6 These activities look at useful vocabulary from the video. Exercises 4 and 5 can be done with the whole class, checking answers and clarifying meanings as you go. For Exercise 6, ask students to work individually and get them to compare answers in pairs before checking with the class.

4 1 blue-sky thinking   **2** innovation economy
   3 open innovation   **4** cutting-edge products   **5** adjacent industry   **6** consumer demand   **7** research staff
5 1 d   **2** e   **3** a   **4** g   **5** h   **6** b   **7** c   **8** f
6 1 innovation economy   **2** blue-sky thinking
   3 staying ahead   **4** consumer demand
   5 look beyond, tap into   **6** rest on our laurels   **7** blind to

7 Put students in pairs or small groups to discuss the questions, then broaden this into a class discussion. Encourage students to elaborate.

8 Depending on the time available, this writing task can be done in class or set as homework. If students write their proposals as homework, step C can be done in the next lesson. Explain the writing task and go through steps A and B with the class. Point out to students that using sub-headings will help them organise their proposal (and their ideas) clearly. In weaker classes, you could let students plan their proposal in pairs. Point out the word limit and set a time limit before students begin. For step C, prompt students with a few questions to think about while comparing their work: Did they have similar ideas? Did they use similar language? Are the proposals organised in a similar way, using (similar) sub-headings? What would they change in their own proposal after reading their partner's?

Model answer
Proposal
Background
We have only one gym at our university campus. The staff are friendly and the quality of the classes is excellent, but there are also negatives as the swimming pool and the changing rooms are very crowded at peak times. The fact that the gym offers a flat membership rate means peak times are always busy. I believe we need to review membership rates. Firstly, we should offer a discount for members who use the facilities at non-peak times. Secondly, we could offer a family membership rate at the weekend to attract more members of the public. Thirdly, we should install cutting-edge technology to avoid non-members entering the gym.
Business case for reviewing gym membership rates
There are positives and negatives in offering a flat membership rate. One advantage is that all members can use the gym at any time. However, there are serious issues with overcrowding at peak times and consumer demand fluctuates greatly, for example, during the holidays, when most students are away, members leave and the gym is underused. Furthermore, non-members sometimes enter illegally, with the help of their friends. There is also a lack of technology at reception and in the changing rooms, compared to other private gyms in the area. As a result, the gym is losing money throughout the year.
In summary, I believe that the university gym needs to be the subject of a management review, offer more attractive rates and incorporate technology to stay ahead of the competition.
Recommendations
I therefore recommend that we offer three types of membership: a discounted rate for non-peak times (mornings), a higher rate for peak times and a family rate to attract members during the summer and the spring break. In addition, I strongly recommend that we install facial recognition at turnstiles when members enter to avoid non-members using the facilities. There could be sensors and cameras installed in the pool area to detect swimmers who are in trouble. There could also be electronic lockers installed in the changing rooms to avoid theft. Finally, we should consider dividing some of the swimming lanes early in the morning so that swimming is more attractive to older members and families.

Vocabulary: Innovation

Students look at vocabulary related to innovation.

5A Students could do this individually or in pairs. Check that they understand the meaning of *complicated* and *accepted* in the definitions before they begin. Check answers with the class, clarifying the meanings of the words and phrases in the box as necessary.

> **1** innovator **2** code **3** well thought out **4** insight
> **5** prototype **6** gadget **7** unorthodox **8** researcher
> **9** high-tech

5B Depending on the strength of your class, you might like to go through the words in the box with students before they attempt the exercise or let them use their dictionaries, then clarify meanings during feedback. If time is short, you could also do this as a whole-class activity, checking answers and clarifying meanings as you go. Encourage students to record the collocations in their vocabulary notebooks.

> The verbs *address*, *face*, *overcome* and *tackle* collocate with *problem*.
> The adjectives *out of date*, *unorthodox* and *well thought out* collocate with *solution*.
> We can say *a digital problem/solution*.

5C Check that students understand the meanings of the words in the box, then elicit the answer. Again, encourage them to record the collocations in their notebooks.

> innovative (adj.)

6 Explain the task and get students to match the sentence halves individually or in pairs, depending on the level of your class and the time available. Check answers with the class, clarifying meanings as necessary. After class feedback, give students 3–4 minutes to discuss the questions in pairs, then invite different students to share their answers with the class.

> **1** i **2** h **3** f **4** d **5** g **6** c **7** b **8** e **9** a

7 This activity practises vocabulary from Exercises 5–6, so students could do it individually. Get them to compare answers in pairs before class feedback.

> **1** read up on, get into it **2** innovative **3** approach
> **4** out of date **5** Innovations, overcome

Extra activities 1.1

C–D These activities practise key vocabulary from the lesson. They are consolidation exercises, so you may prefer students to complete them individually and then compare answers in pairs before class feedback.

> **C 1** came **2** high-tech **3** overcome, gadgets, tackle **4** innovation, researchers **5** unorthodox, innovative, insight, innovator
> **D 1** dreamt up, digital **2** innovative, well thought out **3** coding, technology

Project: The impact of technology

Students discuss the impact of innovative technology on a community.

8A Put students in pairs, explain the task and check that they understand the words and phrases in the box. You could offer some ideas and suggestions to get them started, e.g. high-tech prosthetics, Google Maps™ for the visually impaired, sensors in the home to monitor movements of elderly or disabled people living alone. It might also help if you told an anecdote describing a specific problem or challenge that a friend, relative or colleague has and asked for ideas, e.g. *My friend has mobility issues and going to work every day on public transport is a nightmare because …* Alternatively, if time is short, you could focus on a specific problem known to students, to reduce the time spent on brainstorming ideas.

8B Give pairs time to choose their best idea and research or discuss similar existing products/innovations. It does not matter if the innovation already exists; they may want to improve on it, make it more accessible, offer it to a different target market, reduce production costs, etc. In stronger classes, you could also ask students to think about possible negative effects of the innovation, e.g. it may have high production costs; it might break easily and be difficult to repair or replace parts; it may be limited in that it is a solution for a single person; the device might be harmful to the environment or have a high carbon footprint.

8C Before students share their ideas with the class, give them some time to discuss the questions here in their pairs and think about a) how they are going to present their idea, b) questions they may be asked about their idea and how they could answer them and c) questions they might ask while listening to other pairs' ideas. When they are ready, they could present their ideas directly to the entire class or you could get them to mingle instead.

8D In this final step, students rate the different innovations. You could list the innovations on the board and ask students to score them from 1 to 5, with 5 being the highest score. Encourage them to give reasons for their answers and remind them to consider the questions in Exercises 8B and 8C as they rate each innovation. It might be helpful to write some key words/phrases on the board following up on these questions, e.g. *innovative? easy to use? accessible? cheap to produce? high social impact?*

MyEnglishLab: Teacher's Resources: extra activities; Reading bank
Teacher's book: Resource bank Photocopiable 1.1 p.142
Workbook: p.4

1.2 ❯ How innovators think

GSE learning objectives

- Can get the gist of specialised articles and technical texts outside their field.
- Can understand definitions of technical terms presented in a linguistically complex academic text.
- Can check and correct spelling, punctuation and grammar mistakes in long written texts.
- Can describe reactions to different work-related scenarios in detail.

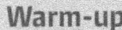

Warm-up

Ask the class these questions, eliciting answers from different students: *Do you consider yourself a creative person? Why/Why not? Can you think of different ways in which a person can be creative? Do you think creativity comes with time and work or are we born with creative talent? In which professional field do you think people are more creative? (e.g. science?) Why?*

Lead-in

Students talk about creativity.

1 Put students in small groups and give them 3–4 minutes to discuss the questions. For question 1, you could briefly explain the meaning of *eureka* before they begin or let them use their dictionaries instead and then clarify as necessary during class feedback, sharing the information in the Note below. At the end of the activity, invite students from different groups to share their answers with the class.

Note

A 'Eureka!' /jʊˈriːkə/ moment is when you suddenly have an innovative idea or solve a difficult problem. It comes from the story about Archimedes (the mathematician of ancient Greece) who shouted, 'Eureka! Eureka!' ('I have found (it)!') after he had stepped into a bath and noticed that the water level rose. At this moment he realised that the volume of water displaced must be equal to the volume of the part of his body he had submerged. This has helped our understanding of the formula for density in physics.

Reading

Students read an article about creative thinking.

2 Tell students that they are going to read an article about creative thinking, refer them to the title of the article and teach or elicit the meaning of *make someone tick*. Explain the task, give students a minute to read ideas 1–4 and check that they understand *cognitive* and *filter* (in this context, a mechanism for selecting or removing a particular type of information). The terms *divergent thinking* and *cognitive flexibility* are explained in the text, so if students ask about them, reassure them that they will understand their meaning when they read the article. Before they begin, reassure them that they do not need to worry about unknown words at this stage; they should focus on understanding the main ideas in order to decide which of the options (1–4) are expressed in the text. Allow plenty of time for students to read the text and complete the task and encourage them to underline the parts of the text that help them decide which of the ideas are expressed in it. Check answers with the class.

Students should tick ideas 1 and 3:
1 *Two qualities that define creativity are divergent thinking – thinking beyond normal boundaries – and cognitive flexibility, which is the capacity to restructure ideas and see connections.* (para 2)
2 Not mentioned in the article.

3 ... *During moments of insight, cognitive filters relax momentarily and allow ideas that are on the brain's back burners to leap forward into conscious awareness.* (para 4)
4 Not mentioned in the article; it says: *If businesses are to encourage innovation, they need to tolerate a degree of uncertainty.* (para 8)

3 Give students 2 minutes to read the questions and options and ask you about any they do not understand. Get them to complete the task individually and then to compare answers in pairs before class feedback. Again, encourage them to underline the parts of the text that give them the answers.

1 a **2** a **3** b **4** a **5** a **6** b

4 Put students in pairs or small groups and go through the questions with them. Give them 3–4 minutes to discuss in their pairs/groups, then elicit answers around the class.

Extra activities 1.2

A This activity provides students with extra reading practice. Let students complete the exercise individually and get them to compare answers in pairs before class feedback. Check answers with the class, clarifying the meanings of the words in the boxes as necessary.

1 industry, creative individuals
2 Scientific research, bipolar disorder
3 psychologist, cognitive disinhibition
4 receptors, brain's
5 frontal lobes, thoughts
6 high IQ, attention
7 creativity, workplace
8 creative idea, organisation

B This activity looks at useful vocabulary from the reading text. Get students to complete it individually or, in weaker classes, in pairs, using their dictionaries if necessary. Check answers with the class, clarifying meanings as necessary. If time is short, you could also do this as a whole-class activity, checking answers and clarifying meanings as you go.

1 ... creativity and schizotypal personality features often go hand in hand ...
2 Understanding their unique ways of thinking is essential to getting the best out of them.
3 What are we doing that's getting in the way of innovation?
4 This can be frustrating when the process of innovation goes against the grain of businesses that demand productivity and efficiency.

Spoken English
p.112: We need a chain of command

1 🔊SE1 Explain to students that they are going to hear people discussing question 3 from Exercise 4 and refer them to the question. Remind them of the meaning of *hierarchical* – elicit a brief explanation – then explain that students should make notes in answer to questions 1–5 while listening. Give them a minute to look at the questions so they know what to listen for, then play the recording, twice if necessary. Check answers with the class.

> **1** The people at the top are less creative (and have less of an idea what their organisation needs).
> **2** The CEOs of some of the biggest tech firms have had very creative ideas (that have changed the world).
> **3** Creative ideas need to reach senior people within the company before they are approved and implemented and they are often filtered out on the way up.
> **4** Smaller companies tend to be more creative in the earlier stages of development. Larger companies tend to take fewer risks.
> **5** Putting certain content out on social media (that can generate interest and influence stocks and market share) / Offering a range of products.

Expressing opinions

2 🔊SE1 Explain the activity and give students time to quickly read through the extracts before they listen. Play the recording again, then check answers with the class, clarifying meanings as necessary. Note that in weaker classes, students may need to listen twice or you may need to pause the recording for them to complete their answers.

> **2** I do think
> **3** dare I say it
> **4** I'm not too sure about that
> **5** I'm not sure if I necessarily agree
> **6** Can I just say
> **7** maybe I'll qualify what I say
> **8** And isn't it also the case
> **9** what about
> **10** I'm not saying
> **11** I'm just saying

3 Tell students that the phrases they wrote in Exercise 2 have two functions. Explain the activity and write the functions on the board: *Expressing an opinion* and *Disagreeing politely*. Give students time to complete the exercise, individually or in pairs, then check answers with the class. Encourage students to record the phrases in their notebooks.

> **a** 1, 2, 3, 6, 7, 8, 9, 10, 11
> **b** 4, 5
> The phrases *dare I say it*, *maybe I'll qualify what I say* and *I'm not saying … I'm just saying …* are phrases that might be used when a speaker expects someone to disagree with their ideas.

Grammar

Students study and practise the definite, indefinite and zero article.

5 Before students do the exercise, write this sentence from the text on the board: *Strictly adhering to a plan restricts the creative process*. Underline *a* and *the* and ask students what type of words they are (articles). Explain that they will be looking at how to use articles, refer them to the Grammar reference on page 116 and go through it with them, clarifying any points as necessary. Then get them to complete the exercise individually and compare answers in pairs before checking with the class.

> **Possible answers**
> **1** If you want to kill <u>a</u> creative idea; have <u>an</u> organisation …
> **2** Dr Shelley Carson, <u>a</u> lecturer in psychology; Gary Klein, <u>a</u> cognitive psychologist
> **3** Ø creative individuals, Ø Technology and Ø industry are increasingly reliant on Ø innovation; Ø awareness; Ø research; a degree of Ø uncertainty
> **4** at <u>the</u> heart of … <u>the</u> brain's … <u>the</u> thalamus to <u>the</u> frontal lobes
> **5** in <u>the</u> chain (referring to the idea of a hierarchical organisation)
> **6** Ø Dr (Shelley) Carson; Ø Harvard University; Ø Sweden
> **7** getting <u>the</u> best

6 Go through the instructions with students and elicit a brief definition and examples of *countable* and *uncountable nouns*. Then do the exercise with the whole class, checking answers as you go. To extend the activity, you could ask students to write one example sentence for each of the highlighted nouns; they could do this in class or as homework.

> *technology*: both; uncountable use in text; countable example: *We can assume that amazing new technologies will emerge to enable steep cuts in greenhouse gas emissions.*
> *businesses*: both; countable use in text; uncountable example: *We've been working with them for years; they are great people to do business with.*
> *thinking*: uncountable
> *people*: countable (*person* = singular). *People* is the usual plural. However, we can also say *peoples* to mean 'national or ethnic groups', as in *the peoples of Europe*, although this is a formal use of the word.
> *disinhibition*: uncountable (Note that *inhibition* can be both countable and uncountable, e.g. *She had no inhibitions about saying what she felt. I was amazed at his lack of inhibition about speaking in public.*)
> *awareness*: uncountable
> *research*: uncountable
> *thoughts*: both; countable use in text; uncountable example: *I've been giving your proposal a lot of thought.*
> *flexibility*: uncountable
> *psychologist*: countable

7 Get students to complete the exercise individually and remind them that they can refer to page 116 if they need help. Check answers with the class.

> **1** Ø **2** the **3** Ø **4** Ø **5** a **6** an **7** Ø **8** the

Extra activities 1.2

C This activity gives further practice of the definite, indefinite and zero articles. It is a consolidation exercise, so it would be better for students to do it individually. Before they begin, explain that this is the last part of the article they read on page 10 of their Coursebook and encourage them to read it quickly before attempting the exercise. If there is time, get them to compare answers in pairs before class feedback.

1 Ø 2 Ø 3 Ø 4 Ø 5 the 6 Ø 7 the / Ø 8 Ø
9 the / a 10 the 11 an / Ø 12 the 13 a

Writing

Students practise using articles by writing an intranet post.

8A–B Explain the scenario: students are managers at their organisation and have just received an email from their CEO, who has recently attended a training course on creativity and innovation and would like to know more about what her managers understand to be creativity in the workplace. Give students a minute to read the email, then explain the writing task. Point out the word limit and if necessary, explain what a company intranet is (a private network for exchanging information within an organisation, accessible only to the organisation's staff). Set a time limit for the writing task and tell students they should check their post for correct use of articles; again remind them that they can refer to the Grammar reference on page 116 if they need help. While students are writing, monitor and help as necessary.

Model answer

What is creativity?

Most people associate creativity with artistic tasks such as writing **a** novel, painting or composing music. While these are creative activities, not all creative thinkers are artists. Certain jobs involve artistic creativity, such as writing advertising copy or designing **a** logo. Many jobs in business and science also require creative thinking.

Some people think of science and engineering as **the** opposite of creativity. However, **the** field of STEM (Science, Technology, Engineering and Maths) is highly creative. Writing **an** innovative computer program or designing **a** video game, for instance, requires **a** lot of creativity.

I am often creative at work. I'm particularly good at problem-solving and last year, when I was working in **the** engineering division, I redesigned machinery on **an** assembly line to improve productivity. That saved our company **a** lot of money.

Creativity simply means being able to come up with something new. Therefore, creative thinking is **the** ability to consider something differently. This could be what we do with **a** set of new data, **a** conflict between employees, or **a** group project.

MyEnglishLab: Teacher's Resources: extra activities
Grammar reference: Articles: *a/an, the,* no article p.116
Spoken English: p.112
Teacher's book: Resource bank Photocopiable 1.2 p.143
Workbook: pp.5–7

1.3 ➤ Communication skills
Pitching your ideas

GSE learning objectives

- Can use persuasive language to convince others to agree with their recommended course of action during a discussion.
- Can make a clear strong argument during a formal discussion.
- Can critically evaluate the effectiveness and appropriateness of a presentation.
- Can compare and evaluate different ideas using a range of linguistic devices.

Warm-up

Ask students to think of a time in their life when they had to persuade someone to do (or not do) something or accept an idea, which they are comfortable sharing with the class. Put them in pairs or small groups to tell each other about their experiences, then invite brief feedback from the class. How easy/difficult was it to persuade that person? What did they do in order to be more persuasive? Invite students from different pairs/groups to share their experiences with the class.

Lead-in

Students are introduced to the concept of, and some persuasive techniques for, pitching.

1A Do this as a quick whole-class activity. Go through the definition with the class and check that students are familiar with the concept of 'a pitch' and 'pitching'. You may also wish to teach these related words and common collocations: *pitch* (n.) (the things someone says to persuade people to buy something, do something or accept an idea); *make a pitch for something*; *pitch (an idea) to someone*; *pitch for a contract, business,* etc.; *business pitch* (a presentation by one or more people to an investor or group of investors; it can also be an email, letter or even an impromptu conversation); *elevator pitch* (a short description of an idea, product or company that explains the concept in such a way that any listener can understand it in a short period of time). Ask for a brief description of the photo, then elicit the answer.

The person is pitching a (business) idea to a group of people.

1B Depending on the size of your class and the time available, you could ask students to discuss the question in pairs or small groups first, then invite different students to share their experiences with the class. Alternatively, if time is short, discuss the question with the whole class. Note that the answers can include many formal and informal examples of pitching (e.g. pitching an idea to your manager to get funding for projects in your department or pitching an idea for a holiday to your family). To get students started, you could give one or two examples of your own.

1C Go through the instructions and techniques for pitching with students, then give them 3–4 minutes to brainstorm ideas; they could do this individually or in pairs. Encourage them to make notes. When they are ready, elicit and discuss their ideas and encourage them to give reasons: why do they think these techniques would be effective? Point out that a pitch is essentially a performance and getting a 'yes' can come down to how well you 'perform' as well as how good the idea may be. You may also wish to tell students that, according to research, 'the catcher' (a manager, an investor, etc.) will be more open to a pitcher's idea if they are made to feel that they are participating in the idea's development.

> **Possible answers**
> - Be friendly and likeable.
> - Show passion and enthusiasm.
> - Listen actively to your audience.
> - Encourage the audience to participate in the idea's development.
> - Be fun and entertaining.
> - Use interesting visuals.

Preparation: Pitching an idea for a magazine section

Students read and think about the scenario for a roleplay.

2 Go through the instructions with students and check that they understand *readership* and *target audience*. Give them a minute to look at the reader profile and ask you about anything they do not understand, then ask them to think about the question. If there is time, get them to discuss their answers in pairs or small groups first, then get brief feedback from the class. Encourage students to give reasons for their answers.

3 Explain to students that *Lifestyle* is looking for ideas for a new section and go through the instructions with them. Give them time to read the email and notes and decide whether the journalist's idea would make a good section for the magazine. Remind them to think about the reader profile they read in Exercise 2 and to give reasons for their answers. If there is time, get them to share their ideas and reasons with a partner first and then with the class.

Roleplay

Students make a business pitch.

4A Explain that students are now going to pitch the idea from Exercise 3 to *Lifestyle*'s Editor-in-Chief. Point out that they need to bear in mind all the points in his email; you could list these on the board for students to refer to during the activity:

- *no more than 2 minutes*
- *short description of concept*
- *why include in magazine?*
- *why appropriate for target audience?*

Also, remind students to use some of the techniques they discussed in Exercise 1C and to think about how they can make their pitch more persuasive. Allow plenty of time for them to prepare for the roleplay, while you monitor and provide help as necessary.

4B Put students in pairs or small groups and ask them to take turns to make their pitches. While listening, they should make brief notes to give their partner(s) feedback afterwards.

Ask them to think about what they think the pitcher did well each time and what could be improved. Remind them of the time limit and either stop them after every 2 minutes and ask the next student to take a turn, or ask the other student(s) in each pair/group to keep time. During the activity, monitor and note down any points to highlight during feedback, but do not interrupt students' pitches.

4C Students should do this in the same pairs or groups as for Exercise 4B. Look at the questions with them and give them plenty of time to discuss them. They should also share any points they noted while listening to their partners' pitches. Finally, go over any points you noted while monitoring.

Video

Students watch a video of a business pitch.

5A ▶ 1.3.1   Explain to students that they are now going to watch a journalist at *Lifestyle* pitching the idea from Exercise 3 at an editorial meeting. Explain the activity and go through the questions with students before they watch. For questions 2 and 3, refer students back to the examples in Exercise 1C. Encourage them to make notes in answer to the questions while watching and play the video. You could get students to compare answers in pairs before discussing them with the class.

> 1 Students' own answers, but in general, it is a very successful pitch; the speaker is well-prepared and convincing.
> 2 Yes, she has an upright, open posture, makes eye contact around the table, smiles and looks enthusiastic, uses hand gestures to emphasise what she says.
> 3 Yes, she speaks clearly, sounds enthusiastic and pauses for effect as necessary.
> 4 By getting them to contribute ideas and responding positively to these ideas. By becoming involved, the editors have a sense of ownership of the idea and get excited about it.
> 5 Students' own answers.

5B Put students in pairs or small groups to discuss the question, then broaden this into a class discussion. Encourage students to elaborate.

6 Students could do this individually or in pairs, depending on the time available. You could encourage them to refer to the videoscript on page 145 and read the text 'around' the words in bold, to help them work out their meanings. Check answers with the class, clarifying meanings as necessary.

> **1** c **2** a **3** g **4** f **5** b **6** d **7** e

> **Extra activities 1.3**
>
> **A** This activity gives further practice of the vocabulary students looked at in Exercise 6. Ask them to complete it individually and remind them that they can refer to the definitions in Exercise 6 if they need help. Check answers with the class.
>
> > **1** jargon **2** appealing **3** gizmo **4** monetise
> > **5** upbeat **6** unbiased **7** tech-savvy

Reflection

Students reflect on the conclusions from the video and their own approach to the pitch.

7A Allow students to work individually first so that they can reflect on their own approach. Remind them to think about the ideas in Exercise 1C and their answers to Exercise 5A and encourage them to make notes in the table. To help them, you could discuss the pros and cons of the approach used in the video with the whole class.

Approach used in the video

Pros: encourages audience to participate in the development of the idea; shows enthusiasm; demonstrates knowledge by explaining interest in tech gadgets

Cons: some audiences might not expect to participate and be reluctant to do so; audience interaction can take more time

---

7B Put students in pairs or small groups to compare their reflections, then get brief feedback from the class.

MyEnglishLab: Teacher's Resources: extra activities; Interactive video activities

1.4 ➤ Business skills

Engaging presentations

GSE learning objectives

- Can recognise rhetorical questions in a linguistically complex presentation or lecture.
- Can make an effective introduction and opening to a presentation.
- Can make an effective summary and conclusion to a presentation.
- Can compare and evaluate different ideas using a range of linguistic devices.

Warm-up

Ask students to think about an *engaging* presentation they have attended at their place of work or study. Ask them to share with the class what they think made that presentation engaging. What do they think helped capture and hold the audience's attention? If there is time, you could let students discuss in pairs or small groups first, then share their experiences with the class.

Lead-in

Students talk about engaging presentations and read an article on the psychology of learning.

1A–B If time is short, briefly discuss the questions with the whole class, nominating a few different students to answer. Alternatively, let students discuss in pairs or small groups first, then get feedback from the class.

1B Possible answers

- use of highly relevant/interesting information/content
- dynamic delivery with use of an enthusiastic voice and good body language
- use of language which is not too complicated to understand
- integration of graphics and visuals which are appealing and communicative
- stories and/or humorous anecdotes

2A Explain the task, give students a minute to read the questions and teach or elicit the meanings of *primacy* and *recency*. Then refer students to the article on page 124 and give them time to read it and answer the questions. Get them to compare answers in pairs before checking with the class.

---

1 The Primacy and Recency effect indicates that we tend to recall information presented to us from the beginning and end of a presentation.
2 (Refer to the bullet point list in the article.)
3 There is a need to focus on making the opening and close of a presentation highly effective so that they are as memorable as possible and memorable for the right reasons.

2B Discuss the question with the whole class, inviting different students to contribute. Make sure they give reasons for their answers: why do they think these techniques would be effective? How could they help make the information more memorable? List the techniques students mention on the board.

Possible answers

- Quote a (famous) person or statistics.
- Tell a story of success or failure.
- Ask the audience a challenging question.
- Test the audience's knowledge with a quiz.

Listening

Students listen to the opening and close of a presentation.

3 ◀) 1.01 Play the recording and ask students to make notes in answer to the questions. In weaker classes, students may need to listen again to check/complete their answers. After checking answers with the class, you could play the recording again and ask students to note which techniques for opening a presentation discussed in Exercise 2B the speaker used.

---

1 He starts by discussing some customer feedback and suggesting how to think differently about customers.
2 Ask questions in order to determine the needs of the customer before suggesting solutions.
3 How to shift from a 'telling' culture to an 'asking' culture, a service culture.

---

4 ◀) 1.02 Explain the task and play the recording, twice if necessary, then check answers with the class. You could then ask students to listen again and note which techniques for closing a presentation discussed in Exercise 2B the speaker used.

---

1 a two-month process including training for everyone and a new electronic feedback process in stores
2 increased sales of over five percent

---

5 Put students in pairs and, if necessary, let them listen to both recordings again and make notes before they discuss the questions. Give them 2–3 minutes to exchange ideas in their pairs, then get feedback from the class.

Useful language

Students look at techniques and useful phrases for high-impact openings and conclusions to presentations.

6A Before students complete the exercise, go through the headings and sentences in the table with them and check that they understand each one. Then get them to complete the exercise individually and check answers with the class, clarifying meanings as necessary. In weaker classes or if time is short, you could also do this as a whole-class activity, eliciting the correct category for each phrase as you go.

> **1** c **2** e **3** h **4** g **5** a **6** j **7** d **8** i **9** b **10** f

6B Go through the questions with students, then put them in pairs and give them 3–5 minutes to discuss. Alternatively, if time is short, you could discuss the questions with the whole class, inviting different students to contribute. During feedback, feed in information from the answer key below as appropriate. When discussing question 3, you may wish to tell students that a three-step structure is often used for rhetorical questions: the speaker asks a question, then comments briefly before moving on to explore solutions to the problem, e.g.

Question: *So how will we build a new culture?*

Comment: *That's the key question. And I think there's a simple answer.*

Solution: *What we need to do is listen to the customer more often.*

> **Possible answers**
> 1 Different speakers will find different techniques more difficult or easier depending on their own communication style and preferences. It is important to use techniques that you are comfortable with and feel natural. At the same time, it is important to stretch your skills and learn new techniques.
> 2 Audiences often like to 'participate' in a presentation, either by asking questions or talking to one another, which also helps people to build networks. However, it is important to understand the audience's expectations; many might consider this approach ineffective as they believe presenters should present and not facilitate discussion.
> 3 Rhetorical questions can allow interesting and relevant questions which are in the minds of the audience to be asked and to link the 'answers' presented to real concerns, helping audiences to feel engaged. Asking questions also breaks the usual flow of a presentation and avoids 'talking at' an audience for extended periods.
> 4 Some people may feel that using powerful language overdramatises situations and that it lacks professionalism and balance.

> **Extra activities 1.4**
>
> **A** This activity looks at a 'three-part formula' for making presentations more engaging. Ask students to look at the table and remind them that rhetorical questions can be particularly effective in opening and closing presentations. Draw their attention to the three headings and explain that it is a good idea to follow rhetorical questions with a comment and then a solution/answer statement. Do the first item as an example with the class to illustrate the 'three-part formula', then ask students to complete the rest of the exercise individually. Check answers with the class.
>
> > **1** f, iv **2** c, i **3** d, vi **4** a, v **5** e, iii **6** b, ii
>
> **B** This activity looks at another three-part strategy for making presentations more engaging: using an opening statement, followed by three repetition statements and then a closing demand. Go through the instructions and example with the class, then ask students to complete the exercise individually. If there is time, get them to compare answers in pairs before class feedback.
>
> > **1** b, vi **2** f, i **3** e, ii **4** d, iii **5** a, v **6** c, iv

Task

Students plan and deliver the opening and close of a presentation.

7A Refer students to page 124 and give them time to read the professional context and ask you about anything they do not understand. You might like to ask a few check questions, e.g. *What type of organisation is DAPRA?* (a German coffee machine manufacturer) *What are its main values?* (team, freedom, entrepreneurialism) *What does it say it can offer young people?* (quick training and [international] career development) *What is DAPRA's 3:3 programme?* (a graduate programme offering working locations in three continents in three years) *What challenges is it experiencing at the moment?* (in recruiting young international talent to the organisation). Make sure students are clear about their audience – international students – and the main objective of the presentation: to *inspire* and *engage* people to consider DAPRA as a future employer.

7B Put students in pairs and explain that they are going to plan the opening and close of the presentation. Go through the list of points to include with them and point out the time limit for each part. Encourage them to make notes for each point and remind them to think about the different techniques they looked at in the lesson. During the activity, monitor and help students as necessary.

7C Do this part in three stages. First, get students to practise their opening and close in their pairs until they feel confident. Remind them to refer to their notes from the previous stage and to use phrases from Exercise 6A. When they are ready, put them in groups and get them to take turns to give their presentations to their group. Remind them of the time limit for each part and make sure both students in each pair have a chance to present, e.g. Student A could present the opening, then Student B the close. Explain that while listening, the rest of the group (the 'audience') should make notes about

a) the clarity and impact of the presentation, b) the techniques used to engage the audience and c) the effectiveness of these techniques: did the presenters manage to make their opening and close engaging? You may also wish to suggest that each group appoints a student as a timekeeper. Finally, when all pairs have given their presentations, the 'audience' should give feedback on points a–c above.

During the activity, monitor and note down points to highlight during feedback, but do not interrupt the presentations or feedback sessions. When they have finished, you could invite different students to share their experiences with the class: which techniques do they think were effective? Which were easy to use? Did students use phrases from Exercise 6A? If yes, were they helpful? What do they think went well? What would they do differently next time? Finally, highlight any points you noted while monitoring.

MyEnglishLab: Teacher's Resources: extra activities; Useful language bank

1.5 ❯ Writing
Investment research

Warm-up

Discuss these questions with the class: *Have you ever read/ written an investment research report? Who was it for and what was it about? How important is it for companies to have such reports prepared before deciding for or against an investment?*

Lead-in

Students read and discuss the main content and structure of a research report.

1 Explain the task and give students time to read the statements before they look at the report. Check that they understand what a venture capital company is (a company which looks for start-up businesses to invest in) and elicit or give a brief explanation of *SWOT analysis* (a strategic planning technique used to identify an organisation's *Strengths*, *Weaknesses*, *Opportunities* and *Threats*). Refer students to page 136 and give them time to read the report and complete the task, then check and discuss the answers with the class. During feedback, highlight the structure of the whole report; you could put it on the board for students to refer to later: *Background → SWOT analysis → Report (strengths, weaknesses, opportunities and threats) → Recommendation.*

1 T **2** T **3** T

2 Put students in small groups and give them 2–3 minutes to write their follow-up questions, then elicit ideas around the class. As an extension, you could ask two students to come to the front of the class and take on the role of representatives of the web development company. Have a short Q&A, with the class asking their follow-up questions and the two students inventing information for their answers.

Useful language

Students look at useful vocabulary for research reports and how to use topic sentences and cohesion to structure paragraphs in a research report.

3 This is best done as a whole-class activity, checking and discussing answers as you go. After discussing the answers for each item, go through the relevant section of the Useful language box with the class. For question 1, you could let students discuss the words/phrases in pairs first, using their dictionaries if necessary, then clarify meanings with the class. For question 2, point out that when two ideas are contrasted within the same paragraph, we do not usually include a topic sentence at the start (see answer key below).

1 *barrier to entry* = something that prevents a company from entering an industry or market; *have a track record* = have a reputation, based on what you have done; *hook somebody* = succeed in making somebody interested in something; *leverage something* = use something again in a different way to have a bigger effect; *risk/reward* = the balance of probability between losing money and gaining money; *scalable* = easy to make bigger; *showcase a concept* = show that a new idea works and is good; *UX* = the experience a user has on a website or interacting with an app, e.g. ease of finding information

2 Yes, for paragraphs 1, 2, 3, 5 and 7 there is a topic sentence. That is, the first sentence introduces and summarises one main idea and the other sentences in the same paragraph give more detail. However, paragraphs 4 and 6 are different. Paragraph 4 does begin with a topic sentence that introduces the next three sentences, but in the final sentence there is a second main idea (the 'scalable site') and so, many people would start a new paragraph here. Note that there is no rule about 'new idea = new paragraph' as long as the ideas are connected and the text is clear and easy to understand. Paragraph 6 has a contrast of ideas within the same paragraph and so does not use a topic sentence at the start.

3 a *First, … ; The second issue is …*
  b *On the other hand, …*
  c *while*

4 a Repetition of *academics* and *web company.*
  b *Businesses* avoids using *company.*
  c *This* in the second sentence refers to *their concept.*
  d *This* in the second sentence refers to *subscription-based, with a free element to hook potential subscribers.*
  e *Ones* in the third sentence refers to *their usual projects.*
  f Repetition of *teams.*
  g *Committed* changes to *commitment.*
  h *Here* in the first sentence refers to the whole business, not specific words in the text.

4 Get students to do this individually and then compare answers in pairs before class feedback. You could list the words/phrases students choose on the board for them to refer to during the writing task.

Extra activities 1.5

A This activity gives further practice of the useful language for research reports students looked at in Exercise 3. Explain that the pairs of paragraphs are from different research reports; they should choose the one which is better each time, giving reasons. Remind them to think about their answers in Exercise 3 and also refer to the Useful language box in their Coursebook if they need help. During class feedback, elicit or explain what makes the correct answer a better option each time.

1 b: In **b**, there is a topic sentence to introduce the paragraph, whereas in **a** there is not – it just goes straight in. Also, there's repetition of the word 'issue'.

2 b: In **b**, the linking words make sense – 'in addition' adds another related point; 'on the other hand' makes a contrast. In **a**, the linking words make no sense – 'whereas' is used for a contrast, but here there is no contrast; 'in addition' is used to make a second supporting point, but here there is a contrast not a supporting point.

3 a: In **a**, it is clear and unambiguous what chart 1 and chart 2 show. In **b**, the words 'it' and 'them' in the second sentence could refer to several things.

4 b: In **b**, the words 'one' and 'another' clearly and unambiguously refer to 'one problem' and 'another problem'. The simplicity helps the reader. In **a**, the longer phrases 'one of the problems' and 'another of them' make the text repetitive and more difficult to read.

5 a: In **a**, the word 'choice' is substituted by the synonym 'option' and then 'choice' is used again at the end. Also, the phrase 'wait for a year' is substituted by 'waiting'. This makes the text more interesting and easier to read. In **b**, the word 'choice' is used three times with no substitution, and the phrase 'to wait for a year' is used twice with no substitution.

Optional grammar work

The report in Exercise 1 contains examples of substitution of nouns and noun phrases, so you could use it for some optional grammar work. Refer students to the Grammar reference on page 116 and use the exercises in MyEnglishLab for extra grammar practice.

Task

Students write a research report.

5A Put students in pairs and refer them to page 136. Give them time to read the SWOT analysis and notes and ask you any questions they may have. Then encourage them to add more details of their own in their pairs, while you monitor and help them as necessary.

5B Explain the writing task and allow students plenty of time to plan their answer. Remind them to think about the content

and structure points discussed in Exercise 1 and to use the useful language and writing strategies from Exercise 3. Set a time limit and while students are writing, monitor and help as necessary. If time is short, the writing task can be set for homework.

Model answer

Research report: biotech company with drug to treat Alzheimer's in phase 2 trials

Background

This is a company that was spun out of a university biochemistry department. Its founders are three PhDs. They are working on a drug that slows the progression of Alzheimer's by slowing the production of beta-amyloid, a protein associated with the disease. Further research may lead to even better treatment and this lab is certainly a world leader in the field.

Our exit strategy would be the normal one for a biotech start-up – selling the business to a large pharmaceutical company. There will be no shortage of potential buyers if the drug gets to phase 3 trials.

SWOT

Strengths
- CEO/CFO are external appointments
- Phase 1 trials passed, phase 2 looking good

Weaknesses
- Running out of cash
- Lack of other drugs in the pipeline

Opportunities
- Gap in the market
- Ageing population

Threats
- Other labs doing similar research

Report

This company is strong in human resources. Not only are there excellent scientists working in the lab, but the management team is also very good. They used their first-round venture capital funding to appoint a CEO and CFO from outside the company. Both managers impressed us when we went to visit the company. And the drug itself looks very promising. It has passed phase 1 trials and is now in the middle of phase 2. The initial results show very positive clinical outcomes.

On the negative side, the company does have difficulties with cashflow. Their early funding has supported them until now but they only have three months' working capital to pay salaries and overheads. They need more funding to continue. There is also cause for concern as they have no other drugs in their pipeline. We would be betting on just one drug.

There is a lot of money to be made by whoever can find a successful treatment for Alzheimer's. There is a big gap in the market as there is currently no drug to treat this condition and the world's ageing population gives a huge potential market. There are many other labs working in this area and so competition is strong. However, there will be room in the market for more than one drug.

Recommendation

My recommendation is that we do make an investment. The risk is no more than usual for a biotech start-up, but the potential profits if the drug proves successful are huge.

5C If students write their report as homework, you could do this activity in the next lesson. Put students in the same pairs as for Exercise 5A and get them to read each other's reports and give their partner feedback. Remind them to think about the content and structure of the report, as well as the language and strategies from Exercise 3. What did their partner do well? What could be improved? Students could then rewrite their reports based on their partner's feedback; they could do this in class or as homework.

MyEnglishLab: Teacher's Resources: extra activities; Writing bank; Interactive grammar practice

Grammar reference: Substitution of nouns and noun phrases p.116

Workbook: p.8

Business workshop ❯1

Innovative thinking

> **GSE learning objectives**
>
> - Can follow presentations on abstract and complex topics outside their field of interest.
> - Can synthesise information from different sources in order to give a written or oral summary.
> - Can participate in discussions using linguistically complex language to compare, contrast and summarise information.
> - Can suggest pros and cons when discussing a topic, using linguistically complex language.
> - Can confidently argue a case in writing, specifying needs and objectives precisely and justifying them as necessary.

Background

Students read about a passenger lift manufacturer looking to take a more innovative approach to product design.

1 Put students in pairs and give them time to read the background and discuss the questions. Check answers with the class and clarify any unknown vocabulary.

> 1 It is a passenger lift company.
> 2 It is known for its safety record, precision engineering and high quality of service and support.
> 3 There is a lack of innovation which has led to slow growth and falling profits.
> 4 Some employees are going to an innovation conference to get some ideas.

Product innovation

Students listen to a presentation on product innovation.

2 🔊 BW1.01 This activity is best done in two stages. Start by putting students in pairs and giving them time to discuss the first two questions in the rubric. Encourage them to make notes. Get brief feedback from the class, then move on to the listening task. Explain that students are going to listen to the first part of a presentation on product innovation and compare their ideas with the presenter's, referring to their

notes. Give them 1–2 minutes to discuss in their pairs before class feedback.

> The presenter mentions examples of mobile phones, light bulbs and refrigerators. She says that businesses that don't innovate will be by-passed by their competitors, will lose their competitive advantage and eventually go out of business.

3 🔊 BW1.02 Explain that students are now going to hear the next part of the presentation and give them time to read the questions before they listen. Play the recording, twice if necessary, then check answers with the class.

> 1 incremental innovation
> 2 They made the camera casings more durable and colourful and made them much simpler to use.
> 3 People no longer buy or rent DVDs and are no longer restricted to watching what is on TV at a specific moment. This has led to DVD rental shops and shops that sell DVDs closing and to terrestrial TV channels creating their own streaming services.
> 4 It is innovation of new technology into a market that previously did not exist. For those reasons, it has the potential to greatly affect society and history.

4A Explain to students that the words in the box are the different types of innovation described in the presentation and check understanding of each one. Ask students to complete the exercise individually and get them to compare answers in pairs before class feedback.

> 1 Incremental 2 Structural 3 Disruptive 4 Radical

4B If time is short, you could do this as a whole-class activity, checking answers as you go. Alternatively, ask students to do it individually, giving them time to read the examples and ask you about any unknown words before they begin. Check answers with the class.

> **a** 2 **b** 1 **c** 3 **d** 4

5 🔊 BW1.03 Remind students of the scenario in the Background: a team from Attollus are attending a conference in order to learn new techniques on how to approach innovation within the organisation. During the conference, they attended Daniela's presentation and are now discussing it. Ask students to listen and answer the question and play the recording. Check the answer with the class, then invite different students to tell the class what they think the team should do.

> There isn't consensus in the group over the type of innovation the company needs.
> Initially, Julia seems to be advocating incremental innovation and some new models of their popular products. One of the male speakers disagrees and favours some disruptive innovation. The second male speaker suggests that that level of innovation would take too long to develop and they should instead try to launch some existing technology into new markets. Julia builds on this and suggests a radical approach to passenger transport.

A This activity looks at useful vocabulary from the listening. You could get students to complete it individually and clarify meanings as necessary during class feedback. Alternatively, if you are short of time, do this as a whole-class activity, checking answers and clarifying meanings as you go.

1 e **2** d **3** b **4** a **5** f **6** c

Ideas for innovation

Students read about and discuss different ideas for innovation within an organisation.

6A Divide the class into four groups, A–D. Explain that each group is going to read about an idea for the next big innovation at Attollus. They should first decide what type of innovation it is, referring to Exercise 4A, and then discuss the pros and cons of the idea. Refer students to their respective information and give them time to read the information and discuss. Encourage them to make notes. During the activity, monitor and provide help as necessary.

Group A: Disruptive innovation
Group B: Incremental innovation
Group C: Radical innovation
Group D: Structural innovation

6B In their groups, students now prepare to present the idea they read about in Exercise 6A. Explain that they should review the information they have on Attollus and the different innovation types and make notes in preparation for their presentation. Set a time limit and, again, monitor and provide help as necessary during the activity.

Extra activities Business workshop 1

B This activity practises useful vocabulary from the lesson. Students could do it individually or in pairs, using their dictionaries to check unknown words if necessary. Alternatively, you could go through the words in the box with them before they begin. Check answers with the class, clarifying meanings as necessary.

1 competitive advantage / market share
2 revenue and profit, competitive advantage / market share **3** quick wins **4** categorise **5** consequences
6 techniques

Task: Create a set of recommendations

Students discuss and decide on the best innovation idea for an organisation and write an email explaining their choice.

7A Put students in groups of four comprising one student from each of the groups in Exercise 6 (A–D). If your class does not divide up into fours, some roles can be doubled. Explain the scenario: students are the product design team at Attollus and are discussing the ideas from Exercise 6 in order to decide on the best one for the company. Go through the instructions with them and make sure they are clear about the different steps they need to follow. They should a) take turns to present the

idea they read about to their group, b) discuss the pros and cons of each idea as a group and c) decide on the two best options: the one the company should adopt and a 'Plan B' one, the next best option. Point out that while listening to their partners' presentations, they should make notes of the pros and cons of each idea, so that they can discuss them with the group later. Also remind students that they should discuss the *reasons* for each of their choices. You could ask them to make notes of their final choices and reasoning if they like, to help them with the writing task in Exercise 8.

7B Students now explain their choices (and reasons) to another group. If there is time, at the end of the activity you could invite one or two groups to explain their choices to the whole class and encourage brief class discussion.

8 Depending on the time available, students could do the writing task in class or as homework. Explain that they are now going to write an email to the Head of Strategy at Attollus, describing and explaining their choice for the next big innovation. Point out the word limit and remind students that they can refer to their notes from Exercise 7B if they have them. In weaker classes, you could let students plan their emails in pairs. If they do the writing task in class, set a time limit before they begin and while they are writing, monitor and provide help as necessary.

Model answer

Dear Nina,

Thanks again for agreeing to our attendance at the recent Innovation Conference. We all found it to be a very insightful and useful experience. We learnt about a number of industry case studies and also about different types of innovation.

After the conference, the team came together to discuss various ideas for innovation and the pros and cons of each suggestion. Given our current market position, we feel that it would be best to plan the following range of innovations in the short to medium term.

Our first choice would be to use an incremental approach to innovating our current models. It should be possible to launch a new model of our currently best-performing passenger lift, the PX52, using some of the ideas the research team has already been working on, for example, better lighting, faster motors and internal TV screens showing the news or advertisements. It would be good if it were possible to accelerate that development.

We also believe that the second-best option is some structural innovation into the medical sector. This would give us a much-needed revenue boost, as well as unlocking a whole new market for the company. We could redevelop our existing motor technology for use in medical instruments and partner with a medical equipment manufacturer if necessary. The growth potential would be up to fifteen percent over the first year. However, this innovation requires more time and investment than innovating our current models to reach fruition.

These two innovations would not only address our falling profits, but could also raise further capital to make larger investments in innovation over the coming years.

I look forward to discussing this with you further in our follow-up meeting.

With kind regards,

Nicole

Review ◀ 1

1 **1** address/tackle   **2** try out   **3** overcome
   4 come up with   **5** address/tackle   **6** get round
2 **1** read up on   **2** catch on   **3** get into   **4** go about
   5 tinker with   **6** dream up
3 **1** Ø   **2** a   **3** a/the   **4** Ø   **5** Ø   **6** a/the   **7** a/the   **8** an
   9 Ø   **10** a   **11** the   **12** a
4 **1** d   **2** a   **3** f   **4** e   **5** b   **6** c
5 **1** b   **2** c   **3** a

2 Life cycle

Unit overview

	CLASSWORK	FURTHER WORK
2.1 **A circular economy**	**Lead-in** Students talk about the life cycles of products they own. **Video** Students watch a video about circular economy approaches in the electronics sector. **Vocabulary** Students look at vocabulary related to products, their life cycles and circular economies. **Project** Students choose a potential partner for their university or company.	**MyEnglishLab:** Teacher's resources: extra activities; Reading bank **Teacher's book:** Resource bank Photocopiable 2.1 p.144 **Workbook:** p.9
2.2 **Product life cycles**	**Lead-in** Students talk about the different stages of the product life cycle. **Listening** Students listen to a podcast on manufacturing and the environment. **Grammar** Students study and practise passive structures. **Speaking and writing** Students talk and write about a plan for waste management.	**MyEnglishLab:** Teacher's resources: extra activities **Grammar reference:** p.116 Additional passive structures **Teacher's book:** Resource bank Photocopiable 2.2 p.145 **Workbook:** pp.10–12
2.3 **Communication skills:** Reformulating and clarifying	**Lead-in** Students talk about communication difficulties and misunderstandings in the workplace. **Roleplay** Students roleplay a meeting where they have to reformulate what they want to say and clarify points they want to make. **Video** Students watch a video about reformulating and clarifying. **Reflection** Students reflect on the conclusions from the video and their own approach to reformulating and clarifying misunderstandings.	**MyEnglishLab:** Teacher's resources: extra activities; Interactive video activities
2.4 **Business skills:** Effective meetings	**Lead-in** Students discuss the findings of research into the effectiveness of communication in business meetings. **Listening** Students listen to conference calls discussing the details of a new business partnership. **Useful language** Students look at strategies and useful phrases for getting their message across more effectively in meetings. **Task** Students hold a meeting where they put forward problems and proposals.	**MyEnglishLab:** Teacher's resources: extra activities; Useful language bank **Spoken English:** p.112
2.5 **Writing:** Minutes of a meeting	**Lead-in** Students talk about the main principles of writing meeting notes and minutes. **Useful language** Students look at strategies and useful language for writing meeting notes and minutes. **Task** Students expand meeting notes to write full minutes.	**MyEnglishLab:** Teacher's resources: extra activities; Writing bank; Interactive grammar practice **Grammar reference:** p.117 Ellipsis **Workbook:** p.13
Business workshop 2 Achieving a circular economy	**Listening** Students listen to a talk on implementing a circular economy. **Reading** Students read reports on an initiative for the implementation of a circular economy. **Task** Students make an action plan for the implementation of a circular economy in a city.	**MyEnglishLab:** Teacher's resources: extra activities

Business brief

The main aim of this unit is to introduce students to the concept of **product life cycles** and what happens to manufactured products at the end of their life span. It also looks at sustainable alternatives to the **throwaway culture** such as the **circular economy**.

Most manufacturers today make money by continually producing updated versions of their products to persuade people to keep buying them, not by investing in designs to make products last longer. In some cases, products or product features are actually designed to become ineffective after a time (**built-in obsolescence**), forcing customers to then buy new versions or other products. This business model, known as the **linear economy**, is based on the idea of manufacturing increasing volumes and types of goods to generate profits and relies on the availability of cheap resources. It is also sometimes called the 'take, make and dispose' model since **raw materials** are taken from the environment, used to make a particular product and then thrown away when the product becomes obsolete.

The product life cycle covers different aspects of marketing and production, but essentially follows four standard stages: **introduction**, **growth**, **maturity**, **decline**. When a new product is launched, the manufacturer needs to **build product awareness** and **develop a market** for it. Once it's established in the market, the focus shifts to **increasing market share** and **growing the brand**. In order to keep the volume of sales high, the manufacturer will then **add value** to it to maintain its appeal. This often takes the form of a new look, extra accessories/services or product extensions. After a period of growth, the product will then **mature** and competitors will launch similar products onto the market to compete with it. At this point, the manufacturer may decide to add further new features or services to increase the product's **competitive edge**. The final stage of the life cycle comes when sales start to decline. To prolong its life in this last stage, the manufacturer may try to find new uses for it, new markets, ways to reduce the cost of producing and marketing it, or simply decide to discontinue making it altogether.

The 'take, make and dispose' model may be good for sales, but it's bad for the environment. Most obsolete products end up in landfill sites, an outcome which is not only wasteful of resources but also damaging for the environment. Society as a whole is now beginning to realise that this model is not **sustainable** in the long term. In addition, as the world's **non-renewable resources** are depleted to manufacture more and more products, their prices are actually rising, resulting in increased production costs and lower profit margins. Manufacturers are therefore looking for alternative models to combat the problem and one of these is the **circular economy**.

In the circular economy model, the focus is on keeping resources in use for as long as possible, extracting the maximum value from them while they are in use and recovering and regenerating components and materials from them at the end of their life cycle. The process is designed so that materials and components can 'flow' within it in a way that is **long-lasting**. Rather than 'take, make and dispose', the idea is to 'take, make and feed back' into the economy for as long as possible. Instead of a product disappearing into the waste stream when its useful life ends, the materials it is made from can be **remanufactured** and **resold**, thereby prolonging its value over time and reducing its long-term impact on the environment.

Product life cycles and your students

All students will have experience of consumer goods they've bought which have quickly become obsolete, such as mobile phones or computers. They will probably be aware of the strategies some manufacturers use to maintain their products' appeal in the mature phase of the product life cycle, such as adding storage capacity to a phone, or improving the quality of its camera. They may also be aware of schemes in their area to recycle domestic batteries, electrical appliances or computer hardware to avoid these ending up in landfill. In-work students may already be aware of the different stages of the standard life cycle of a product and have participated in devising strategies for it.

Unit lead-in

Elicit a brief description of the photo (empty [reusable] glass milk bottles on a doorstep, with a note in one of them) and ask students how they think it might be related to the unit title *Life cycle* (e.g. extending the lifetime of products by reusing them). Elicit a few ideas, then refer students to the quote. Check understanding of *global marketplace* and elicit or explain that *ecological truth* refers to the environmental impact of products. Briefly discuss the quote with the class. What do students think it means? How likely do they think it is for prices of products to 'tell the ecological truth' in future? Would this be a good thing? Why?

2.1 > A circular economy

GSE descriptors

- Can follow presentations on abstract and complex topics outside their field of interest.
- Can extract specific details from a TV programme on a work-related topic.
- Can contribute to a group discussion using linguistically complex language.
- Can contribute ideas in a panel discussion using linguistically complex language.

Warm-up

Put students in small groups and ask them to tell each other about something they own which they replaced recently. How long did the previous item last? Why did they replace it? Was this the first time they'd had to replace this item? Give them 2–3 minutes to discuss in their groups, then invite different students to share their answers with the class.

Lead-in

Students talk about the life cycles of products they own.

1 Put students in pairs and give them 3–4 minutes to discuss the questions, then invite students from different pairs to share their answers with the class. Encourage them to elaborate where appropriate. You could use this Lead-in as an opportunity to teach some useful vocabulary from the Video section which follows, e.g. *waste stream, landfill, dispose of a product, consumerism, obsolescence.*

Video

Students watch a video about circular economy approaches in the electronics sector.

2A You could do this as a whole-class activity, eliciting ideas from a few different students and then checking the answer with the class. Alternatively, let students discuss in pairs or small groups first, then get brief feedback from the class and check the answer.

b

2B ▶ 2.1.1 Explain the activity and before students watch, go through the factors in the box and check understanding of each one. Play the video, then check the answer with the class.

planned obsolescence

3 ▶ 2.1.1 Give students a minute to read the questions before they watch again and check understanding of *throwaway culture* in question 2. Ask them to make notes in answer to the questions while watching and play the recording, twice if necessary. Check answers with the class.

1 They make up the largest waste stream and are disposed of in landfills.
2 It is not good for the environment, but it helps the economy because it encourages innovation.
3 It is called 'take, make, dispose' and refers to taking natural resources from the ground, manufacturing something with them and then disposing of them.
4 Products are designed from the start to flow through a system and when they are older, they are fed back into the same system. They can be remade and resold as new products.
5 They need to sell many products, not design those which will last a long time.
6 The resources prices were decreasing but since 2000 they are increasing.
7 They save the valuable elements in electronic goods.
8 It is suggested that finding new materials will help to create an economy where we no longer just throw things away.

4 Put students in pairs or small groups and give them 2–3 minutes to discuss the question. When the time is up, invite different students to share their answers with the class; encourage them to give reasons.

Extra activities 2.1

A ▶ 2.1.1 Explain the activity and tell students that the completed sentences paraphrase information from the video. Ask them to complete the exercise individually and if necessary, let them watch the video again before they begin. Point out that they need to underline the information in the videoscript which has been paraphrased and if necessary, do the first item as an example with the class. If there is time, get students to compare answers in pairs before class feedback.

1 d (*Electronics is the fastest growing waste stream in the UK. Every year around five hundred thousand ... end up in landfill.*)
2 h (*Our throwaway culture is bad for the environment but it's good for the economy. And it also drives innovation because new stuff tends to be better than old stuff.*)
3 a (*When we talk about a circular economy, we design the economy from the outset for the materials, components and products to flow within a system, so rather than take, make and dispose, you take, you make and then those products feed back into the economy.*)

4 e (*So rather than something that goes off into the waste stream and you have no value from, it keeps coming back to you. You can remanufacture it and resell it as a new product.*)

5 b (*That model of creating profits by manufacturing more and more stuff ... relied really on the idea that we had not just cheap resource prices but resource prices that were continually coming down.*)

6 f (*Many broken and unloved gadgets end up here where their precious contents are salvaged, but what if our electronic goods didn't break in the first place?*)

7 c (*If we made our gadgets from self-healing plastics, they would always look shiny and new ...*)

8 g (*It's new materials like this which are a key part of creating a circular economy, to change electronics from a throwaway item to something that's central to our lives, something we can still love and be enthusiastic about and we don't have to feel the guilt of throwing them away.*)

Alternative video worksheet: The circular economy

1 If there is time, get students to discuss the questions in pairs or small groups first, then get feedback from the class. Otherwise, discuss the questions with the whole class. List students' ideas on the board so that they can refer to them later, when they do Exercise 8.

2 ▶ ALT2.1.1 Tell students that they are going to watch a video about the circular economy. Put them in pairs, give them 1–2 minutes to discuss the question, then invite students from different pairs to share their ideas with the class. Play the first part of the video (0:00–0:52) for students to check their ideas, then elicit the answer.

Possible answer

In the circular economy, products are made and used but are then remade or reused, thus exploiting resources for as long as possible.

3 ▶ ALT2.1.1 Explain that students are going to watch the second part of the video and answer some questions; give them time to read the questions before they watch. Encourage them to make notes in answer to the questions while watching, then play the video (0:53–3:26), twice if necessary, and check answers with the class.

Possible answer

1 Greensole make shoes by taking the (rubber) soles of discarded shoes and trainers and refurbishing/turning them into sandals.
2 Oat shoes are unusual because they are designed to biodegrade in six months. This means flowers can grow out of your old sneakers!
3 BMW and Tata Motors
4 Local Motors are making 3D-printed cars using recyclable body parts (that can be reprinted into another car).
5 Carbon-fibre-reinforced ABS plastic

4 ▶ ALT2.1.1 Explain the activity and give students time to read the questions and options before they watch. Play the last part of the video (3:27–4:55) and check answers with the class.

1 b **2** c

5 This exercise looks at useful vocabulary from the video. Students could do it individually or in pairs, using their dictionaries if necessary. Alternatively, you could go through the words in the box with them before they begin. Check answers with the class, clarifying meanings as necessary.

REUSE: recycle, refurbish into sth, remanufacture, reprint
THROW AWAY: dispose of, (get) rid of sth
MATERIALS: bio-cotton, cork, hemp
WASTE: landfill, leftovers, peel
CHEMICAL PROCESS: biodegrade, decompose

6 You could do this as a whole-class activity, checking answers and clarifying meanings as you go. Alternatively, let students complete the exercise individually, then check answers and clarify meanings with the class.

1 resources **2** recyclable **3** biodegradable
4 decompose **5** viable **6** footprint

7 This exercise looks at vocabulary from Exercises 5 and 6, so students should be able to do it individually. Before they begin, point out that all the options fit grammatically, so they should think carefully about *meaning* in order to choose the correct one. If there is time, get students to compare answers in pairs before class feedback.

1 d **2** c **3** b **4** a **5** b **6** c

8 Put students in pairs or small groups and go through the questions with them. For question 3, remind them to refer to the list of products on the board. Give them 3–4 minutes to discuss the questions in their pairs/groups, then get feedback from the class.

Vocabulary: Circular economies

Students look at vocabulary related to products, their life cycles and circular economies.

5 Get students to complete the exercise individually. Check that they understand the meaning of *flow* and *disposal* in the definitions before they begin. If time allows, encourage them to find the words in videoscript 2.1.1 on page 145 before matching them with their definitions, so that they can see them used in context; this will help them work out their meanings. Check answers with the class, clarifying meanings as necessary.

> **1** e **2** c **3** g **4** a **5** d **6** h **7** f **8** b

6 Explain the activity and get students to complete the exercise individually or in pairs, using their dictionaries if necessary. Check answers with the class, clarifying meanings as necessary.

> **1** new products **2** get rid of
> **3** 'take, make and feed back' into the system
> **4** eagerly accept **5** valuable contents
> **6** to repair a circuit which is not working
> **7** produce state-of-the-art products
> **8** not to be reused

7 This activity looks at vocabulary covered in Exercises 5 and 6, so students should be able to do it individually. Point out that the words in the box are different parts of speech and encourage students to look carefully at the words around each gap to help them choose the correct word. Check answers with the class.

> **1** waste stream **2** outset **3** modular **4** precious metals
> **5** commodities **6** upgrade **7** salvage
> **8** recycling plants
> (Not used: forward-thinking and throwaway)

8 Put students in pairs and, before they begin, elicit brief definitions of *linear economy* and *circular economy*. Give them a few minutes to discuss the question in their pairs, then invite students from different pairs to share their views with the class.

Extra activities 2.1

B This activity practises key vocabulary from the lesson. Explain to students that the sentences are definitions of the words in bold; some of them are correct and some incorrect. Students should decide which definitions are correct and then correct the incorrect ones. If necessary, do the first item as an example with the class and, if there is time, get students to compare answers in pairs before class feedback.

> **1** F (A waste stream occurs when products are disposed of completely.)
> **2** T
> **3** F (When something is salvaged, it is saved.)
> **4** T
> **5** T
> **6** F (When we do something from the outset, it means that we do it from the start.)
> **7** T
> **8** F (The throwaway economy drives innovation as manufacturers look for new goods to produce).

Project: Replacing electronic devices

Students hold a panel discussion about the life cycles of products they own.

9A Put students in small groups and explain that they are going to set up and then hold a panel discussion. Start by eliciting or giving a brief explanation of what a panel discussion is and how it works (a situation in which a selected group of people discuss a specific topic or issue in front of an audience; panel discussions are often used in meetings, conferences and conventions). Give students a few minutes to decide which electronic device they replace most often. Then explain that they need to come up with ideas on how to avoid replacing this device. Draw their attention to the words in the box and encourage them to use these, as well as other words from the lesson. Point out that each member of the group will later present one idea to another group in a brief panel discussion. Give groups time to prepare for their discussion and encourage students to make notes. During the activity, monitor and help as necessary.

9B Students now hold their discussions. Join groups together into new, bigger groups and explain the activity: students will take turns to present and explain their ideas; they each have 1–2 minutes. It is a good idea to appoint someone as a timekeeper, to keep the group focused and give a feeling of responsibility for time management to the group. The timekeeper can also give feedback at the end, saying how long each person spoke for. Explain that there will be a Q&A session at the end, so while listening to each presentation, the group can note down questions about each idea. During the activity, monitor and note down points to highlight during feedback, but do not interrupt students' discussions.

9C Groups now choose one spokesperson to give a one-minute summary to the whole class. The class could then vote for the best idea(s).

My English Lab: Teacher's Resources: extra activities; Reading bank
Teacher's book: Resource bank Photocopiable 2.1 p.144
Workbook: p.9

2.2 > Product life cycles

GSE descriptors

- Can understand most of a linguistically complex podcast.
- Can follow a group discussion on complex, unfamiliar topics.
- Can use persuasive language to convince others to agree with their recommended course of action during a discussion.
- Can make detailed notes of the key action points from feedback.

Warm-up

Put students in pairs or small groups and ask them to tell each other about products they own which they think can be repurposed and reused. Give them 2–3 minutes to discuss in their pairs/groups, then elicit a few ideas around the class.

Lead-in

Students talk about the different stages of the product life cycle.

1 Put students in pairs, explain the activity and check that they understand the meanings of *decline* and *maturity*. Give them 2–3 minutes to exchange ideas in their pairs, then invite different students to share their ideas with the class. Do not confirm answers yet, as students will check them in the next activity.

> **1** the introduction phase **2** the growth phase
> **3** the maturity phase **4** the decline phase

Listening

Students listen to a podcast on manufacturing and the environment.

2 ◀ 2.01 Tell students that they are going to hear the first part of a podcast on the effect of manufacturing on the environment and go through the instructions with them. Ask them to listen and check their ideas from Exercise 1 and play the recording. Discuss the answers with the class, feeding in information from the Note below as necessary.

> *Note*
>
> **Product life cycle**
> • The introduction phase: branding/quality established; copyrights obtained; low price to enter market; promotion aimed at early adopters; limited distribution; awareness of product increased; product development continued.
> • The growth phase: product quality and pricing kept at same level; more distribution channels added; promotion broadened; change from awareness building to brand building; product often extended through new products in same line.
> • The maturity phase: increased competition in market; may be necessary to lower prices; incentives offered to customers; promotion looks at product differentiation; sales growth stops or stagnates; profit margins decline; find way to produce more cheaply; make small changes to product; add new features; increase funds for promotion and discounts.
> • The decline phase: product can be maintained (possibly through new uses of product), harvested (costs reduced, sell to niche or other new market; try to harvest as much as possible from brand) or discontinued; unit sales fall due to various reasons, e.g. outdated products or change in buying habits.

3A ◀ 2.02 Tell students that they are going to hear the next part of the podcast and give them a minute to read the questions. Play the recording, then check answers with the class.

> **1** She feels it can be better exploited and that companies should start doing this.
> **2** She feels it can be considered as something sustainable by looking at the way raw materials are sourced, that the manufacturing process should be looked at carefully and that packaging and disposal need to be taken into account. She thinks everyone needs to change the way they think about product life.

3B ◀ 2.02 Before students listen again, give them time to read the statements and options and check that they understand *equality*, *fairly sourced materials* and *planned obsolescence*. To check answers, you could play the recording again and get students to ask you to pause when an answer is heard.

> **1** b **2** a **3** b **4** c **5** a **6** c

> **Extra activities 2.2**
>
> **A** ◀ 2.01 ◀ 2.02 This activity provides students with extra listening practice. Give them time to read the statements before they listen, then play the recording and check answers with the class. In weaker classes, students may need to listen twice: once to decide if the statements are true or false and then a second time to correct the false statements. You may also need to pause the recording to give students time to write their answers.
>
> **Track 2.01**
> **1** F (In the introduction stage of a product, companies work at building product awareness and developing a market. It is during the growth stage of a product that companies work at increasing their market share.)
> **2** T
> **3** F (In the growth stage, it is necessary to grow the brand. / In the growth stage, the level of quality must be maintained and other services or extensions can be added to the product.)
> **4** T
> **5** T
>
> **Track 2.02**
> **6** F (Maja agrees that in many cases companies are too interested in profit and don't look after the planet when they consider the life cycle of products.)
> **7** F (Maja feels that the market for recyclable goods is not being exploited at the moment.)
> **8** T
> **9** F (Maja feels that the product life cycle can be sustainable by looking at the sourcing of raw materials, the manufacturing process and packaging and disposal.)
> **10** T

Grammar: Additional passive structures

Students study and practise passive structures.

4A Start by briefly revising the form and main use of the passive. Write this sentence from the podcast on the board: *Companies are usually driven by profit and the planet comes second.* Underline *are usually driven* and ask students what structure the verb is in (the passive). Elicit the form (*be* + past participle) and main use of the passive (when the 'doer' of an action is not important, not known or obvious), then continue with the exercise. You might like to do it as a whole-class activity, checking answers and explaining the uses as you go. Alternatively, you could get students to complete it individually and clarify any points as necessary during feedback. When discussing sentence 2, elicit the form of the causative *have*, which is different from the other structures here (*have* + object + past participle).

1 c 2 e 3 b 4 d 5 a

4B At this point, you may wish to refer students to the Grammar reference on page 116, go through it with them and answer any questions they may have. Alternatively, students can look at the Grammar reference after Exercise 4C. Explain the activity and refer students to audioscripts 2.01 and 2.02 on page 151. Get them to complete the exercise individually and then to compare answers in pairs before checking with the class. During feedback, clarify any points or queries as necessary.

is looked at (a), is often based on (a), needs maintaining (d -*ing* form), need to be removed (d infinitive), having been given (b), can be thought of (a), are considered to be (c), were hoping to be recognised (c), can be applied to (a), have a new kitchen installed (e), have other work done (e)

4C Do this as a whole-class activity. For sentence 4, point out that when *need* is followed by -*ing*, we do not use a form of *to be* (e.g. *This type of thinking needs implementing sooner rather than later*).

Possible answer
They all indicate that the action is more important than the agent or that the agent is unknown or irrelevant.
All but one use a form of *to be* and the past participle. The exception is the causative *have* (sentence 2), which does not use a form of *to be*.
(Note also that when the -*ing* form follows the modal verb *need*, no form of *to be* is used.)

5–6 Both of these activities give further practice of the passive structures covered in Exercise 4, so students should be able to complete them individually. Remind them that they can refer to the Grammar reference if they need help and, if there is time, get them to compare answers in pairs before checking with the class.

5 1 b 2 c 3 b 4 a 5 c 6 a 7 b 8 c 9 c 10 b
6 1 The marketing campaign was aimed at a young demographic.
   2 We don't like being / to be criticised for no reason.
   3 It was the first recycling initiative to be tried / to have been tried by our city.
   4 The campaign needs promoting / to be promoted at universities.
   5 I had my car tested for emissions.
   6 It was the last idea to be tested / to have been tested.

Extra activities 2.2

B This activity gives further practice of passive structures. Get students to complete it individually and then, if time allows, to compare answers in pairs before class feedback. As an optional extension, you could ask them to match the examples of the passive with the uses in the box in Exercise 4A of the Coursebook (see answers in brackets below).

1 are looked at / are being looked at (a)
2 being told (b)
3 have/get, certified (e)
4 taking / to be taken (d)
5 to have been incorporated (c)
6 have been affected by (a)
7 to have been concerned / to be concerned (c)
8 to be reused / reusing (d)
9 being forced (b)

Speaking and writing

Students talk and write about a plan for waste management.

7A Students should do this individually. Explain the activity, go through the phrases in the box with students and check that they understand each one. Point out that they can use their own ideas as well as those in the box.

7B Put students in pairs and get them to compare and discuss their lists. They should decide on the action points, prioritise them and then discuss the steps that need to be taken. Encourage them to make notes so that they can refer to them when they do the writing task in Exercise 7C. Allow 4–6 minutes for pairs to discuss their ideas, while you monitor and help as necessary.

7C Explain the writing task and before students begin, go through the tips for memo writing with them. Remind them that memos go to more than one person; this increases the use of the passive as the agent is not important but the actions are. Also remind students of the format/structure: today, memos come as emails so the format is the same, but it is different from a personal email due to the way it is worded. Depending on the level of your class and the time available, you could let students plan their memos individually or in the same pairs as for Exercise 7B. If short of time, students could plan their memos in class and write them as homework.

Model answer
Dear all,
It has been determined that we have too much waste in the department. Waste materials need to be separated so that they can be recycled. We realise that reading online is not always simple and no one likes being told that they are not allowed to print documents; however, please consider using paper that has been printed on one side already. Non-energy-efficient light bulbs need replacing by the facilities department; we ask you to inform them if you still have old bulbs at your desk. Lights which are switched on during the day are often not necessary; please give this some thought. In addition, the plastic cups and cutlery have been replaced in the kitchen with glass and china; make sure you wash them and put them away when you are finished.
Thank you.

MyEnglishLab: Teacher's Resources: extra activities
Grammar reference: Additional passive structures p.116
Teacher's book: Resource bank Photocopiable 2.2 p.145
Workbook: pp.10–12

2.3 ⟩ Communication skills
Reformulating and clarifying

GSE descriptors

- Can reformulate what they want to say during a conversation or discussion using linguistically complex language.
- Can follow an animated conversation between two fluent speakers.
- Can infer meaning, opinion, attitude, etc. in fast-paced conversations between fluent speakers.
- Can compare and evaluate different ideas using a range of linguistic devices.

Warm-up

Ask students to think about a situation in which they or someone they know had to deal with a communication difficulty or misunderstanding. Put them in pairs or small groups to share their experiences, then invite different students to share them with the class. How did students overcome this difficulty? How did they clear up the misunderstanding?

Lead-in

Students talk about communication difficulties and misunderstandings in the workplace.

1A Go through the instructions and look at the example with the class; check that they understand *jargon*, *acronym* and *distracted*. Students could then complete the activity individually or in pairs, depending on the time available and the level of your class. You may also like to ask students about some difficulties that can occur specifically when working in international teams (see Note below for examples). Do not conduct class feedback at this point; move straight on to the next activity.

1B Put students in small groups (if they did Exercise 1A in pairs, join pairs together into groups of four) and explain the activity. Encourage students to add to their lists from Exercise 1A and also to note down the possible solutions they discuss in their groups. When they are ready, invite students from different groups to share their ideas from both exercises with the class, then broaden this into a class discussion, feeding in information from the Note below as necessary.

Possible answers

The speaker:
- can't express what isn't yet a clear thought in their own minds.
- knows what they want to say but struggles to communicate it clearly and effectively, e.g. can't find the right words, leaves out important information ('underexplaining') or is long-winded and repetitive ('overexplaining').
- doesn't want to say something, e.g. doesn't challenge or disagree with the manager due to personal habit or cultural norms.

The listener:
- is unable to mentally process the information conveyed by the speaker for any number of reasons, e.g. it's an unfamiliar or complex topic, difficulties with the speaker's accent, poor internet/phone connection.
- has misconceptions, biases or is not open to other perspectives/viewpoints.
- misinterprets what is said and jumps to conclusions.

Note

Communication can be difficult and misunderstandings at work occur for many reasons. For example, you could be listening to some highly complex information or a speaker is mumbling, rambling, being unclear by giving a long-winded explanation or, conversely, by giving too short an explanation and omitting key details. Listeners are also distracted by multi-tasking, which blocks understanding and is more prevalent than ever today due to smartphones.

International teams can have another layer of difficulty: cultural differences, e.g. unfamiliar accents, not openly disagreeing or saying you do not understand something, not saying 'no', not admitting there is a problem or that you cannot do something.

In general, listeners do not ask for clarification when they do not hear or understand something. So a speaker has to be able to use clarifying and reformulating strategies to get their message across clearly and help get the listener's attention and understanding. Simply asking, 'Do you understand?' can be perceived as too direct and impolite to some people. In addition, the answer may be 'Yes,' which could mean, 'No, not really'.

Other barriers that can inhibit communication include making incorrect assumptions. The speaker needs to be able to clear up inevitable misunderstandings and respond politely to any incorrect assumptions/ interpretations made by the listeners.

It is also important to remember that listeners will generally 'filter' what they hear and only pay attention if they consider what is being said is worth processing and relevant enough to them. 'Mindful listening' is a method that helps someone be fully in the present moment by reducing internal and external distractors and paying full attention to what another person is saying without preconceptions or judgements.

Preparation: Meeting to discuss an article

Students read and think about the scenario for a roleplay.

2A If you have already done Lesson 3 in Unit 1, ask students what they remember about *Lifestyle* magazine and its target audience. If this is the first Communication skills lesson for your class, briefly set up the context and tell students that they are going to read about *Lifestyle*'s target readers in the text that follows. Go through the instructions with them and check they understand *pitch*; if they did Lesson 3 in Unit 1, they should be able to give a brief explanation themselves. Give students time to read the guidelines and pitch and think about the question; remind them that they will need to give reasons for their answers.

Tell them not to worry about any unknown words in the texts for now. If there is time, give them 2–3 minutes to share their opinions in pairs or small groups first, then get brief feedback from the class. Finally, help students with any unknown vocabulary from the texts.

> **Possible answer**
>
> It probably would interest this profile of readers, many of whom would be concerned about environmental problems and aware of plastic pollution. Many would be curious about how someone was able to adapt their life to reduce their plastic consumption. They'd also be interested in ideas showing them how they could take simple steps to do the same.

2B Explain the activity and give students time to read the email and think about the question, then elicit the answer.

> Donna thinks Susan plans to interview someone about their life without plastic, but the idea she pitched was actually to do the experiment herself. Susan needs to make it clear she's going to carry out the experiment.

Roleplay

Students roleplay a meeting where they have to reformulate what they want to say and clarify points they want to make.

3A Put students in pairs and tell them that they are now going to roleplay a meeting between the Assistant Editor and Journalist from Exercise 2, to discuss the feature article for *Lifestyle*. Assign roles (or let students choose), refer students to their relevant information and give them plenty of time to read it while you monitor and help them with any vocabulary or other questions they may have. Set a time limit for the preparation stage and point out that students should think carefully about any misunderstandings which may occur and how they are going to clear them up. Remind them of the reasons and strategies they discussed in Exercise 1. During the activity, monitor and provide help as necessary.

3B Students now roleplay their meetings. Set a time limit before they begin and remind them that they should try to reformulate and clarify any points they are trying to make in a polite, professional way. During the activity, monitor and note down any points to highlight during feedback after Exercise 3C, but do not interrupt the meetings.

3C Students now reflect on their roleplays. Go through the questions with them and give them 3–4 minutes to discuss in their pairs, then invite different students to share their answers with the class. During feedback, you may also wish to ask students for examples of language they used to reformulate and clear up misunderstandings (e.g. *What I'm saying is …* , *What I mean (to say) is …* , *(To put it) In other words, …* , *To put it another way, …*). You could list any phrases students mention on the board and use them as a point of reference/ comparison when discussing question 3 in Exercise 4A below. (More practice of this language is available in Interactive video activities on MyEnglishLab.) Finally, highlight any points you noted during the roleplays.

Video

Students watch a video about reformulating and clarifying.

4A ▶ 2.3.1 Remind students of the scenario from Exercise 2 and tell them that they are now going to watch Donna and Susan discuss the feature article. Ask them to make notes in answer to questions 1 and 2 while watching, play the video and check answers with the class. Then refer students to question 3 and the videoscript on page 146; remind them of the useful language they looked at in Exercise 3C and refer them to the list on the board. Give them time to find the phrases in the script, then elicit them and add them to the list on the board. Encourage students to record the phrases in their notebooks.

> 1 This follow-up meeting seemed to be very successful.
> 2 If they hadn't cleared up the misunderstandings, then the article Susan produced wouldn't have been what Donna was expecting; it may not have been in keeping with editorial policy, there would have been extra costs incurred and delays.
> 3 *(I think) there's been a misunderstanding … ; I just wanted to clarify what I actually meant by … ; What I'm saying is … ; That's to say … ; If you don't mind, I'll just go over … ; In other words, … ; The point I'm trying to make is … ; I mean …*

4B Put students in pairs or small groups to discuss the question, then broaden this into a class discussion. Encourage students to elaborate.

5 Students could do this individually or in pairs, depending on the time available. You could encourage them to refer to the videoscript on page 146 and read the text 'around' the words in bold, to help them work out their meanings. Check answers with the class, clarifying meanings as necessary.

> 1 are   2 gets   3 completely   4 simple   5 throw it away
> 6 simple   7 agree

> **Extra activities 2.3**
>
> **A** This activity practises useful vocabulary from the video. Students practised the vocabulary in Exercise 5 of the Coursebook, so they should be able to complete this exercise individually. Remind them that they can refer to the definitions in the Coursebook if they need help.
>
> > **a** on the same page   **b** got the wrong end of the stick
> > **c** chuck out   **d** undivided attention   **e** hook
> > **f** dumbing it down   **g** in layman's terms

Reflection

Students reflect on the conclusions from the video and their own approach to reformulating and clarifying misunderstandings.

6A Allow students to work individually first so that they can reflect on their own approach. Go through the headings with them before they begin and remind them to think about the ideas in Exercise 1 and their answers to Exercises 3C and 4. To help them, you could give them one or two examples from the model answers below.

Model answers

<u>What I've learnt about my performance when reformulating and clarifying misunderstandings in a discussion</u>

I probably don't reformulate and clarify what I mean often enough. This can contribute to communication difficulties and misunderstandings when working with others.

<u>What I will remember about reformulating and clarifying in the video</u>

That both women spent time reformulating and clarifying what they wanted to say in a polite and professional way.

<u>What lesson(s) I will take away from the video</u>

Reformulating and clarifying our meaning are essential elements of clear communication and good teamwork.

<u>Where I can apply this knowledge in future</u>

I can spend more time listening actively, checking understanding, clarifying and reformulating in order to make sure I'm understood and that I understand others when working together.

6B Put students in pairs or small groups to compare their reflections, then get brief feedback from the class.

MyEnglishLab: Teacher's Resources: extra activities; Interactive video activities

2.4 ❯ Business skills
Effective meetings

GSE descriptors
- Can follow a work-related discussion between fluent speakers.
- Can put forward a smoothly flowing and logical structured argument, highlighting significant points.
- Can describe a business proposal in detail.
- Can use persuasive language to convince others to agree with their recommended course of action during a discussion.
- Can compare and evaluate different ideas using a range of linguistic devices.
- Can participate in extended, detailed professional discussions and meetings with confidence.

Warm-up
Discuss these questions with the class: *How often do you have to attend meetings at your place of work/study? What kind of meetings are they, e.g. Training meetings? Briefings? Meetings to discuss your academic progress/ goals? How effective do you think time spent in meetings is? What do you think makes a meeting effective?*

Lead-in
Students discuss the findings of research into the effectiveness of communication in business meetings.

1A Put students in pairs, explain the activity and give them time to look at the quiz. Check that they know what a status update meeting is (a meeting where team members with a common goal communicate updates on different tasks and discuss progress, challenges and next steps). Ask pairs to try and guess

the answers to the quiz by selecting one of the options in italics for each statement. As feedback, invite students from different pairs to share their ideas with the class, but do not confirm answers yet, as students will check them in the next activity.

> **a** 25m **b** 15% **c** 4 **d** 50% **e** 67% **f** $37bn **g** 92% **h** 69% **i** 80% **j** true

1B 🔊 2.03 Explain that students are going to listen to a podcast and check their ideas from Exercise 1. Play the recording and go over the answers to the quiz with the class. Then refer students to the questions and discuss questions 1 and 2 with the whole class, inviting different students to contribute. Students could brainstorm ideas for question 3 in pairs or small groups first if there is time. Elicit ideas around the class and write them on the board in three columns headed *Before*, *During* and *After*. You could then do a class vote on which idea(s) students think would be the most effective for each stage.

3 Possible answers

Before a virtual meeting, particularly with people you don't know very well, it can be useful to send an email to clarify your expectations of the meeting and align with their expectations. Sometimes a telephone call in advance of the meeting can be useful to discuss the agenda and possible outcomes to ensure that the actual meeting is as efficient as possible.

During the meeting, it is important for individuals to avoid speaking too much as others might stop listening or simply find it difficult to follow. At the beginning of the meeting, allow time for some small talk to create a professional but relaxed atmosphere; asking people general questions about work and everyday life to help build relationships, introducing people who do not know each other at the beginning, etc. Asking check questions to gauge opinions and allowing people to respond and clarify is useful.

After the meeting, if necessary, it can be useful to email or call people to check their feelings about the meeting, for example, to ensure the discussion was clear and the decisions were acceptable; sometimes people do not admit that they do not understand or disagree.

❯ **Spoken English**
p.112: I do wonder, really, where it all ends up, I mean

1 🔊 SE2 Go through the instructions with students and give them time to read the questions so they know what to listen for. Play the recording, then check answers with the class. Note that students may need to listen twice in order to check/complete their answers. As an extension, you could ask students if they have ever repurposed any of their electrical appliances and if not, if they would consider doing so. What electrical appliances could they repurpose and how?

Possible answers
1 She's been downsizing.
2 Charity shops
3 You put things in different categories so they can be recycled.
4 Waste Electrical and Electronic Equipment
5 She put lights in it and created a little lamp.

Using adverbs for effect

2 ◆ SE2 Explain the activity and give students time to quickly read through the extracts before they listen. Point out that all the gapped words are adverbs. Play the recording again, then check answers with the class, clarifying meanings as necessary.

> **1** literally **2** absolutely **3** Exactly **4** apparently
> **5** actually **6** necessarily **7** pretty

3 This exercise looks at using different types of adverb for effect. Ask students to complete it individually and get them to compare answers in pairs before class feedback. Alternatively, if time is short, you could do it as a whole-class activity, checking answers as you go.

> **a** absolutely **b** apparently **c** literally
> **d** exactly **e** pretty **f** actually **g** necessarily

Listening

Students listen to conference calls discussing the details of a new business partnership.

2A ◆ 2.04 Go through the instructions with the class and give students a minute to read the questions before they listen. Play the recording, twice if necessary, then check answers with the class.

> **1** WinGreen is formally in talks to form a strategic partnership with BioGrad.
> **2** It will help WinGreen innovate with new green materials for its products and so meet their strategic sustainability goals.
> **3** Tammy is worried that the partnership may result in the use of more expensive materials with a negative reaction from customers to possible price rises and a negative impact on sales.
> **4** Paula decides to reschedule the meeting to the next day and send some information in advance. She also promises to speak to Tammy and Frank one-to-one in advance of the second meeting.

2B ◆ 2.05 Again, after explaining the task, let students read the questions before you play the recording. If necessary, especially in weaker classes, let students listen again in order to check/complete their answers.

> **1** Paula has talked individually to the meeting participants and sent a summary report explaining in more detail the planned relationship with BioGrad.
> **2** The fact that the relationship with BioGrad has actually been driven by WinGreen's customers, with the top 50 customers approving of this new focus on sustainable materials.
> **3** The first consulting milestone is when beginning a new product development project. The second is when preparing marketing materials before product launch. What is important is that WinGreen keeps control and takes the final decision on materials used and the final marketing message.
> **4** Frank wants to share a summary of the BioGrad information with key customers as some of them are nervous.

2C ◆ 2.05 Point out that in the second meeting, Paula took a number of steps in order to handle the meeting more effectively. Ask students to listen again and note anything they think helped make the meeting more effective. Elicit answers around the class and list them on the board, adding any points from the key below which students do not mention. If there is time, you could play the recording again, pausing (or telling students to ask you to pause) each time Paula does one of the things listed on the board.

> **Possible answers**
> Paula:
> - decides to slow down and talk through the issues more carefully.
> - prepares people for the second meeting by talking to them individually in advance.
> - signals clearly her respect for the other participants' opinions to create an engaged atmosphere.
> - sends a summary document to support the second meeting.
> - clearly signals which pages of the document people should read during the discussion, checking as she goes if people can see the relevant pages.
> - involves people by name to better control who is speaking about what.
> - makes clear the assumptions and implications of her ideas so people can follow.
> - structures and signposts her explanation explicitly to make her message clearer.
> - says what she is *not* saying to avoid confusion and misinterpretation.
> - offers a translation of the key document to help people locally understand.

Useful language

Students look at strategies and useful phrases for getting their message across more effectively in meetings.

3A Go through the tips and principles with the class, clarifying any vocabulary as necessary. Do not discuss these in detail yet, as students will do this in the next activity. Instead, move straight to the exercise, getting students to complete it individually or, in weaker classes, in pairs. Check answers with the class.

> **1** c **2** i **3** d **4** b **5** f **6** g **7** j **8** a **9** e **10** h

3B Put students in pairs and give them plenty of time to brainstorm ideas and expressions. Monitor and help them with any vocabulary they may need. During feedback, elicit ideas around the class and list them on the board. Encourage students to record the expressions from question 2 in their notebooks.

> **Possible answers**
> **1** Principle 1: Find points you have in common quickly to establish rapport.
> Principle 2: Check from time to time if people feel the process used is appropriate and/or meeting expectations.
> Principle 3: Suggest starting the discussion with other people's priorities rather than your own.
> Principle 4: Take meeting minutes using a computer screen visible to everyone in order that people can follow the discussion and decision-making.

2 Motivating people to collaborate (*Paul, you're expert in this. What do you think? Jon, do you agree with Paul?*) Asking for feedback at the close of meetings (*How effective was our meeting today? What can we do to improve our meetings in the future?*)

3 People may feel they lack the English to participate fully. Relationships between participants may not be optimal. Some participants may feel they lack expertise or confidence in the subject matter and so are reluctant to speak.
Possible ways to overcome the challenges: agreeing ground rules and 'good behaviours' collaboratively is an effective method. Giving one-to-one feedback after the meeting to confirm good behaviours and to challenge less effective behaviours is also necessary.

Extra activities 2.4

A Explain to students that sentences a–j are extracts from different meetings and students need to match them to the tips in Exercise 3A of the Coursebook. Give them time to read the tips again if necessary and ask them to complete the exercise individually. Check answers with the class.

> **a** 4 **b** 9 **c** 10 **d** 2 **e** 6 **f** 3 **g** 1 **h** 8 **i** 7 **j** 5

B This activity looks at useful signposting expressions from the lesson. Explain the activity, ask students to complete it individually, then check answers with the class. If there is time, students could then practise the dialogue in groups of three.

> **1** d **2** c **3** f **4** a **5** g **6** b **7** h **8** e

Task

Students hold a meeting where they put forward problems and proposals.

4A Put students in groups of four and give them time to read the professional context and ask you about anything they do not understand.

4B Explain that students are going to hold a meeting where they each present a problem, lead a short discussion about it and then, with the rest of the team, decide on a proposal to present to the CEO for implementation. Assign roles A–D (or let students choose) and refer students to their respective role cards. If your class does not divide up into fours, any one of the roles can be omitted or doubled. If you have to double one of the roles, the students with two roles can take turns to present the problem and lead their part of the discussion. Give students time to read their role cards while you monitor and help them with any questions they may have. When they are ready, they should prepare to present their problem and then lead a discussion on it. If there is time, you could group students with each role together to brainstorm ideas and then get them to regroup to hold their meetings. Encourage them to make notes and remind them to try to follow the four-principle model and use phrases from Exercise 3. Set a time limit for this preparation stage and also for each part of the meeting.

4C Students now hold their meetings. Point out that they may choose to have a face-to-face meeting or simulate a virtual meeting; for the latter, they can sit back to back so they cannot see the other participants. Remind students of the time limit and ask them to begin. During the activity, monitor and note down points to highlight during feedback, but do not interrupt the meetings.

4D Let students discuss in their groups first, then broaden this into a class discussion. Remember to highlight any points you noted while monitoring.

MyEnglishLab: Teacher's Resources: extra activities; Useful language bank
Spoken English: p.112

2.5 ❯ Writing
Minutes of a meeting

GSE descriptors

- Can write detailed minutes of a meeting.

Warm-up

Discuss these questions with the class: *Do you usually take notes during meetings? If so, what are your notes like: do you use full sentences? Abbreviations? Symbols? Formal language? Visuals, e.g. mind maps? Have you ever had to write (or been given) the full minutes of a meeting? What are they like? What kind of information do they include? What style are they written in?*

Lead-in

Students talk about the main principles of writing meeting notes and minutes.

1 Explain the activity, go through the list with the class and clarify any vocabulary queries. Ask students to work individually and then compare their lists in pairs before class feedback. Note that the answer key below includes a *suggested* answer; the order of some of the stages may vary. Write the answer in a numbered list on the board.

Possible answers

Welcome
Apologies for absence
Approval of previous minutes
Matters arising (from previous minutes)
Review of agenda
Agenda items (numbered)
Summary of decisions and action points
AOB (Any other business)
Date and time of next meeting

2 Do this as a whole-class activity, checking answers as you go. It is a good idea to record the answers on the board, so students can refer to them when they do the writing task in Exercise 6. For question 1, you could mark the numbered list from Exercise 1, adding ticks and crosses for each item.

1 Welcome, Review of agenda, Summary of decisions and action points (the summary is a verbal reminder; the decisions themselves will be under each agenda item).

2 date, time and location of meeting, attendees (= participants), name of meeting chair, apologies, name of minute-taker

3 what action is required, who will do it and the timeframe

3 Refer students to page 137 and give them time to read the notes and minutes and answer the questions. Get them to compare answers in pairs before checking with the class. During feedback, point out that the extent and amount of information in meeting minutes obviously varies according to company style.

1 The factual information in the full minutes is the same as in the notes. There are just a few extra, unimportant points that come from the minute-taker's memory: LD will write a short report for the client, not just contact them verbally; the light fixtures were discussed at the last meeting.

2 Possible answer: The notes and minutes are effective and contain just about the right information.

Useful language

Students look at strategies and useful language for writing meeting notes and minutes.

4 This is best done as a whole-class activity, checking and discussing answers as you go. After discussing the answers for each item, go through the relevant section of the Useful language box with the class. For question 1, you could let students discuss the words/phrases in pairs first, using their dictionaries if necessary, then clarify meanings with the class. For question 2, list any additional abbreviations students suggest on the board and encourage them to record them in their notebooks. Point out that when taking notes for themselves, students may choose to abbreviate other words to help them write faster; this is perfectly OK as everyone has their own, personal note-taking style. For question 3, again point out that when taking notes, students may choose to omit more/ fewer/different words.

1 cost overrun: when there is an unexpected increase in actual costs compared to budgeted costs; **milestone:** a significant point in a project that marks the completion of a major phase of work; **scope:** the goals of the project in terms of deliverables, features, functions and tasks (in other words, what needs to be achieved to deliver a project)

2 PE, NW, LD, RK, VY, CJ, KN (i.e. all names); cc = carbon copy; qual = quality; mats = materials; elec = electrical; re = regarding; Feb = February; lab = labour; Jan = January; AOB = any other business; regs = regulations; doc = document; info = information; mtg = meeting

3 The architect wants to be included in the email cc list for project updates. This was agreed.
The client is worried about the thermal insulation in the roof. He/She wants pricing on better qual mats. LD will investigate then get back to the client with options. NW spoke to the elec subcontractor re the revised schedule that is due to the problem with the light fixtures. All okay; they can deal with it. There are no manpower issues as the elec crew can be reassigned to other work until things are sorted out.

PE reported that the actual costs to the end of Feb are 2.6% over budget. The main reason is extra lab costs in Jan and Feb due to overtime.
CJ believes the April 20th milestone is still achievable. He/She also says that the small cost overrun can be absorbed within the profit margin, but we will have to monitor it closely.
There are new fire regs coming next year. See the doc on Sharepoint for more info; search for 'fire regulations'. Let CJ know if you can't come.

4 Use short sentences. Use semi-formal vocabulary and grammar structures. Overall, the writing style is simple, direct and concise. You can use some ellipsis, but not as much as in the notes taken during the meeting.

5 Passive forms make the verb impersonal and the style more formal, but are not common in modern business and can look old-fashioned if overused. Many English language learning coursebooks focus on these constructions but in reality, they are only used when it is really necessary to not mention the subject (quite rare in real life) or for reasons of cohesion, simplicity and not repeating a subject too many times (more common).

5 Ask students to do this individually and then get them to compare answers in pairs before checking with the class. During feedback, encourage different students to contribute and elicit or suggest ways to help students with any language they find difficult.

Extra activities 2.5

A This activity gives further practice in writing full meeting minutes from notes. Explain the task, look at the example with the class and ask students to complete the exercise individually. Point out that they need to use the words in brackets and remind them that they can refer to the Useful language box on page 26 of the Coursebook if they need help. Check answers with the class.

1 GW stated that he was very pleased with the positive impact of the new marketing strategy.

2 AH thanked the sales team for their hard work at the Seoul trade fair.

3 The legal department is not happy about the penalty clause because it has the potential to be very expensive for us if we have / there are production problems.

4 There are still big quality issues in relation to the new electric motor currently under test.

5 It was suggested that we reduce the marketing spend on Facebook as our target customers no longer use this platform.

6 Please remind all team members that the IT system will be down for routine maintenance next weekend.

7 The HR department has produced a preliminary job specification for the position of Regional Sales Manager; please send any comments or suggested changes to the HR Director by 4 November.

Optional grammar work

The notes and minutes in Exercise 1 contain examples of ellipsis, so you could use them for some optional grammar work. Refer students to the Grammar reference on page 117 and use the exercises in MyEnglishLab for extra grammar practice.

Task

Students expand meeting notes to write full minutes.

6A Put students in pairs and refer them to the notes on page 138. Give them time to read through the notes and identify the abbreviations, short forms and ellipses. Check all the abbreviations during feedback and elicit a few examples of short forms and ellipses.

HK, YD, SW, ML, PG, DS, PA (names); Obvs (= Obviously); w/ house (= warehouse); re (= regarding); prev (= previously); > (= more than / over); AOB (= any other business); Fri (= Friday); mtg (= meeting)

6B Students should do this individually. Explain the writing task, point out the word limit and set a time limit. Before students begin, remind them to refer to the Useful language box and also to the information on the board (if appropriate). If there is no time to do the writing task in class, it can be set for homework. Note that in places, the model answer below extends the information in the minutes and uses different vocabulary. Students can be encouraged to do the same.

Model answer
Meeting: Project review – construction of new plant.
Client: RELAY
Date: 9 November
Time: 14.00
Present: HK, YD, SW, ML
In chair: PG
Apologies: DS, PA

1 Matters arising
The shortage of workers is now solved because we are working with a new agency. However, we will need to pay higher wages to attract workers from abroad. YD asked how much more money this will be. There was some discussion but no clear figures.

2 Scope
The client now wants the warehouse to be 20% larger. There are some planning issues still to be resolved with City Hall (see next item). Obviously, the client expects to pay an increased cost, but the meeting questioned whether this should be 20% directly in proportion to the extra warehouse area. HK to investigate this and prepare a new budget for the larger warehouse.

3 Schedule
Mr Stephens at City Hall says that he expects to have a response to the planning application in the next few weeks. SW stated that there are no immediate issues regarding the schedule as work on the warehouse has not started yet.

4 Cost
The issues of higher wage costs and an expanded warehouse area were discussed previously. In relation to wages, the meeting felt that there will be some hard negotiations with the employment agency as they will want to keep a big percentage of the fee for themselves.

5 Overall progress
The project is on track. It was pointed out that the client's business is obviously going well, given their request for extra warehouse space. Therefore, we may be able to charge them more than 20% for the extra construction costs involved in building the larger warehouse.

6 AOB
Maria in the Accounts department is leaving the company after 20 years. There will be a small leaving party this Friday, 12 November, at 13.00 in meeting room 3.

7 Next meeting
The next meeting will be on 2 December at 14.00 in meeting room 1.

6C If students write their minutes for homework, you could do this exercise in the next lesson. Put students in the same pairs as for Exercise 6A and ask them to read their partner's minutes and discuss the questions. You could then ask them to write a final, improved version of their minutes, in class or for homework.

MyEnglishLab: Teacher's Resources: extra activities; Writing bank; Interactive grammar practice
Grammar reference: Ellipsis p.117
Workbook: p.13

Business workshop ➤2
Achieving a circular economy

GSE descriptors

- Can follow presentations on abstract and complex topics outside their field of interest.
- Can infer opinions in a linguistically complex presentation or lecture.
- Can identify a speaker's point of view in a linguistically complex presentation or lecture in their field of specialisation.
- Can discuss the information presented in a complex diagram or visual information.
- Can contribute to a group discussion using linguistically complex language.
- Can discuss a plan of action for dealing with a work-related task.
- Can make a detailed, formal, evidence-based argument in a presentation or discussion.

Background

Students read about an initiative for the implementation of a circular economy in a city.

1 Go through the questions with the class and teach or elicit the meanings of *initiative* and *address* (v.). Then put students in pairs and ask them to read the background and answer the questions. During class feedback, help students with any unknown words from the text and also feed in information from the Note below as required. Note that the term *district heating* in the text refers to a method of providing heat and hot water from a central source to a number of buildings or a district. It uses pipes under the streets and eliminates use of fossil fuels such as coal or wood being burned in individual buildings and homes.

1 To continue using products and materials for as long as is possible, to reduce waste and to reuse products at the end of their life cycle.
2 They separate their rubbish and bring items to a recycling centre.
3 Repair rather than replace appliances, use district heating and use alternative forms of transport.
4 Production and sourcing of raw materials that are environmentally friendly, how food is handled and what happens to food wastage, how critical materials are recovered from products, dealing better with the waste from constructing and tearing down buildings and using bio-products for energy rather than fossil fuels.
5 People are aware of the issues, but it means having to change the way they live or do business to comply with the initiatives.

Note

As the European Union is concerned about the impact of products and waste on the economy, they have produced a document to help a transition to a more circular economy. The goal is to keep products, materials and resources for as long as possible and to reduce the amount of waste produced. This, in turn, will lead to a more sustainable and competitive economy. Through this initiative, it is hoped that industry can be protected from a lack of resources, that they will find more efficient ways to produce goods and that new jobs will be created. In addition, being more careful about both production methods and waste treatment will help the environment and cut back on air and water pollution. The report notes that industry, consumers and governments must all work together towards the same goals in order to achieve a more balanced system for production, use and recycling of products.

Dealing with the problems

Students listen to a talk on implementing a circular economy.

2 ◄) BW 2.01 Explain the activity, go through the topics in the box with students and check they understand each one. Point out that students need to identify *two* topics which are not mentioned in the talk. Play the recording, then check answers with the class.

innovative new materials, overflowing landfills

3 ◄) BW 2.01 Give students a minute to read the questions so that they know what they need to listen for. Encourage them to make notes in answer to the questions while listening and play the recording, twice if necessary. Check answers with the class.

1 Scientists feel that the damage being done will be irreversible.
2 The speaker feels it will help the economy, create jobs and encourage innovation, sustainability and competition between companies.
3 The speaker says that they are a step in the right direction but we must continue working towards sustainability as much as possible.

4 The speaker says that what they buy or don't buy has a huge impact on the economy.
5 The speaker feels that governments are responsible for providing information, creating initiatives, setting up regulations and raising awareness of environmental issues among citizens.

Sharing perspectives

Students read reports on an initiative for the implementation of a circular economy.

4A Put students in groups of three and explain that they are each going to read a report on the initiative they read about in the Background. Explain that each report is written from a different perspective: that of an industry representative, a citizens' representative and the municipal government. Tell students that they should work individually for this stage, refer them to their respective reports and give them time to read the texts and make notes on the most important points.

4B Explain the task and before students begin, give them 2–3 minutes to think about the questions and decide what they are going to say about each one. Tell them that they can refer to their notes but should not read them aloud; they should use their own wording as far as possible. Conduct a brief feedback session discussing the answers to the questions with the whole class.

Possible answers

A Industry representative
1 follows regulations and has reduced emissions; treats waste properly; is looking for ethical sourcing of resources
2 finds it difficult to create longer product life cycles without significantly increasing costs; also difficult to check the supply chains all along the line because price, quality and reliability are priorities; difficult to find natural resources without long-distance transportation
3 says that there will always be questions about repairing or replacing products, as it is difficult to say when it is worth doing so; also says they need to make a profit to pay their employees, their taxes, etc.
4 they are looking into finding more natural resources locally and working on products with longer life cycles

B Citizens' representative
1 is able to use the waste management system for everyday waste and uses public transport
2 has found it difficult to use the subsidy programme to repair an appliance as it was too complicated; also a problem to recycle certain goods if you don't have a car
3 says that public transport is generally very good, but it doesn't run frequently enough in the evenings and at weekends; thinks solar-powered parking ticket dispensers are positive; welcomes the various initiatives from the government, but thinks the city needs to do more to raise awareness
4 provide more public transport for those who come from outside the city so they don't need to use their cars; people can reduce waste by not buying items in plastic packaging

C Government representative
1 has begun a programme of subsidies for repairing rather than replacing household goods; has introduced funding for solar panels; has tried out a programme for installing photovoltaic panels; offers electric taxis, a car-sharing service, reduced-price annual ticket for public transport

2 is finding it difficult to get more people to use the 'repair' subsidies or put up solar/photovoltaic panels; there are still too many cars in the city centre

3 feels that most people are careful about waste separation; it's good that shops take back batteries, etc.; also says that the reduced-price annual ticket for public transport is a success

4 they are looking into ways to expand the subsidy for repair initiative and the initiatives for solar/photovoltaic panels; also to find a way to have car-free days and keep cars out of the city centre

---

Extra activities Business workshop 2

A This activity looks at useful vocabulary from the listening. You could get students to complete it individually and then clarify meanings during class feedback or, if time is short, you could do it as a whole-class activity, checking answers as you go.

> **1** d **2** h **3** a **4** f **5** c **6** b **7** e **8** g

---

An infographic

Students summarise and discuss information from the listening and reading texts.

5 Students should do this activity in the same groups as for Exercise 4. Explain the task and look at the examples with them, then give them time to complete the infographic in their groups. Check answers with the class. Depending on the resources you have available, you could project the infographic onto the board and write (or invite students to write) the answers there.

> **Possible answers**
>
> **Industry** (left to right): ethical sourcing of raw materials; treating waste; repair of goods; *reducing emissions*
>
> **Citizens** (top to bottom): waste separation; public transport; shopping with reusable bags; *drop-off centre*
>
> **Government** (top to bottom): electrical charging station(s); *food bank*; recycling centre(s); solar energy

6 Depending on the time available and the size of your class, students could do this in the same groups as the previous two activities or as a whole class. If they work in groups, allow plenty of time for them to discuss and rank the priorities, then invite students from different groups to share their answers (and reasoning) with the class.

Extra activities Business workshop 2

B This activity practises collocations from the listening. Explain the task and point out that there is one extra word in each box. You could go through the words in the boxes with students before they begin or let them attempt the exercise individually, using their dictionaries if necessary, then clarify meanings during class feedback.

> **1** raise awareness **2** district heating **3** fossil fuels
> **4** household appliances **5** inconsistent labelling
> **6** planned obsolescence **7** waste management
> **8** food banks **9** photovoltaic panels
> (Not needed: supply chain)

Task: Make an action plan

Students make an action plan for the implementation of a circular economy in a city.

7A Put students in groups of three, assign roles A–C and explain the scenario: in their roles as citizen and industry representatives and the municipal government, they are going to hold a meeting in order to create an action plan for future initiatives for the Austrian city. Go through the instructions with them and point out that they each need to make notes on specific points for the meeting. Refer them to their respective information and give them time to read it and prepare for the roleplay, while you monitor and provide help. In weaker classes, you could group students with each role together for this preparation stage and then get them to regroup to hold their meetings.

7B Students now hold their meetings. Go through the instructions with them and set a time limit for the meetings before they begin. To help them, you could list the points they need to discuss and bear in mind on the board, for them to refer to during the activity:

Action plan satisfying ALL three groups:
• how to raise awareness?
• how to work together?
• time frame?

7C In the same groups, students now create a poster outlining their action plan. Encourage them to be creative and remind them that they can use the infographic in Exercise 5 and the information on their role cards to help them express their ideas. Set a time limit and during the activity, monitor and provide help as necessary. Then, they take turns to give a brief presentation to the class (or, in larger classes, to one or two other groups). The class then vote for the best action plan(s).

MyEnglishLab: Teacher's Resources: extra activities

Review ◀ 2

> **1** 1 a 2 b 3 b 4 a 5 b 6 b 7 a 8 b 9 b 10 a
> **2** 1 throwaway culture 2 dispose of 3 upgrade
>   4 from the outset / precious metals (although it is unknown whether the company products contain precious metals) 5 forward-thinking
> **3** 1 having 2 been 3 to 4 being 5 have 6 to
>   7 had 8 been
> **4** 1 appreciate 2 make 3 concerns 4 through
>   5 relevant
> **5** **Possible answers**
>   1 Adi Ratna reported that sales figures were up 6% year on year and (that) the product line launched in January is (was) selling very well.
>   2 The chair suggested that the Marketing Director should prepare a report for the next meeting on the effectiveness of our social media promotions.
>   3 There could be problems in the future in relation to (regarding) production volumes unless we build another assembly line in the Bratislava plant.
>   4 Vanya Bakshi said that the technical specifications for the new website are (were) not ready yet, but the full details should be available for the web development team by the end of next week.
>   5 The next meeting is in conference room (number) 2 on Thursday 18 May at 9 a.m. Please let Christine Jones know if you can't come.

3 > Finance and investment

Unit overview

	CLASSWORK	FURTHER WORK
3.1 > **How traders work**	**Lead-in** Students talk about finance and investment. **Video** Students watch a video about financial traders in London. **Vocabulary** Students look at vocabulary related to finance and investment. **Project** Students give a presentation on different types of investment.	**MyEnglishLab:** Teacher's resources: extra activities **Teacher's book:** Resource bank Photocopiable 3.1 p.146 **Workbook:** p.14
3.2 > **Financial investments**	**Lead-in** Students talk about investment and different financial products. **Listening** Students listen to people giving investment advice. **Grammar** Students study and practise different expressions to talk about the future. **Speaking and writing** Students write an email about personal investments and then discuss the ideas in it.	**MyEnglishLab:** Teacher's resources: extra activities; Reading bank **Spoken English:** p.113 **Grammar reference:** p.117 Expressing attitudes to the future **Teacher's book:** Resource bank Photocopiable 3.2 p.147 **Workbook:** pp.15–17
3.3 > **Communication skills:** Challenging facts politely	**Lead-in** Students talk about fact-checking. **Roleplay** Students roleplay a discussion in which they have to challenge facts and data for an article. **Video** Students watch a video of a discussion during which the participants challenge facts and data. **Reflection** Students reflect on the conclusions from the video and their own approach to challenging facts and data.	**MyEnglishLab:** Teacher's resources: extra activities; Interactive video activities
3.4 > **Business skills:** Exploring options	**Lead-in** Students talk about using questions as a negotiation strategy. **Listening** Students listen to people negotiating an office lease. **Useful language** Students look at strategies and useful language for using questions in negotiations. **Task** Students hold a meeting to negotiate time involvement on a project.	**MyEnglishLab:** Teacher's resources: extra activities; Useful language bank
3.5 > **Writing:** Budget report	**Lead-in** Students read and discuss the main content and structure of a budget report. **Useful language** Students look at useful language for, and the main structure of, a budget report. **Task** Students write the executive summary of a budget report.	**MyEnglishLab:** Teacher's resources: extra activities; Writing bank; Interactive grammar practice **Grammar reference:** p. 118 Modal verbs: possibility **Workbook:** p.18
Business workshop 3 > Financial strategy	**Reading and Listening** Students read a text and listen to a conversation about different financing options for a company. **Task** Students discuss different financing options and write an email requesting financial support.	**MyEnglishLab:** Teacher's resources: extra activities

Business brief

The main aim of this unit is to familiarise students with the way financial markets work, the functioning of the international stock market and the role of traders. It also looks at different ways investors can choose to invest their money and how these various types of investment work.

Finance is the science of money management and is chiefly concerned with the allocation or **investment** of **assets** over time, usually with an element of **risk** or uncertainty involved. **Trading** is the buying and selling of **securities**, tradeable financial assets such as **currencies** on the New York Stock Exchange. Someone who buys and sells currencies, **bonds** or other **commodities** in this way is known as a **trader**.

Traders in financial markets aim to buy and sell financial assets based on three factors: their risk level, their fundamental value and the expected **rate of return** (**return on investment** or **ROI**), the profit gained from the investment. Traders in the global foreign exchange market trade in the change in value of one currency against another (**ForEx trade**). For example, a trader who believes the euro will increase in value against the U.S. dollar will buy euros with U.S. dollars. If the exchange rate rises, he or she will then sell the euros, making a profit.

The **commodities market** trades in raw materials and agricultural crops rather than manufactured products. Examples are coffee beans, cocoa, fruit and sugar and **hard commodities** such as gold, rubber and oil. Commodities traders can access about 50 major commodity markets worldwide although nowadays, physical trades in which goods are actually delivered are increasingly rare.

Stocks and bonds are two of the most traded items. Stocks are **shares** in a publicly traded company and investors who invest in a particular company are **shareholders** who receive a share of the company's profits in return. **Bonds** are a form of **fixed-income loan** the investor makes to a government or corporate entity. Securities sold on the bond market are essentially all various forms of debt; the investor is effectively lending money for a set period and charging interest, like a bank.

Financial advisers recommend investment options for clients that are best suited to their individual circumstances and financial goals. This may be in currencies, commodities, stocks and bonds or other securities like commercial property. Since no one can consistently predict which assets or sectors will perform best, most advisers recommend spreading investment across a range of different assets in an **investment portfolio**. This avoids an investor having too much money invested in one type of asset. Many also recommend investing regular **premiums** (regular small amounts) as a more sensible way to invest, rather than making large, high-risk, one-off payments and always keeping a **cash buffer** to avoid being forced to sell stocks when prices are falling and thereby losing money.

Two of the most frequent ways investors lose substantial sums are when there's an **economic bubble**, or if they invest in a **Ponzi scheme**. The first starts when there's a sudden surge in the price of a group of stocks which isn't linked to any real increase in the value of the commodity. Driven by rumour, more and more people are encouraged to buy the stocks, which continues to push up the price. Eventually the number of new investors wanting to buy the stocks at the higher prices diminishes and a massive **sell-off** occurs, leaving most of the investors with worthless stocks.

The second is actually a form of **fraud** which often attracts inexperienced investors. They are tricked into believing that a rise in the value of their stocks is the result of product sales, when actually it's due to new investors being persuaded to finance the scheme. The stocks appear to do well as long as there are always new investors to put money into the scheme and investors don't demand repayment. As soon as investors do demand repayment, however, the whole scheme collapses.

Finance and investment and your students

Some students may have made investments in stocks, commodities or currencies themselves, or have family members who have done so. Older students may have invested in a property. Most will know something about international exchange rates if they have ever gone abroad on holiday and so be able to understand something of currency markets. In-work students may also have received financial advice about pension investments, or how best to maximise their savings.

Unit lead-in

Draw students' attention to the unit title, teach or elicit the meaning of *finance* and elicit a brief description of the photo. Then look at the quote with the class and check that they understand the meaning of *interest*. Briefly discuss the quote with the class: What do students think it means? (A possible answer might be that educating yourself will repay you in the future.) Do they agree? Why / Why not?

3.1 ❯ How traders work

GSE descriptors

- Can extract specific details from a TV programme on a work-related topic.
- Can contribute fluently and naturally to a conversation about a complex or abstract topic.
- Can make a linguistically complex business presentation with the help of notes.
- Can evaluate the strength of a speaker's assumptions in a linguistically complex presentation or lecture.

Warm-up

Discuss these questions with the class: *What springs to mind when you hear the word 'investment'? How important is it to invest? Would you consider hiring a professional to help you manage your investments / achieve your financial goals?* If there is time, you could get students to discuss the questions in pairs or small groups first, then get brief feedback from the class.

Lead-in

Students talk about finance and investment.

1 Depending on the size of your class and the time available, you could ask students to discuss the questions in pairs or small groups first, then invite different students to share their ideas with the class. Alternatively, if time is short, discuss the questions with the whole class.

> **2** City of London, Frankfurt, Wall Street (New York City), Singapore, Tokyo, etc.
> **3** They buy and sell financial products, e.g. stocks, bonds, shares and assets, on behalf of investors, i.e. investment banks, large organisations, wealthy individuals, etc.

Video

Students watch a video about financial traders in London.

2 ▶ 3.1.1   Tell students that they are going to watch a video about financial traders in London and give them time to read the questions so they know what they need to watch for. Check they understand the meaning of *infer* and *hunch* and if desired, teach this vocabulary from the video: *(the) noughties, level* (adj.), *foreign exchange (market), read between the lines, high-risk* (adj.). Play the video, then check answers with the class.

> **1** Trading used to be done on the physical trading floor of a stock exchange. It is now mostly done from a computer.
> **2** It's an announcement about interest rates in the Eurozone.
> **3** That the European Central Bank was closer to cutting rates this month than they were the month before.
> **4** Eight thousand dollars.

3 ▶ 3.1.1   Explain the activity and give students time to read the summary before they watch again. Play the first part of the video (0:00–1:54), then get students to compare answers in pairs before class feedback.

> Traders these days work from computers rather than a trading floor or 'pit'. Piers Curran has been a trader since the early ~~nineties~~ noughties. He says he wouldn't do the job if he ~~didn't have physical presence~~ was ten years older. Markets are more competitive than before. Piers ~~works for~~ works with / runs his business together with Will De Lucy. They trade the ~~European Central Bank~~ (global) foreign exchange market, trying to make money from the change in value of currencies. A global event today might affect the ~~British pound~~ euro and it's a ~~wasted~~ perfect opportunity for traders to make money. The main issue is whether the Central Bank will lower ~~lending~~ interest rates or not in order to help the Eurozone.

4 ▶ 3.1.1   Explain to students that they are now going to watch the second part of the video and complete the extract with two to four words in each gap. Point out that contractions count as one word. Give them time to read the extract, then play the video (1:54–end) and check answers with the class.

> **1** reading between the lines
> **2** their every word
> **3** what moves markets
> **4** the euro to drop
> **5** we haven't cut rates
> **6** 5.3 trillion dollars
> **7** than physically exists
> **8** bought more back
> **9** tiny timeframe
> **10** risk and reward

5 Put students in pairs or small groups and give them 3–4 minutes to discuss the questions. Then invite students from different pairs/groups to share their answers with the class.

Extra activities 3.1

A ▶ 3.1.1   This activity provides students with extra listening practice. Give them time to read the statements and options before they watch again, then play the video. Alternatively, if you think your students may remember some of the information, you could ask them to answer as many of the questions as they can before watching again, then play the video for them to check/complete their answers. Check answers with the class.

> **1** b **2** c **3** a **4** b **5** a **6** b **7** a **8** b **9** c

Alternative video worksheet: Investing in art

1 ▶ ALT3.1.1 Students could discuss the questions in pairs first, if there is time, or you could discuss them with the whole class. Elicit a few ideas around the class, then play the first part of the video (0:00–1:34) for students to check their answers.

> **1** Students' own answers
> **2** Possible answer: Sotheby's and Christie's are international auction houses. At auction houses, paintings, antiques, etc. are sold to the person who offers the most money for them.

2 ▶ ALT3.1.1 Explain the activity and give students time to read the statements. Play the first part of the video again (0:00–1:34), then check answers with the class. Students may need to watch the video twice for this activity: once to decide if the statements are true or false, then a second time to correct the false statements.

> **1** T
> **2** F (*Van Gogh's painting* L'Allée des Alyscamps *sold for $66.3 million in 2015.*)
> **3** T
> **4** F (*There has been a rise in the number of investors investing in art recently.*)

3 ▶ ALT3.1.1 Tell students that they are going to watch the rest of the video and explain that the sentences here are summaries of what they will hear. Point out that they need to fill the gaps with numbers or units and give them time to read the summaries before they watch. Play the video (1:35–4:38), twice if necessary, then check answers with the class.

> **1** thousand, $400 million **2** ten **3** 5 percent / 5%
> **4** 120,000, two million

4 This exercise practises useful vocabulary from the video. Explain the task and point out that the first letter of each missing word is given. Also point out that students may need to change the form of some words. In weaker classes, you could give students the missing words in a list on the board and ask them to complete the sentences with them. Check answers with the class, clarifying meanings as necessary.

> **1** sum **2** investors **3** houses **4** artworks **5** asset
> **6** stream **7** haven **8** gain

5 This activity looks at useful collocations from the video. You could do it as a whole-class activity, checking answers and clarifying meanings as you go. Alternatively, let students attempt the exercise individually, then check answers with the class, clarifying meanings as necessary.

> **1** art **2** investment **3** money

6 Get students to complete the exercise individually and then to compare answers in pairs. You could play the video again for students to check their answers or go over them with the class, clarifying meanings as necessary.

> **1** money, investment funds **2** liquid asset
> **3** income, dividends, rental **4** hedge, currency
> **5** rocket, profit **6** monetary gain

7 Put students in pairs or small groups and give them 3–4 minutes to discuss the questions. When they are ready, invite students from different pairs/groups to share their answers with the class.

8 Put students in new pairs or groups and give them 3–4 minutes to discuss their ideas. Remind them that they need to give reasons for their choices. When the time is up, invite different students to share their decisions (and reasons) with the class.

Vocabulary: Finance and investment

Students look at vocabulary related to finance and investment.

6 This is best done as a whole-class activity: elicit the answers, then go through the collocations in the box with students, clarifying meanings as necessary. During feedback, highlight that the noun *trade* usually collocates with *make*.

> **a trade:** a high-risk trade, a trade-off, put a trade on
> **b make:** make a good trade, make a return, make five trillion

7 Draw students' attention to the dials and explain that they need to turn them in a clockwise or anti-clockwise direction in order to form collocations using all the words. Refer them to the example and make sure they understand that the dial has moved in an anti-clockwise direction once. Say: *We have moved the dial one place, so what is the collocation with*

'business'? (*business partner*). Elicit the remaining three collocations (*human resources, stock market, mobile phone*). You may wish to point out that turning the example dial in a clockwise direction three times is also possible. Give students 2–3 minutes to find and write the collocations in each dial; they could do this individually or, in weaker classes, in pairs. They can turn the dials in either direction to form the correct collocations (see answer key below). Check answers with the class, clarifying the meanings of the collocations as necessary and encourage students to record the collocations in their vocabulary notebooks.

> **Dial 1:** Turn the dial one segment in a clockwise direction or three segments in an anti-clockwise direction.
> financial instruments, trading floor, interest rates, foreign exchange
> **Dial 2:** Turn the dial two segments in a clockwise or anti-clockwise direction.
> be in profit, reap rewards, make an investment, deal in currencies

8 Depending on the level of your class and the time available, you could go through the words in the box with students before they attempt the exercise or let them use their dictionaries to complete the sentences, then clarify meanings during feedback.

> 1 backer 2 yield 3 ROI (= return on investment)
> 4 ballpark figure 5 level playing field

9 This activity practises vocabulary from the previous exercises in this section, so students should be able to do it individually. Give them 3–4 minutes to complete the task, then check answers with the class. During feedback, check that students understand the meaning of the incorrect options as well.

> 1 level playing field 2 reap the rewards
> 3 financial instruments 4 interest rate
> 5 make a trade-off 6 on the trading floor

10 Explain to students that they are going to prepare and give a presentation on how best to invest 100 million euros. Before they do, they need to discuss the questions here. Give them a minute to read the questions and check that they understand *dedicated software* in question 2. Put them in pairs and give them 3–4 minutes to discuss the questions, then get brief feedback from the class.

Extra activities 3.1

B This activity gives further practice of key vocabulary from the lesson. Students should do it individually, as consolidation. Encourage them to read the whole text quickly first, before they attempt the exercise. Check answers with the class.

> 1 investment 2 makes 3 rewards 4 investors
> 5 make 6 profit 7 rate

Project: Presentations: types of investments

Students give a presentation on different types of investment.

11A Put students in pairs or small groups and go through the instructions and steps 1–3 with them. When going through step 1, draw their attention to the chart and check that they understand how these products work:

- bond: an official document promising that a government or company will pay back money that it has borrowed, often with interest.
- mutual fund: an arrangement managed by a company, in which you can buy shares in many different businesses (note that the British English synonym is *unit trust*).
- collectible (or collectable): something that is likely to be bought and kept as part of a group of similar things, especially because it might increase in value.

You could give students the definitions above as a glossary or write them on the board for students to refer to during the activity. Alternatively, you could read out the definitions and ask students to call out the correct product from the chart.

Encourage students to look up information on the returns and risks of various investment options. They will probably have parents or relatives who have made investments and have given them advice about investing in property or stocks. For example, real estate and art are usually considered safe options, depending on the market. Investing in a precious metal such as gold has also traditionally been considered a good investment. There might be a higher risk in investing in antiques, depending on the quality of the items or students' knowledge of the market. Buying stocks/shares will depend on the kind of company students want to invest in, e.g. an exciting new tech start-up might have higher risk but also higher returns over ten years, although there could be a strong possibility of it failing in the first few years. A more established company might be considered a safer investment as it will have lower risk; however, it will probably also have a larger number of investors and lower returns. Refer students to internet sources on investing, if you have them, and encourage students who are less interested in finance to research real estate, local farmland, art works or antiques (e.g. vintage cars), to see what they could buy for their money.

Allow plenty of time for this preparation stage and encourage students to make notes for their presentations. Point out that the students in each pair/group should take turns to give their presentations in the next stage and remind them that they should try to use 'financial language' from Exercises 6–8 as far as possible.

11B Pairs/Groups now take turns to give their presentations. The rest of the class should listen and make notes for (a) and (b) in the instructions. When all pairs/groups have given their presentations, students as a class should discuss the best investment plans and choose the pair/group that will probably make the highest returns, but have chosen high-risk investments and those whose choices sound like safer, low-risk investments but will yield a steady ROI.

Model answer

For our second investment, we plan to put 10,000 euros into an exciting new start-up that has been created by classmates. It's a tech business that … We're hopefully looking at a ROI of between 30 and 40% in five to ten years' time.

Finally, we'd like to invest the remaining 10,000 euros into collectible art. Our preferred artwork is a painting/ photograph by … We consider it a wise investment. We hope to hold on to this investment for many years and we are sure it will give us a steady rate of return of 20% over ten years.

MyEnglishLab: Teacher's Resources: extra activities
Teacher's book: Resource bank Photocopiable 3.1 p.146
Workbook: p.14

3.2 ❯ Financial investments

GSE descriptors

- Can follow the main points in a linguistically complex interview, if provided with written supporting material.
- Can use 'on the point of' (Br Eng) to talk about imminent events in the present and past.
- Can develop a written case to persuade others about the advantages or disadvantages of a course of action.
- Can describe reactions to different work-related scenarios in detail.

Warm-up

Put students in pairs or small groups and draw their attention to the photo. Ask: *Do you think it's better to save or invest your money? Why?* Give students a few minutes to discuss the question in their pairs/groups, then invite different students to share their views with the class. Encourage them to give reasons.

Lead-in

Students talk about investment and different financial products.

1A In a strong class, let students attempt the matching task individually first, then check answers with the class, clarifying the meanings of the financial terms in the box as necessary. In weaker classes, you could do this as a whole-class activity, checking answers and clarifying meanings as you go.

> **1** volatility **2** assets **3** liquidity **4** equities **5** bonds

1B Give students a minute to read the questions and teach or elicit the meanings of *diversify* and (*investment*) *portfolio*. You could then get them to discuss the questions in pairs or small groups and get brief feedback from the class after 2–3 minutes, or, if time is short, you could discuss the questions with the whole class.

Possible answers

2 It's wise to invest in a range of different stocks and sectors to diversify an investment portfolio. This is because some stocks/shares are riskier, but may offer high returns in the short term (e.g. high-tech companies), while other safer stocks/shares (e.g. in 'blue-chip' and more established companies) may offer lower returns, but provide good opportunities for investments to grow steadily over time.

3 Information technology, or tech companies, healthcare and financials (financial companies) are the most popular with investors (at the time of writing).

❯ **Spoken English**
p.113: Don't put all your eggs in one basket

1 ◀》SE3 Explain to students that they are going to hear people discussing question 1 from Exercise 1B and refer them to the question. Explain the activity and give students time to read the questions so they know what to listen for. Encourage them to make notes in answer to the questions while listening, then play the recording, twice if necessary. If you think your students will find this activity hard, you could get them to compare answers in pairs before class feedback, referring to audioscript SE3 on page 162.

Possible answers

1 A personal pension plan; an individual savings account, a house with the mortgage paid off.
2 The interest is tax free.
3 Advantage: a return is guaranteed. Disadvantage: the return on investment is low.
4 A low rate of return won't keep up with inflation.
5 You have to be patient and invest over the long term; the stock market always goes up over time.

Expressing attitudes to risk taking

2 ◀》SE3 Explain the activity and give students time to quickly read through the extracts before they listen. Again, if you think your students will find it hard to make out what the speakers are saying, you could let them refer to the audioscript to check/complete their answers. Check answers with the class.

> **1** You know where you are.
> **2** it's fairly stable
> **3** that's not a guarantee
> **4** I can't take those kind of risks
> **5** that's not going to work
> **6** I don't want to test that
> **7** it's still a chance you're taking

3 This activity looks at different ways to refer to safer and riskier investments. Get students to complete it individually, then check answers with the class. Alternatively, if time is short, you could do it as a whole-class activity, checking answers as you go.

> **a** 1, 2
> **b** 3, 4, 5, 6, 7

Listening

Students listen to people giving investment advice.

2 ◀)) 3.01   ◀)) 3.02   ◀)) 3.03   Explain the task and give students time to look at the list of topics a–c. Play the recordings, then check answers with the class.

> **1** b   **2** c   **3** a

3 ◀)) 3.01   Explain that students are going to listen to the first speaker again and answer the questions. Give them time to read the questions and check that they understand *making the most of*. Encourage them to make notes in answer to the questions while listening and play the recording, twice if necessary. Check answers with the class. During feedback, you may wish to elaborate on these financial terms mentioned in the recording: *bull run*: a period of time when prices rise on a financial market; *equity markets*: the markets where equities are traded (stocks/shares are also called equities).

> **1** no (*... nobody can predict which assets or sectors will perform best*)
> **2** diversification (*... spreading your money across different assets, such as equities, commercial property and cash*)
> **3** It has changed and bonds are linked to equities (*In the past, bonds and equities were less linked and investors expected to do well in bonds when equity markets were falling. But bond markets are not the same anymore. The likelihood is that you will lose money less slowly with bonds in a downturn but you are very unlikely to make any money.*)
> **4** investing regular premiums (*... regular small amounts ... rather than making large and high-risk one-off payments*)

4 ◀)) 3.02   Tell students that they are going to listen to the second speaker again and explain the task. Give them a minute to look at the four topics and check that they understand *cash buffer* and *inflation*. Play the recording, twice if necessary, then check answers with the class.

> **Possible answers**
> **1** An investor won't be forced to sell when prices are falling.
> **2** Everybody should have some cash savings to deal with emergencies. Have enough cash to live on for at least 3–6 months.
> **3** Inflation eats away savings. If there is annual inflation of 2.5 percent, savings of 10,000 euros will end up with the purchasing power of 5,310 euros.
> **4** Low-interest savings accounts won't leave you with enough money when you retire. If your goal is to retire comfortably, cash is unlikely to deliver.

5 ◀)) 3.03   Explain the task and give students time to read the statements before they listen. Play the recording, then check answers with the class. In stronger classes, you could ask students to correct the false statements; they may need to listen again for this.

> **1** T
> **2** T
> **3** F (*I plan on having a diversified portfolio based on <u>my retirement age</u>.*)
> **4** F The third error is mental accounting. (*It... is how we evaluate our finances depending on the money's source ..., whether it's from an inheritance, our monthly pay, or a credit card. And how you plan to use it.*)
> **5** T
> **6** F (*... gamblers refer to any money they have just won as 'house money'. They then use this money to gamble again and, inevitably, lose it.*)
> **7** T

Extra activities 3.2

A ◀)) 3.01   ◀)) 3.02   ◀)) 3.03   This activity provides students with extra listening practice. Explain the task and point out that students need to use between two and four words in each gap. Give them time to read the extracts quickly before they listen, then play the recording, twice if necessary. Check answers with the class.

> **1** effect on your finances
> **2** equities, commercial property
> **3** comes with extra costs   **4** less liquidity
> **5** a better chance of   **6** a smaller amount of
> **7** a short-term goal   **8** loss aversion
> **9** low-interest accounts
> **10** enough money in retirement
> **11** can beat the market   **12** disposable income

6 Refer students to the two statements in the boxes, ask them which one is true for them and ask for a show of hands for each statement. Then put students in pairs; ideally, students who chose the same statement should work together. Give them 3–4 minutes to discuss their questions in their pairs, then invite students from different pairs to share their ideas with the class.

Grammar: Expressing attitudes to the future

Students study and practise different expressions to talk about the future.

7A Draw students' attention to the heading *Expressing attitudes to the future* and ask them what tenses/verb forms they know of to talk about the future. Briefly elicit a few answers around the class (e.g. *going to, will*, Present Continuous, Present Simple), then explain to students that we can also use certain words and phrases to refer to the future. Explain that the words/phrases in bold in statements 1–8 are examples of such phrases and explain the activity. Ask students to do the exercise individually or, in weaker classes, in pairs, then check answers with the class. During feedback, clarify meanings as necessary, pointing out the *-ing* form after *plan on* and *be on the point of* and the infinitive after *be about to* and *be bound to*. Refer students to the Grammar reference on page 117 and go through the explanations and examples with them, clarifying any points as necessary.

> **1** b   **2** d, e (both apply to 'expected to do')   **3** c   **4** d   **5** d
> **6** a   **7** d   **8** a ('on the point of spending'), b ('are bound to lose')

7B Ask students to do this individually and, if time allows, get them to compare answers in pairs before class feedback.

> **Track 3.01**
> *... if you have too much money in one area that doesn't do well, it's **bound to** have a negative effect on your finances. (b)*
> *... but you **are very unlikely to** make any money. (b)*
> ***... the expectation is that** diversification comes at extra costs such as more volatility, higher charges and less liquidity. (c)*
>
> **Track 3.02**
> *Any good investor **is likely to** have some cash to protect them against falling stock markets. (b)*
> *... they can **expect** to have a better chance of avoiding losses. (d)*
> *I **expect** many people will have more cash than that ... (d)*
> *However, if your goal is to retire comfortably, then cash **is unlikely to** deliver. (b)*
>
> **Track 3.03**
> *I **plan** to invest in the middle ground. (d)*
> *... we're more **likely to** see it as disposable income ... (b)*

8 Again, students should work individually. Remind them that they can refer to the Grammar reference on page 117 if they need help and if necessary, do the first item as an example with the class. Check answers with the class and clarify any errors as necessary.

> **1** predict **2** strong possibility **3** on the point of
> **4** planning **5** envisage that **6** expecting **7** unlikely
> **8** likelihood **9** hope to **10** 'm expected

9 Put students in pairs and give them 3–4 minutes to discuss the questions. Encourage them to use words/expressions from Exercises 7A and 7B. As feedback, invite students from different pairs to share their answers with the class.

Extra activities 3.2

B This activity gives further practice of phrases and structures for expressing attitudes to the future. Ask students to complete it individually, as consolidation. If there is time, get them to compare answers in pairs before class feedback.

> **1 a** diversify **b** diversifying
> **2 a** expect / are expecting **b** is expected to
> **3 a** are likely to **b** is unlikely (that)
> **4 a** is bound to **b** are about to
> **5 a** on the point of **b** on the verge of
> **6 a** anticipates / anticipated / was anticipating
>   **b** (an) anticipation
> **7 a** is a likelihood (that) **b** is a (strong/slight)
>   possibility (that)

C Students should do this exercise individually. Encourage them to read the text quickly first, before attempting to complete the gaps. Remind them that for each answer, they should think about meaning as well as the type of word needed to complete the sentence. To help them, you could tell them that all the gapped words complete phrases or structures which express attitudes to the future. Check answers with the class and for questions 6 and 7, try to elicit both possible answers.

> **1** verge **2** to **3** possibility **4** to **5** anticipate
> **6** bound /likely **7** likelihood/chance **8** expected

Writing and speaking

Students write an email about personal investments and then discuss the ideas in it.

10A Explain the scenario and writing task and if time allows, brainstorm some possible investments with the class. Encourage students to be creative; the more unusual the investment, the better, e.g. investing in government bonds in a failing economy; purchasing a derelict hotel in a remote region; buying an expensive work of art by an unknown artist or a broken-down vintage car; investing in the failing business of a friend or relative; financing a personal film project. Point out that students are writing to a friend, so they should write an informal email and encourage them to use words/ expressions with future meaning from Exercises 7A and 7B. In weaker classes, you could let students plan their emails in pairs. While they are writing, monitor and offer help as necessary. If time is short, you could assign the writing task as homework and then do Exercise 10B in the next class.

> **Model answer**
>
> Dear Pete,
>
> How are you? I hope you're well. I'm writing to you because I recently inherited €1,000,000 from a distant relative. I know you know something about investments, so I wanted to ask what you thought. I'm thinking of investing as follows:
>
> Property: I'd like to invest €500,000 in purchasing commercial property. I'm currently looking at an office near the city centre. Do you think I should buy a larger property that needs further investment or a smaller property that has already been refurbished? It is likely to be used for my new venture: a film production company – see below!
>
> Equities: I understand that equity markets are having a bull run at the moment and I hope to invest approximately €200,000 in stocks. I'd prefer to invest in companies in the information technology or energy sector. Have you got any advice as to which one(s) you envisage doing well over the next ten years? In addition, I'm not sure whether to invest by paying one-off payments or premium payments. What do you recommend?
>
> Film project: My third investment of €200,000 is for a film project. Some of my university friends have written a screenplay and they need backers to produce the film. I realise this is a high-risk investment, but we expect it will be a low-budget movie filmed in our hometown. What do you think is the likelihood of a return on investment for this project?
>
> Finally, I plan on keeping the remaining €100,000 in a low-interest savings account.
>
> Maybe we could meet up some time to catch up. I'd love some advice on how best to invest!
>
> Best wishes,
>
> Rachel Spielberg

10B Put students in pairs and explain that they should now read each other's emails and tell their partner what they think of the investments he/she is considering. Remind them to think about the advice in the three interviews they listened to; refer them to audioscripts 3.01–3.03 on pages 152 and 153. Allow 4–6 minutes for students to discuss in their pairs, then get brief feedback from the class.

MyEnglishLab: Teacher's Resources: extra activities; Reading bank
Spoken English: p.113
Grammar reference: Expressing attitudes to the future p.117
Teacher's book: Resource bank Photocopiable 3.2 p.147
Workbook: pp.15–17

3.3 › Communication skills
Challenging facts politely

> ### GSE descriptors
>
> - Can participate in extended, detailed professional discussions and meetings with confidence.
> - Can successfully challenge points made during a presentation.
> - Can evaluate the strength of a speaker's source materials used to make a point in a presentation or discussion.
> - Can compare and evaluate different ideas using a range of linguistic devices.

> ### Warm-up
>
> Do this as a quick whole-class activity. Ask students if they have ever used facts and data (e.g. in a discussion, in a project at their place of work or study) which they later found were inaccurate. What was the situation? Where did they find this information? How did they find out it was inaccurate? Invite volunteers to share their answers with the class.

Lead-in

Students talk about fact-checking.

1 If time is short, discuss the questions with the whole class, nominating a few different students to answer each question. Alternatively, let students discuss in pairs first, then get feedback from the class. When discussing question 4, you may wish to tell students that psychologists refer to this as 'confirmation bias': we tend to undervalue or ignore evidence that contradicts our beliefs and overvalue evidence that confirms them. We believe in 'alternative facts' if they support our pre-existing beliefs.

> 1 Facts are observable, verifiable and provable. Claims, like opinions, though possibly based on facts, are subjective; expressions of individual/personal feelings, beliefs, experiences, tastes, ideas, viewpoints.
> 3 Good: when it is a reliable website such as an official or government site. Not so good: user-generated content such as Wikipedia or forums or blogs may be less reliable.

Preparation: Discussing research for an article

Students read and think about the scenario for a roleplay.

2 If this is the first Communication skills lesson for your class, briefly tell students about *Lifestyle* magazine and the profile of its readership. Otherwise, elicit this information from students before you start. Go through the instructions with students and give them time to read the text. Help them with any vocabulary questions they may have, then invite different students to share their ideas with the class.

> A possible answer is that media companies might face legal consequences if they publish incorrect information.

Roleplay

Students roleplay a discussion in which they have to challenge facts and data for an article.

3A Put students in pairs and explain the scenario: they are journalists working together on an article on money for *Lifestyle*; draw their attention to the infographic and briefly go through it with them. Explain that in preparation for writing their article they have gathered some facts and data for it, which they are going to share with their colleague. As part of the fact-checking process, they should politely challenge information from their colleague and respond to any challenges from him/her. Refer students to pages 126 and 129, give them time to read their information, then set a time limit for the preparation stage. Note that in order to elicit challenges, there are intentional inconsistencies between the two sets of information. Remind students to think about their magazine's fact-checking guidelines from Exercise 2 and also how they are going to challenge facts / respond to challenges politely. During the activity, monitor and provide help as necessary.

3B Students now roleplay their discussions. During the roleplays, monitor and make a note of any difficulties or frequent errors, as well as good use of language and strategies students use.

3C Students should do this in the same pairs as for Exercise 3B. Look at the questions with them before they begin and give them plenty of time to discuss them; encourage them to give reasons for their answers. Finally, go over any points you noted while monitoring.

Video

Students watch a video of a discussion during which the participants challenge facts and data.

4 ▶ 3.3.1 Tell students that they are going to watch a conversation similar to the one they have just had about the joint assignment. Go through the instructions with them and give them a minute to read the questions before they watch. Play the video, then check answers with the class.

> 1 They are calm and relaxed about it.
> 2 They both appeared to be confident.
> 3 Yes, it's an important part of their jobs as journalists and they know their work will be subject to close scrutiny by their editor.
> 4 **Possible answers:**
> - They should use the most up-to-date figures they can get from the Bank of England on credit card debt and state the date and source of the figures in the article. This figure should have a hyperlink to the original official source if possible.
> - They probably shouldn't use the figure for credit-card debt per household because as Yvonne points out, it's a 'statistical fallacy'. It's essentially a meaningless figure, as no two households are the same, nor will they have the same amount of credit-card debt.
> - They should include the figures from the debt charity and include the full name and job title of their source. A hyperlink to the charity's official site would also be a good idea to lend credibility to the source.
> - They could include the interview with Frank Turner, a financial psychologist, making it clear that these are his opinions.

- They definitely need to mention that the source of the millennials' attitude to money survey was an online bank, if they decide to use it, because it's a commercial entity that commissioned the research. It would be a good idea to back this up with some more 'desk research' and find a reputable academic source.

5 Students could do this individually or in pairs. You could go through the words in the box with them before they begin or let them use their dictionaries to look up unknown words and clarify meanings during feedback. To check answers, you could play the video again and pause after each sentence is heard to elicit the correct answer.

> 1 dug up 2 conflicting 3 fallacy 4 oversight 5 verify
> 6 reputable 7 delve into 8 stickler for detail
> 9 fine-tooth comb 10 stand up to scrutiny

Extra activities 3.3

A–B These activities practise useful vocabulary from the video. Students saw the words and expressions in Exercise 5 in the Coursebook, so they should be able to complete both activities individually. When checking the answers to Activity A, write (or invite students to write) the gapped words on the board so they can check their spelling.

> A 1 comb 2 conflicting 3 delve 4 dig 5 fallacy
>   6 oversight 7 reputable 8 stand 9 stickler
>   10 verify
> B 1 delve into 2 Conflicting 3 fallacy
>   4 oversight 5 a fine-tooth comb 6 verify
>   7 stickler for detail 8 stand up to scrutiny
>   9 dig up 10 reputable

Reflection

Students reflect on the conclusions from the video and their own approach to challenging facts and data.

6A Allow students to work individually first so that they can reflect on their own approach in the roleplay. Go through the questions with them before they begin and remind them to think about their answers to Exercise 3C.

6B Put students in pairs or small groups to compare their reflections, then broaden this into a class discussion. Encourage students to elaborate.

MyEnglishLab: Teacher's Resources: extra activities; Interactive video activities

3.4 ➤ Business skills
Exploring options

GSE descriptors

- Can propose a range of different options in a complex negotiation.
- Can compare and evaluate different ideas using a range of linguistic devices.

Warm-up

Ask students to think about a time in their life when they had to negotiate over something. What was the situation? What were they negotiating about? What did they expect from the negotiation? Do they think the negotiation was successful? Get them to share their experiences in pairs or small groups, then invite a few students to share them with the class.

Lead-in

Students talk about using questions as a negotiation strategy.

1 If there is time, get students to discuss the questions in pairs or small groups, then get brief feedback from the class. Otherwise, do this as a quick whole-class activity, eliciting a few answers around the class.

2A Put students in pairs, explain the activity and give them time to read the texts and discuss the questions. For questions 2 and 3, encourage them to give reasons for their answers. Allow 5–7 minutes for the pairwork activity, then check the answer to question 1 with the class and invite students from different pairs to share their answers to questions 2 and 3 with the class.

> 1 asking questions

2B Students could do this in the same pairs as for Exercise 2A, in new pairs or in small groups. Give them a few minutes to discuss in their pairs/groups, then broaden this into a class discussion, feeding in information from the answer key below as required.

> Asking questions in a negotiation can be good:
> - to gain participation. When you ask questions, you encourage the other person to talk. This makes your counterpart like you better and helps you learn more about them than they learn about you. It's especially helpful to get your counterpart to talk when you realise you have said something they didn't agree with or understand. Having a chance to talk it out will have a calming effect on them.
> - to give information. Sometimes you may want to provide information that will help your counterpart understand your goals. For example, you could ask, 'Did you know that the Kelly Blue Book value of your car is only $2,100?' (This type of question can also be used as a test to see whether your counterpart recognises if your information is correct.)

- to get an opinion. Questions that ask for someone's opinion not only provide knowledge, but also indicate that you are interested in what that person has to say. For example, ask, 'Can you tell me why you like living in this neighbourhood?'
- to bring attention back to the subject. Appropriate questions can keep the conversation heading toward your goal. Salespeople often ask personal questions about a prospect to find a starting point for their presentation. This is fine, but eventually, you need to discuss the real reasons for meeting. Asking questions like, 'Can we get back to the salary issue and benefits package once again?' refocuses attention on the important issues.
- to reach agreement. Asking questions can help you find out how far apart your goals are from your counterpart's. For example, suppose a seller is asking $150,000 for his house. You ask whether he is willing to take $140,000, since the house needs landscaping and a new roof.
- to reduce tension. If negotiations start to become tense, it can be helpful to ask questions about your counterpart's viewpoint. Understanding his concerns may help you restructure the negotiation. For example, you might say, 'Every time we talk about mandatory drug testing for all employees, you seem to be adamantly opposed. Can you share a little about why you are opposed to this testing?'
- to give positive strokes. Simply put, positive strokes questions make your counterpart feel important. Suppose your counterpart has received three phone calls from complaining customers during your 15-minute meeting. You might ask, 'Are you having a tough day?'

Asking questions can be perceived in many less than positive ways. For some, being asked questions may be viewed as challenging and as a way of attempting to undermine the person or to tactically destabilise the communication process. For those with seniority, it may be viewed as a form of disrespect. In some cultures, potentially, asking (too many) questions may be viewed as impolite, suggesting that your counterpart is unclear as a communicator and so questions are needed to make things clearer. In the end, no communication strategy can be guaranteed to succeed as it will depend on the other person's perception of the strategy. If you make clear a positive intention, then questions can be extremely useful in building mutual understanding and positive relationships.

Listening

Students listen to people negotiating an office lease.

3A 🔊 3.04 Explain the scenario and activity and give students time to read the questions. Teach or elicit the meanings of *build rapport* and *lease* and ask students to make notes in answer to the questions while listening. Play the recording, twice if necessary, then check answers with the class.

1 Diana shows interest by asking questions about Lina running a marathon and how long she has worked with Ron.
2 Diana wants to reduce the scope of the lease to just one floor.
3 The company doesn't need the space at the moment with so many staff working from home. And there is a need to reduce overheads.
4 Ron is not sure, as he prefers to lease the whole premises to a single client and there's very little demand in the market for one-floor office spaces at the moment.
5 Ron feels challenged by Diana's suggestion that start-ups in the area might be interested in leasing the property.

3B 🔊 3.04 Before students listen again, point out that they should listen for all the questions Diana uses during this first stage of the negotiation: her opening questions to build rapport and the questions she asks later on in the conversation to challenge Ron and Lina's views. Accept any reasonable answers, as long as students can justify them.

Possible answer

Diana's opening questions to build rapport seem successful, but her later questions result in Ron's frustration at her challenging him about whether start-ups would be interested in leasing his office space.

3C 🔊 3.05 Explain the activity, give students time to read the questions and check that they understand the meaning of *win–win* in question 5. Note that students may need to listen twice in order to check/complete their answers. If time allows, get them to compare answers in pairs before class feedback, referring to audioscript 3.05 on page 153 if necessary. Check answers with the class.

1 Diana tries to understand the needs of her counterparts by checking if loss of potential income is the main concern.
2 Diana suggests reducing the scope of the lease – to the one floor – but also extending the timeframe.
3 To lease just one floor, extend the lease for another five years and pay $55,000 for just the one floor.
4 Ron is worried about not finding another tenant for the entire year.
5 The lease is reduced to one floor but extended for another five years. The lease is $75k in year one and $50k for the final four years. This gives security on income on one floor to Ron for a long period. If a new tenant is found quickly and charged $55k for the other floor, Ron can also increase his total lease value.

3D If there is time, get students to discuss the question in pairs or small groups first, before asking them to share their views with the class.

Overall, Diana handled the second phase well. She used a good range of questions to engage her counterparts, listened well to what they said, provided creative ideas and focused the discussion on a win–win solution.

Useful language

Students look at strategies and useful language for using questions in negotiations.

4A This activity is best done in two stages. Start by asking students to find and underline all the questions in scripts 3.04 and 3.05 on page 153 and check answers with the class. Then explain to students that they need to match all the underlined questions with reasons 1–8. Go through the list of reasons with the whole class and allow plenty of time for students to complete the matching task, individually or in pairs. Check answers with the class and encourage students to record the questions (and reasons) in their notebooks.

> **1** To clarify understanding:
> *Was the situation with my company clear and what we're looking for?*
> *You want to reduce the scope of the current office lease from two floors to just one floor, yes?*
> *How do you mean, exactly?*
> *For five years, did you say?*
> **2** To build relationships:
> *Are you the Lina who ran the marathon?*
> *How did it go?*
> *Have you been working with Ron long?*
> **3** To challenge and provide alternative views/information:
> *Is that true? (I heard the opposite.)*
> *Really? Did you know that quite a few start-ups have opened downtown?*
> **4** To elicit opinions and ideas from your counterpart:
> *Could this work for you?*
> *How does that sound?*
> **5** To focus attention on the most important topic:
> *Can we come back to your main issue?*
> *Can I just focus on that key point?*
> **6** To reduce tension and conflict by trying to understand others' needs:
> *Is your problem here the potential loss of income?*
> **7** To propose solutions:
> *Would you be willing to consider reducing … but maybe also … ?*
> *What if you cut … and in return we extended … ?*
> *What if we pay 75,000 in year one and 50,000 for the last four years?*
> *So, if we reduce the contract to one floor, you'll commit to an extended contract with these new payment terms, $75,000 in year one, yes?*
> **8** To highlight and celebrate agreement:
> *Yes. It's win–win, isn't it?*
> *Sounds good?*

4B Start by asking students to write a few example questions for some of the reasons in Exercise 4A. You could let them decide how many and which reasons to choose or assign different reasons to different students around the class, to ensure that all the reasons are covered during feedback. Give them a few minutes to write their questions individually, then put them in pairs to compare and discuss their answers. As feedback, elicit ideas around the class and list them on the board; you could encourage students to record them in their notebooks as example/useful questions for each of the reasons 1–8.

> **Possible answers**
> **1** Do you mean … ? Are you saying … ? Is it clear what we're looking for? Do you understand what we are asking for?
> **2** Is this your first time in … ? How do you like … ?
> **3** Are you sure that's right / about that?
> **4** What do you think about … ? What are your thoughts on … ?
> **5** Can we / Perhaps we could return to … ?
> **6** Is there an issue with … ? Is your problem the … ? Are you worried about … ?
> **7** How about if we … ? Could I suggest that … ? Are you / Would you be prepared to … ?
> **8** Does this sound like a good … ? It's good for both of us, right?

4C Students could do this in the same pairs as for Exercise 4B or in new pairs. In weaker classes or if you think they will struggle to come up with reasons/strategies of their own, you could give them a few ideas on the board (see answer key below) and get them to write the example questions for each one. Give them 3–5 minutes to write their questions in their pairs, then join pairs together into groups of four and get them to compare their ideas. As feedback, elicit strategies/reasons and example questions from different pairs and write (or invite students to write) them on the board. Encourage the class to add more questions for each of the strategies on the board. Again, you could encourage students to record the strategies and example questions in their notebooks.

> **Possible answers**
> Identifying weaknesses in a counterpart's position:
> *Wouldn't it be difficult to … ?*
> *Isn't it problematic to try to … ?*
> Simplifying:
> *Can't we just … ?*
> *Wouldn't a quick solution to this be to … ?*
> Exploring decision-making authority:
> *Do we need to involve our management if we … ?*
> *Who else may need to be consulted on this?*
> Suggesting a summary:
> *Should we recap?*
> *Can we go over this one more time?*

Extra activities 3.4

A–B These activities provide further practice and consolidation of the functional language from the lesson. Ask students to complete them individually and get them to compare answers in pairs before checking with the class.

> **A 1** c **2** a **3** d **4** f **5** b **6** e
> **B 1** If we agreed
> **2** with the proviso that
> **3** Given that
> **4** Just thinking a little out of the box
> **5** To play devil's advocate
> **6** Supposing we
> **7** Just as an idea, how about
> **8** Say we were to agree

Task

Students hold a meeting to negotiate time involvement on a project.

5A Explain to students that they are going to hold a negotiation meeting, put them in pairs and give them time to read the professional context. You may wish to check understanding by asking a few check questions, e.g. *What is Next Gen?* (a research project for a range of sunblock creams, run by Beautifies) *What is Sam Birreg's role and what problem is he facing?* (He is the project lead. He is struggling to meet deadlines.) *What is Jean Piaget's role?* (Head of Sales) *What was the agreement between Sam and Jean about Pierre's role?* (to spend 25 percent less time on his usual sales activities in order to help with the Next Gen project) *Why are Sam and Jean meeting today?* (to negotiate Pierre's time involvement on the Next Gen project; Sam needs Pierre to devote more time to Next Gen but Jean needs him for important and urgent sales activities). Allocate roles (or let students choose) and explain that they are going to roleplay the meeting between Sam and Jean.

5B Refer students to their role cards on pages 126 and 130 and give them time to read the information while you monitor and help them with any questions they may have. Point out the bulleted points in each role card, which students need to make sure are covered during the meeting. Explain that before they meet, they need to plan their strategy for the negotiation; remind them of the questioning strategies they looked at in Exercises 4A–C and encourage them to think about the questions they can use to help them negotiate successfully. Highlight that they should aim to create a *positive process and outcome*. In weaker classes, you may prefer to let students plan their strategy in A–A and B–B pairs, then return to their original pairs to hold their meetings. Allow plenty of time for students to prepare while you go round and provide help as necessary. When they are ready, they hold their meetings; set a time limit before they begin. During the activity, monitor and note down points to highlight during feedback, but do not interrupt the meetings.

5C Let students discuss in their groups first, then broaden this into a class discussion. Remember to highlight any points you noted while monitoring.

MyEnglishLab: Teacher's Resources: extra activities; Useful language bank

3.5 > Writing

Budget report

GSE descriptors

- Can write internal communications about a company's financial status.
- Can write a detailed structured report on work-related topics.
- Can correct structural errors in someone else's written report.

Warm-up

Discuss these questions with the class: *What is a budget report?* (an internal report, usually used by management, which compares the estimated budgeted performance with the actual performance achieved during a specific time period) *Have you ever read or written one? How important do you think it is for companies to have such reports prepared on a regular basis? Why? What purposes might these reports serve (for management)?* (Possible answers may include: they can help managers correct problems in order to bring the company's actual performance more in line with the financial goals in the budget; managers can evaluate how accurate and realistic their predictions were and then adjust their next budget accordingly.)

Lead-in

Students read and discuss the main content and structure of a budget report.

1 Start by eliciting or giving a brief definition of *income statement* (a report showing the income, expenses and resulting profits or losses of an organisation during a specific time period). Then do the activity with the whole class, checking answers and clarifying meanings as you go.

Cost of sales: raw materials + direct labour costs (manual/blue collar)

Operating expenses: salaries (office/managerial/white collar), marketing, distribution costs, office supplies, rent, utilities

2 Refer students to the budget report on page 138 and draw their attention to the table at the top. Check that they understand the meaning of *variance* and highlight that in the table, the figures in brackets are 'unfavourable', i.e. the fact that they were more or less than forecast was bad news for the company. Give students time to read the report and decide if the statements are true or false, then go over the answers with the class. After feedback, point out the main structure of a budget report: there is a heading, a table showing the budgeted figures, the actual figures and the variance between them and finally an executive summary of around 350 to 400 words, organised into clear paragraphs.

1 T   **2** T   **3** F

Useful language

Students look at useful language for, and the main structure of, a budget report.

3 This is best done as a whole-class activity, checking and discussing answers as you go. After discussing the answers for each item, go through the relevant section of the Useful language box with the class. For question 2, point out that *CapEx* stands for 'Capital Expenditure' and *OpEx* for 'Operating Expenditure'. For question 3, you may wish to point out that some verbs, such as *increase* and *decrease*, can be both transitive and intransitive depending on how they are used in a sentence (e.g. *We increased sales* – transitive; *Sales increased* – intransitive). For question 5, you could ask students what other linking words they know which can be used to express reasons (e.g. *as* + clause; *since* + clause; *due to* + noun; *owing to* + noun).

1 future investment plans: paragraph 3; budget forecast: paragraph 4; figures for variance are correct

2 CapEx (Capital Expenditure) is money spent on fixed assets such as equipment, computers, furniture, vehicles, buildings, land. OpEx (Operating Expenditure) is money spent on normal business operations such as salary costs, rent, marketing, insurance, R&D. 'X is offset by Y' has meaning a).

3 **be down:** go down, be lower, fall, reduce sth, decline, drop
be up: go up, be higher, rise, raise sth

4 It would be more difficult to read because exact figures all the way through the report become pedantic, distracting and difficult for the brain to process. Sometimes exact figures are needed, but often it is much easier to say that something is 'a little more', 'considerably less' or 'about the same'.

5 *Because* is followed by a <u>subject + verb</u>; *because of* is followed by a <u>noun phrase</u>. Examples: *Sales increased* **because** we had *a marketing campaign. Sales increased* **because of** the marketing campaign.

6 The three phrases show increasing levels of formality, but all are acceptable in a report and it is good to use a variety of structures.

4 Get students to do this individually and then compare answers in pairs before class feedback. You could list the words/phrases students choose on the board for them to refer to during the writing task.

Extra activities 3.5

A This activity gives further practice of useful vocabulary for budget reports. It is a consolidation exercise, so it would be better for students to do it individually. You could get them to compare answers in pairs before checking with the class.

1 likely 2 revenue 3 expenses 4 budgeted
5 CapEx 6 raise 7 slightly 8 significantly
9 variance 10 rise 11 OpEx 12 due

Optional grammar work

The budget report in Exercise 1 contains examples of modal verbs for possibility, so you could use it for some optional grammar work. Refer students to the Grammar reference on page 118 and use the exercises in MyEnglishLab for extra grammar practice.

Task

Students write the executive summary of a budget report.

5A Put students in pairs, explain the activity and give them time to read the information on page 139. Before they add their own ideas to the notes, check that they understand *consumer confidence*, *regional office* and *catch up with* and point out or elicit that *R&D* stands for 'Research and Development'. If you think your students will struggle to come up with ideas for the notes in their pairs, you could do this as a whole-class activity, inviting volunteers to contribute ideas and listing them on the board for students to refer to during the writing task. If students do work in pairs, monitor during the activity and provide help as necessary.

5B Students now write their executive summaries. Point out the word limit, set a time limit for the writing task and remind students to refer to the model summary on page 138 and the Useful language box. Also encourage them to look again at the questions in Exercise 3 before they write their summaries. If there is no time to do the writing task in class, it can be assigned as homework.

Model answer

Executive summary

In this second quarter ending 30 June gross profit was up by $8,000,000 on the budgeted figure, an increase of around 12% from the forecast $68,000,000. This was due to two factors. First, sales increased significantly, the most likely reason being a general improvement in consumer confidence as a result of the better economic environment. Second, cost of sales went down slightly as we are now seeing the benefits of the investment we made in automation last year and fewer workers are needed in the plant.

This increase in gross profit was partly offset by a small rise in operating expenses. There were two reasons for this, both related to the decision to open a new regional office in the north of the country. First, we recruited new employees to staff the office and this pushed our salary costs higher. Second, there were additional general expenses associated with the opening of the new office, such as rent, utilities and other overheads. These two factors caused our total operating expenses to rise by $4,000,000 on the budgeted figure. Marketing costs were in line with budget and so had no impact on costs. In general, this increase in total operating expenses is modest, given the opening of the new office.

Overall, operating profit rose from a forecast $11,000,000 to an actual $15,000,000. This is a considerable increase and puts us in a very good position with our investment plans. We have CapEx planned for a new factory in Slovakia and this will help us become a major player in the European market. We also have significant OpEx investments that we want to make in the area of R&D; our competitors all have bigger R&D programmes than us and we need to catch up.

After discussions with the senior management team, I am able to give some provisional forecasts for the rest of the year. Revenue is expected to grow strongly as the economy improves and consumer spending rises. We are also in the fortunate situation that costs are likely to be unchanged; we have already budgeted for both the factory in Slovakia and the larger R&D department. Therefore, it is probable that operating profit will continue to go up. I am optimistic about the future, partly because of our expected growth in Europe and partly because the economy looks like it will continue to improve.

5C If students write their summaries for homework, this activity can be done in the next lesson. Put students in pairs and ask them to read each other's summaries, think about the questions and give their partner feedback. You could then ask them to write a final, improved version of their summary, in class or for homework.

MyEnglishLab: Teacher's Resources: extra activities; Writing bank; Interactive grammar practice
Grammar reference: Modal verbs: possibility p.118
Workbook: p.18

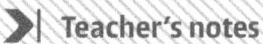

Business workshop 》3
Financial strategy

GSE descriptors

- Can suggest pros and cons when discussing a topic, using linguistically complex language.
- Can understand the details in a linguistically complex audio recording.
- Can participate in discussions using linguistically complex language to compare, contrast and summarise information.
- Can write about complex subjects, underlining the key issues and in a style appropriate to the intended reader.

Background

Students read about a design start-up specialising in handmade sofas.

1 Before students read, you may wish to teach/check understanding of these words from the background and questions: *found* (v.), *source* (v.), *take on*, *working capital*, *feasible*. Put them in pairs and ask them to read the text and discuss the questions, then check answers with the class.

1 They make handmade sofas with international design influences.
2 They have been quite successful. They are two years old and now employ four staff. They more than doubled in their second year. They've just received a single order for 200 sofas, which is substantially larger than the number of sofas they made last year.
3 Their problem is cashflow. They won't get paid for the larger order for seven months, but will need to pay staff costs, rental costs and for materials in the meantime. They don't have enough working capital to cover that much.
4 They've asked their accountant to advise them on their options.

Dealing with fast growth and cashflow

Students read a text and listen to a conversation about different financing options for a company.

2A Explain the activity and look at the four headings in the text with students before they read. Check understanding of *partial*, *up front* and *crowdfund*, then ask students to read the text and answer the question. If there is time, get them to share and discuss their answers in pairs or small groups before class feedback. Elicit a few ideas around the class, then help students with any unknown vocabulary from the text.

Possible answer

One other option could be to look for an investor to give them the £150,000 in return for a share of their business.

2B Put students in pairs, explain the activity and give them time to discuss the pros and cons of each option and make notes in the table. Let them read the text again if they need to. If time is short, you could also do this as a whole-class activity, eliciting ideas around the class.

Possible answers

1 <u>Apply for a short-term bank loan</u>
Pros: Shouldn't be difficult to get.
Cons: Interest will be high.

2 <u>Get partial payment up front</u>
Pros: Saves going to the bank.
Cons: Might be difficult to get. The customer might ask for a discount.

3 <u>Delay supplier payments</u>
Pros: Saves going to the bank or to the hotel customer and could be easier to organise than the previous two options.
Cons: Might be difficult to get and could put relationships under strain. It could put the materials suppliers under financial strain as well.

4 <u>Crowdfunding the £150K</u>
Pros: It could be easy to get the money, given the current growth and brand name. It would save going to any of the other options and would be good advertisement in itself.
Cons: Might look like G&K is not financially secure and that could damage other orders.

Extra activities Business workshop 3

A This activity looks at useful vocabulary from the texts in Exercises 1 and 2A. You could do it as a whole-class activity, checking answers and clarifying meanings as you go. Alternatively, ask students to complete it individually; encourage them to find the words in the texts to help them work out their meanings. Check answers with the class, clarifying meanings as necessary.

1 f 2 h 3 a 4 g 5 b 6 c 7 d 8 e

3A ◀》 BW 3.01 Explain to students that they are going to hear the two founders of G&K discussing the different options. Encourage them to make notes in answer to the questions and play the recording. You could get students to discuss question 1 in pairs before class feedback.

2 A fifth option they think of is to seek an investor to give them the £150k in return for a share of the business. The advantage of doing this is that they could look for an investor who will help advise them on growth and expansion. A disadvantage is that they would have to give up some equity in the business, possibly 15 to 20%, in return for a cash boost for only six months. The benefits of the investor's experience would have to outweigh this disadvantage.

3B Depending on time available, you could do this as a quick whole-class activity or get students to complete it individually and then check answers with the class.

> **A** Payment up front **B** New investor **C** Bank loan
> **D** Delay supplier payments **E** Crowdfunding

Extra activities Business workshop 3

B This activity looks at useful vocabulary from the listening. Ask students to complete it individually or, in weaker classes, in pairs, using their dictionaries if necessary. Point out that *discount* is a noun here and encourage them to look carefully at the words around each gap and think about the type of word/phrase needed (e.g. a verb/verb phrase, a noun/noun phrase); this will help them choose the correct option for each item. Check answers with the class, clarifying meanings as necessary.

> **1** projections **2** interest payable
> **3** pay in instalments **4** discount
> **5** commit to a long lease **6** raise the money
> **7** protect their brand **8** combination of options

Task: Pitch an idea

Students discuss different financing options and write an email requesting financial support.

4A Put students in pairs or small groups. Note that if they work in pairs, they should work with a different partner to Exercise 3. Explain that they are going to discuss the five different options available to G&K and choose the best one(s). Refer them to the flowchart on the right and go through it with them, making sure they are clear about what they need to do for each step. Point out that they can choose a single option or a combination of options, depending on what they think is best for the company. Note that there is no definitively correct answer. There could be a number of legitimate reasons to choose any of the five options. The financial cost of each option is not the only consideration when it comes to building business relationships, easing financial pressure and getting external support and advice, e.g. from an investor. Set a time limit before students begin and during the activity, monitor and note down any points to highlight during feedback, but do not interrupt students' discussions; go through them in a brief feedback session after Exercise 4B.

4B Pairs/Groups now take turns to present and explain their choice(s) to the class. Remind them that they need to give reasons. After all pairs/groups have presented their choices, you could ask them if they now want to change their minds and follow the recommendations of any of the other pairs/groups. That way, they will also be reflecting and learning from each other.

5 Explain the writing task and go through the five options with the class. Depending on the option students choose, it could be useful to follow the structure below; write it on the board for them to refer to while they are planning/writing their emails:

1 *Explain the situation and the background.*

2 *Make the request.*

3 *Address any potential concerns or reasons to reject your request.*

4 *Highlight the benefits of supporting you with this request.*

5 *Close politely and with indication of next steps or a call to action.*

If there is no time to do the writing task in class, it can be assigned as homework.

> **Model answer (for Option 1)**
>
> Dear Ms Smith,
>
> As you know, we have recently received a larger order from a new customer. This is a great opportunity both for the business and its employees.
>
> However, this new opportunity will present us with a cash-flow challenge. We will need to pay for materials and staff costs before we receive full payment from our customer. In order to cover this shortfall, we would like to request a short-term loan of £150,000.
>
> As you will see from our attached customer order, our recent company accounts and our forecast for this year, our revenue is consistently growing and we are on track for further growth this year. This will hopefully convince you of the strength of our company and the low risk to you in approving this request.
>
> If we are able to get this support from you and fulfil this order, it will enable us to grow our business considerably, both in terms of revenue and the creation of new jobs. It will also further enhance our brand and market share in the handmade sofa and furniture business.
>
> I do hope you are able to respond positively. I look forward to hearing from you soon. Please let me know a good time for us to schedule a call to talk about this further.
>
> With kind regards,
>
> Garry Griffin

MyEnglishLab: Teacher's Resources: extra activities

Review ◀ 3

> **1 1** instruments **2** exchange **3** ballpark **4** yield/return
>   **5** return **6** reward **7** rate **8** backer
> **2 1** c **2** d **3** a **4** e **5** b **6** f
> **3 1** b **2** a **3** b **4** a **5** a
> **4 1** d **2** a **3** f **4** c **5** b **6** e **7** g
> **5 1** forecast **2** due **3** significantly **4** figure **5** rise
>   **6** costs **7** less **8** gross

4 ▶ Disruptors

Unit overview

	CLASSWORK	FURTHER WORK
4.1 ▶ **Disruptors in business**	**Lead-in** Students are introduced to the concept of disruption in business. **Video** Students watch a video about disruption in business. **Vocabulary** Students look at vocabulary related to disruptors and disruption in business. **Project** Students hold a panel discussion on products/services that are essential to them.	**MyEnglishLab:** Teacher's resources: extra activities; Reading bank **Teacher's book:** Resource bank Photocopiable 4.1 p.148 **Spoken English:** p.113 **Workbook:** p.19
4.2 ▶ **Disruptive innovation**	**Lead-in** Students talk about the risks involved in disruptive innovation. **Reading** Students read an article about disruption in business. **Grammar** Students study and practise different forms for expressing hypotheses. **Speaking and writing** Students talk and write about possible big disruptors.	**MyEnglishLab:** Teacher's resources: extra activities **Grammar reference:** p.118 Hypothesising **Teacher's book:** Resource bank Photocopiable 4.2 p.149 **Workbook:** pp.20–22
4.3 ▶ **Communication skills:** Finding solutions	**Lead-in** Students talk about brainstorming. **Roleplay** Students roleplay a brainstorming meeting. **Video** Students watch a video of a brainstorming meeting. **Reflection** Students reflect on the conclusions from the video and their own approach to brainstorming.	**MyEnglishLab:** Teacher's resources: extra activities; Interactive video activities
4.4 ▶ **Business skills:** Reporting and planning	**Lead-in** Students talk about different attitudes to change. **Listening** Students listen to a meeting about change management. **Useful language** Students look at useful language for discussing change. **Task** Students roleplay an interview for a change management consultancy role.	**MyEnglishLab:** Teacher's resources: extra activities; Useful language bank
4.5 ▶ **Writing:** Supply chain choices	**Lead-in** Students talk about supply chain management. **Useful language** Students look at useful language for, and the main structure of, an internal recommendation report. **Task** Students write an internal recommendation report.	**MyEnglishLab:** Teacher's resources: extra activities; Writing bank; Interactive grammar practice **Grammar reference:** p. 119 Emphasis using inversion and fronting **Workbook:** p.23
Business workshop 4 ▶ Disruption – planning ahead	**Listening** Students listen to a radio programme about the development and use of autonomous vehicles. **Speaking** Students talk about the impact of a move towards autonomous vehicles on different sectors of the economy. **Task** Students give a presentation on mitigating the impact of a major disruption and write a handout summarising the main points of a presentation.	**MyEnglishLab:** Teacher's resources: extra activities

Business brief

The main aim of this unit is to explore the concept of disruption in business and examine how disruptors and the changes they engender influence consumer behaviour.

The contemporary business landscape is characterised by **disruption**. New ideas and new technologies are continually arriving which are displacing traditional solutions and challenging established companies. A **disruptive innovation** is an innovation that creates a new market for goods and services by disrupting existing markets and causing the **repositioning** of market-leading firms. **Disruptors** are people, companies, or business ideas that radically change the traditional way an industry and its customers operate. Many of these challengers are people who have relatively little experience in the field they are launching into (for example Amazon, Netflix, Dyson, Airbnb). This potentially enables them to see opportunities that more experienced or traditional operators have missed, so they can offer fresh, new perspectives on the business, or identify more effective ways to operate in it.

Disruptive innovations tend to be produced by outsiders and entrepreneurs in **start-ups**, rather than existing market-leading companies, because the business culture of market leaders doesn't allow them to pursue disruptive innovations when they first arise; these are generally not profitable enough at first to be **game-changers** and their development can also take resources away from **sustaining innovations** (the new inventions and modifications established businesses generate in an attempt to stay relevant to customers) which are needed to compete against current competition. A **disruptive process** can also take longer to develop, and the risk associated with it is higher, than with the other more gradual or evolutionary forms of innovation. However, once it's deployed in established markets, it can achieve a much faster penetration and a higher degree of impact.

The pace of disruption in business today is something unprecedented in the history of the world economy, forcing even **long-established companies** to find new strategies to help them adapt in the constantly evolving **business landscape**. The challenge for them is to figure out when disruption is happening to them, understand where it's coming from and have a strategy in place for more than one 'future' for their organisation.

Disruptors emerge most often, not as the result of a single, sudden change in the landscape, but in response to a combination of factors. These include changes in customer behaviour, new types of competition and new methods or technologies. The success of a disruptor also depends on whether the more established operators they're competing with are agile in their response to the disruption or complacent. In most scenarios, products or services developed by market leaders become too sophisticated, too inaccessible or too expensive and customers consequently start to look for other, sometimes radical, alternatives to meet their needs. A disruptor arrives who develops an innovative product that initially is attractive to only a niche segment of customers. However, as it improves through various iterations in response to feedback, its appeal widens. The disruptor then becomes a real threat to existing market leaders.

In recent decades, large companies have developed defensive strategies for confronting this kind of customer-driven disruptive innovation. Most commonly, they'll acquire the disruptor outright, for example by investing in a start-up so that they can 'piggy back' on it, participating in the disruption without having to reform their own organisation. Alternatively, they may try to 'disrupt themselves' by setting up autonomous business units within their organisation to look at potentially disruptive innovations. However, established companies are not always successful at adapting their business culture to accommodate disruption. How quickly they can pivot from one way of doing things to another depends on multiple factors, from the nature of the threat, to the size of the business, to the attitudes of the senior leadership team.

Disruptors and your students

Most students will be aware of well-known industry disruptors who have developed products and services they use in their own lives, such as Apple, Amazon, Skype and Wikipedia. They may also know something about how evolving technologies and customer expectations have engendered innovation in the entertainment industry: for example, allowing Netflix to disrupt the market for video rental stores and cable TV subscriptions, or Spotify and other online streaming services to transform the music industry. Some in-work students may be able to provide examples of disruption in the fields their companies work in.

Unit lead-in

Draw students' attention to the unit title *Disruptors* and write *disrupt – disruption – disruptive* on the board. Teach or elicit their meanings and ask students what they think the term *disruption* might mean in business. Elicit or give a brief explanation (see Business brief on the previous page) and tell students that they will be looking at the concept of disruption in business in this unit. Draw their attention to the quote and briefly discuss it with the class. Can students explain it in simpler words? Do they agree? Why / Why not?

4.1 ❯ Disruptors in business

GSE learning objectives

- Can extract specific details from a TV programme on a work-related topic.
- Can follow a group discussion on complex, unfamiliar topics.
- Can recognise a wide range of idiomatic expressions and colloquialisms, appreciating register shifts.
- Can summarise and reformulate ideas from members of a panel discussion to clarify a point.
- Can contribute ideas in a panel discussion using linguistically complex language.
- Can evaluate arguments in a debate or discussion and justify the evaluation.

Warm-up

Write the word *innovation* on the board and ask for a brief definition of the term in relation to business and the marketplace (the process of introducing a new idea, method, product, service, etc., or the idea/method/product/service itself). Ask students for examples of *innovations* and list their ideas on the board; you could prompt them by adding a few examples of your own (e.g. the TV set, the camera, the car, the World Wide Web, nuclear power, the computer, electric light). Then ask which of the innovations on the board students think has had the biggest impact and why; elicit a few ideas around the class, then move on to the Lead-in questions.

Lead-in

Students are introduced to the concept of disruption in business.

1 Depending on the time available, you could let students discuss the questions in pairs or small groups first, then get brief feedback from the class.

Spoken English
p.113: My time is my own and I can take work or leave it

1 ◀)) SE4 Explain the activity and give students time to read the questions first, so they know what to listen for. Encourage them to make notes in answer to the questions while listening, then play the recording, twice if necessary and check answers with the class.

Possible answers
1 awful and exhausting
2 He wants to work from home on Mondays and Fridays.
3 about nine months ago
4 a nine-to-five job (in an office)
5 money

Attentive listening

2 ◀)) SE4 Refer students to the heading *Attentive listening* and elicit or give a brief explanation: listening attentively means being an active and engaged listener, making a conscious effort to understand not only the words being spoken, but also the message being communicated. It means actively engaging in conversation, responding to what the other person is saying and letting them know you are listening to them. Explain that the gapped words/phrases are different ways in which the speakers show they are listening attentively. Give students time to quickly read through the extracts before they listen, then play the recording, twice if necessary. Check answers with the class.

1 Wow   2 Two hours   3 Yes, without a doubt
4 Not really, no   5 That's a good idea
6 that's me, that's me   7 Oh, I'm sorry
8 I'm so sorry   9 It must be very hard
10 Exactly   11 You do

3 Go through the functions with students before they begin and get them to complete the exercise individually and then to compare answers in pairs before class feedback. Alternatively, do this as a whole-class activity, checking answers as you go.

a 1, 2, 11   b 3, 10   c 4   d 5   e 6   f 7, 8, 9

Video

Students watch a video about disruption in business.

2A Do this as a quick whole-class activity. If you did the Unit lead-in activity, students should be familiar with the term. Elicit from a few different students which option they think is correct but do not confirm the answer yet, as students will check it in the next activity.

2B ▶ 4.1.1 Tell students that they are going to watch the first part of a video about disruption in business and check their ideas from Exercise 2A. Play the video (0:00–0:33) and confirm the answer. If you did not do the Unit lead-in activity, you may wish to also teach the verb *disrupt* and derivatives *disruptive* and *disruptor*.

Possible answer

According to the video: *New ideas and technologies … are continually arriving on the scene, displacing traditional solutions and challenging established companies by creating new markets for goods and services which did not exist before.* Therefore, c is the best definition.

3 ▶ 4.1.1 Give students time to read the questions before they watch and check they understand *business landscape* and *adapt (to)*. Ask students to make notes in answer to the questions while watching, play the video and check answers with the class, encouraging students to use their own words to answer the questions. Note that students, especially in weaker classes, may need to watch the video a second time in order to check/complete their answers.

Possible answers

1 They replace the usual solutions for problems, may cause problems for traditional companies and create new markets which had not formerly existed.
2 It is much faster than it has been before and the lifespan of companies is much shorter than in the past.
3 Smartphones have changed the way people deal with email, social media and the internet, the way they listen to music or look at videos, their money management, how they get around and what they do with photos.
4 Nikon has changed what they now produce and sell, namely, more expensive cameras and lenses. This means they are no longer competing to the same extent with smartphones.
5 Vinyl recordings had already been replaced by CDs before streaming arrived. Once peer-to-peer file sharing platforms became popular, the need for physical recordings stopped and the industry has kept up by charging for digital recordings. The industry is embracing digital distribution.
6 New taxi services have been set up, which challenges traditional taxi companies and more people are living without cars. Companies like BMW and Volkswagen are setting up their own services for people to share cars.
7 They took away customers from traditional airlines, but also created a new market for short-haul passengers. Traditional airlines have reacted by using some of the same ideas such as charging for 'extras'. Low-cost airlines have begun offering services to business people and are now found at major airports. They have also recently expanded into the long-haul market.
8 They have to buy a disruptor and use it, invest in a disruptor, create an organisation that is similar to a disruptor, or change (rewire) their own business culture.

4 Put students in pairs, explain the activity and give them 3–4 minutes to discuss the question. To help them, you could give them examples of specific areas to consider: **university**: getting grades, signing up for classes, receiving a timetable, arranging a meeting with colleagues, arranging meetings with teachers; **work**: clocking in, finding out about training courses and signing up for them, setting up appointments internally and with clients, submitting travel expense forms, receiving pay slips. If your students are uncertain about how to start, this could be done as a brainstorming session to ask what tasks they need to do in order to study or work, such as the ones above. They can then discuss any differences they have noticed. Once they have discussed in their pairs, invite different students to share their answers with the class.

Extra activities 4.1

A ▶ 4.1.1 Explain the activity and give students time to read the statements. Tell them that they should first decide whether the statements are true or false and then find evidence in the video to support their answers. They could do this by referring to videoscript 4.1.1 on page 147 and underlining the relevant parts of the script and/or watching the video again and asking you to pause each time the evidence for one of the answers is heard. If it's necessary to clarify any answers, play any relevant parts of the video again during class feedback.

1 F (*New ideas and new technologies, often introduced by small, agile start-ups, are continually arriving on the scene, displacing traditional solutions and challenging established companies …*)
2 T (*When you look at the pace of disruption, we've actually never experienced this in the history of any of our economies. The average lifespan of companies used to be about 67 years old. Now the average lifespan for the S&P 500 is looking to be 15 years.*)
3 F (*Smartphones have been the biggest game changer … As well as being able to check email, use social media and surf the net, consumers can now do many other things using just one device – listen to music, watch videos, manage their money, find their way and take photos.*)
4 T (*The music industry responded swiftly by embracing the potential of digital music distribution, counting on the fact that consumers would be willing to pay for music.*)
5 F (*Long-established car manufacturers like BMW and Volkswagen are responding by embracing the disruptive model and launching their own car-sharing services. BMW's service, for example, enables drivers in many cities to use an app to locate and hire a nearby car …*)
6 T (*Low-cost airlines have not only seized a huge slice of the market … , but they have also created an entirely new market for short-haul air travel.*)
7 F (*And budget airlines are also applying the low-cost model to the long-haul market.*)
8 F (*But ultimately, no business is safe. Disruption might be just around the corner and it's something no one can predict.*)

Vocabulary: Disruptors and disruption

Students look at vocabulary related to disruptors and disruption in business.

5 Get students to complete the exercise individually. Check that they understand the meanings of *rebranding* and *viable* in the definitions before they begin. Encourage them to find the words in videoscript 4.1.1 on page 147 before matching them with their definitions, so that they can see them used in context; this will help them work out their meanings. Check answers with the class, clarifying meanings as necessary.

> **1** d **2** i **3** b **4** f **5** h **6** a **7** g **8** j **9** c **10** e

6 Go through the collocations in the box with students before they begin and get them to complete the sentences individually. Alternatively, you could let them complete the sentences using their dictionaries and then clarify meanings during feedback.

> **1** game changers **2** digital transformation
> **3** venture fund **4** business landscape
> **5** established companies **6** novel approach
> **7** disruptive innovation

7 This exercise practises vocabulary from Exercises 5 and 6, so students should be able to do it individually. Remind them that they can refer to these exercises if they need help and, if there is time, get them to compare answers in pairs before checking with the class.

> **1** presenting themselves differently to customers
> **2** respond to changes in the market
> **3** market situation
> **4** have been in business for many years
> **5** other companies
> **6** is used for one particular purpose
> **7** the product or service that keeps them in business
> **8** is completely new and becomes popular very quickly

8 Put students in small groups and give them 3–4 minutes to discuss the question, then invite students from different groups to share their views with the class. Encourage them to give reasons. Note that digital natives may find it hard to conceive of life before the digital age. Nevertheless, they will hopefully be able to say something here about how the introduction of computers to the workplace made an immense difference to the earlier generations, as did the arrival of cheap mobile phones in making us constantly reachable and making it harder to disconnect from work at the end of the day. The rapid adoption of the internet will also likely have made substantial differences to the lives of students, their parents or their grandparents (depending on the age of your students). Travelling for work/meetings is also something that has changed in many people's lifetimes as the internet has brought with it a whole range of options for online meetings, rather than face-to-face ones. In the event that you have a mix of older and younger students in your class, this is an ideal moment to draw on the life experience of the older students who will be able to share at first hand how their working lives have been disrupted.

Extra activities 4.1

B This activity gives further practice of key vocabulary from the lesson. It is a consolidation exercise, so it would be better for students to do it individually. Go through the instructions with them and point out that there are two sentence endings that they do not need to use. Check answers with the class.

> **1** i **2** c **3** h **4** a **5** g **6** j **7** e **8** d
> (Not needed: b and f)

Project: Products we rely on

Students hold a panel discussion on products/services that are essential to them.

9A Explain to students that they are going to hold a panel discussion and put them in small groups. If you have not done Lesson 2.1 yet, start by eliciting or giving a brief explanation of what a panel discussion is and how it works (a situation in which a selected group of people discuss a specific topic or issue in front of an audience; panel discussions are often used in meetings, conferences and conventions). Go through the instructions with students, pointing out that they can choose to talk about a product or a service. Give them 1–2 minutes to decide on their products/services.

9B Ask students to work individually for this stage. Go through the instructions and questions with them and explain that they each need to 'defend' the product or service they chose in Exercise 9A, explaining why it is essential to them. Encourage them to make notes.

9C Students now hold their panel discussions, with one student in each group acting as a moderator and also keeping track of time. Each student is allowed one minute to make their point and the moderator then controls the Q&A at the end. If your group is too large for everyone to have a turn, students could be paired up to defend their products/services or the class can be split into two large groups, with each holding their panel discussions and Q&A sessions at the same time. It is also important for a different spokesperson each time to explain why they voted for the defence of a particular device or service, as they should be able to summarise the main points of the arguments set out. During the activity, monitor and note down any points you may wish to discuss during feedback but do not interrupt the discussions. Highlight these in a brief feedback session after Exercise 9D.

9D Students now vote for the best 'defence' of the products/services discussed in their group. Encourage them to give reasons for their choices and point out that they cannot vote for themselves! When they have finished, invite students from different groups to tell the class about the 'winner' in their group, explaining their reasons. Finally, go over any points you noted while monitoring. As a follow-up, you could ask students to a) make a list of reasons for companies to create disruptive products or b) make a list of conditions they think create a market for a disruptor.

MyEnglishLab: Teacher's Resources: extra activities; Reading bank
Teacher's book: Resource bank Photocopiable 4.1 p.148
Spoken English: p.113
Workbook: p.19

4.2 ⟫ Disruptive innovation

GSE learning objectives

- Can understand complex arguments in newspaper articles.
- Can distinguish between facts and opinions in linguistically complex written proposals.
- Can justify a point of view using linguistically complex language.
- Can talk about hypothetical events and actions and their possible consequences.
- Can write a review of a product or service using linguistically complex language.

Warm-up

Write *Pros and cons of innovation* on the board, put students in pairs and ask them to brainstorm ideas for the benefits innovation might bring to a business and the risks it involves. (Possible answers: **benefits**: better quality products/services which meet changing customer needs and, as a result, higher sales and profits; a broader product/service range; improved brand recognition and value; increased competitiveness; improved staff retention and morale; **risks**: can be costly and time consuming; failing to meet quality or cost requirements; failing to attract enough customers for the product/service; ending up wasting resources by developing a product/service that doesn't sell.) Give students a few minutes to discuss in their pairs, then invite students from different pairs to share their ideas with the class.

Lead-in

Students talk about the risks involved in disruptive innovation.

1 Put students in pairs, explain the activity and go through the questions with them. Check that they understand the meaning of *buzzword* before they begin. Give them 3–4 minutes to discuss the questions in their pairs, then get feedback from the class. List students' ideas for question 1 on the board.

Possible answers

1 An innovation may create a new market for goods and services that wasn't there before or change the way consumers use particular products or services. This can cause problems for established or market-leading companies who may need to change their products or the way they operate in order to remain competitive.
2 The digital revolution has led to a great many innovations which are regarded as disruptive. Therefore, disruption has come to be seen as popular and forward-thinking, breaking the old rules and challenging the established order of business. It is also a way for small companies to describe themselves in order to get funding from investors.

Reading

Students read an article about disruption in business.

2 Tell students that they are going to read an article about the definition of the term *disruption* in business and if desired, teach these key words from the text: *entrenched* (adj.), *upmarket* (adv.), *usher* (v.), *label* (v.), *fleeting* (adj.), *battle* (v.), *plough* (*money*) *into* (phr. v.), *nimble* (adj.), *complacent* (adj.). Give students plenty of time to read the text and identify the points from Exercise 1 it mentions; remind them to refer to the list on the board and underline the parts of the text where they think each point is mentioned. Discuss the answers with the class.

3 Students could do this individually or, in weaker classes, in pairs. Make sure they understand that, in addition to deciding whether the statements are true or false, they need to find evidence in the text for the true statements and correct the false ones. If they work individually, get them to compare answers in pairs before class feedback. As a follow-up, you could put them in pairs and ask them to discuss a company they feel has introduced disruptive ideas that have made important changes in the way we interact with others, deal with our everyday needs and do business, giving reasons to support their ideas.

Possible answers

1 F (Some businesses making use of fundraising strategies brand themselves as 'disruptive'.)
2 T (para 2)
3 F (The phrase was used in a specific way; it was about a product or service that was fairly simple, but moved into an important place in the market and displaced a business that was already there.)
4 F (Many of them are only successful for a short time. For example, although Encarta was labelled as a disruptor and was on the way to replacing Britannica, it was itself replaced by Wikipedia.)
5 T (para 7)
6 F (Being able to cause a disruption may depend on whether or not the established companies react quickly or ignore the presence of the disruptor.)
7 F (Sometimes a disruption is just based on luck, something that cannot be predicted.)

Extra activities 4.2

A This activity provides students with extra reading practice. It can be done individually or, in weaker classes, in pairs. Encourage students to underline the parts of the text where they find the answers and elicit these during class feedback.

1 used by 2 better known
3 made personal computers popular
4 competitors had not entered the market
5 may 6 quickly competitors react

Grammar: Hypothesising

Students study and practise different forms for expressing hypotheses.

4 Depending on the level of your class, you could do this as a whole-class activity, checking answers as you go, or let students attempt the matching task individually and then clarify any points as necessary during feedback. After checking answers, refer students to the Grammar reference on page 118 and go through it with them, clarifying any points as necessary. For the section on fixed expressions, point out the different verb forms to refer to different situations (present, unreal or possible future situations). For the section on inversion, remind students of the patterns for second and third conditional sentences

> **1** c **2** a **3** b

5A Students could do this individually or, in weaker classes, in pairs. Remind them that they can refer to the Grammar reference on page 118 if they need help and point out that there may be more than one possible answer for some items. You may also wish to point out that they need to find and correct *one* mistake per sentence. If students work individually, get them to compare answers in pairs before class feedback. Check answers with the class, clarifying any points/difficulties as necessary. For item 4, note that students may ask whether *will* changes to *would* in the second part of the sentence. Make clear that both *will* and *would* are possible here; as this is not technically a mistake, it is not included in the answer key to this exercise.

> **1** If only I <u>had</u> …
> **2** Given that he <u>lived</u> / <u>has lived</u> …
> **3** <u>Had we gone</u> to see / <u>If we had gone</u> to see …
> **4** <u>Let us imagine</u> that …
> **5** Suppose you <u>were</u> presenting …
> **6** Were we <u>to</u> open … / <u>If we were to</u> open …
> **7** If only they listened to their customers, they <u>might grow</u> the company. / If only they <u>had listened</u> to their customers, they might have grown the company.
> **8** What if I <u>had</u> spoken to her …

5B You could get students to complete the exercise individually and then check answers with the class or, if time is short, do this as a quick whole-class activity, eliciting answers as you go.

> **1** If <u>only</u> **2** Given that **3** *no word* (*inversion*) [or *if* if corrected with a third conditional] **4** Let us imagine (that) **5** Suppose **6** *no word* (*inversion*) [or *if* if corrected with a second conditional] **7** If <u>only</u> **8** <u>What</u> if

6 Ask students to do this individually. Remind them again to refer to page 118 if they need help and encourage them to look carefully at the verb forms after the gaps. Check answers with the class.

> **1** Let us **2** If only **3** Supposing **4** I wonder if **5** Were
> **6** I wish **7** Speculating for a moment **8** Had

B This activity gives further practice of hypothetical sentences. It should be done individually, as a consolidation exercise. Check answers with the class, clarifying any points or errors as necessary.

> **1** c **2** g **3** i **4** a **5** d **6** j **7** e **8** b **9** h **10** f

Speaking and writing

Students talk and write about possible big disruptors.

7A Put students in pairs and explain the activity. Encourage them to be creative and if necessary give them an example for the first item (e.g. *I wonder if an app which showed us where we could relax between classes and automatically beeped to remind us to go to our next class would be popular with people our age*). During the activity, monitor and check that they are using the forms/expressions for hypothesising correctly. As feedback, invite students from different pairs to share some of their ideas with the class.

7B In their pairs, students now discuss their ideas from Exercise 7A, choose one as the next 'big disruptor' and write a short blog or intranet post presenting it. Remind them to use forms/expressions for hypothesising and point out the word limit before they write their posts. The writing task could be done individually or in pairs and if there is no time to do it in class, it can be assigned as homework.

Model answer

My colleagues and I wonder if most students wish they had more free time between classes. Supposing we invented an app which showed us where we could relax and meet friends between classes and automatically beeped to remind us to go to our next class? Given that we all use electronic devices, we could certainly use an app to better organise our university life. Let's imagine that we didn't need to keep track of our timetables and social arrangements ourselves, but just relied on a device instead. Had we used this app last month, we wouldn't have been late for our last exam. It seems like something that would be of interest to a number of students and something we could investigate further.

MyEnglishLab: Teacher's Resources: extra activities
Grammar reference: Hypothesising p.118
Teacher's book: Resource bank Photocopiable 4.2 p.149
Workbook: pp.20–22

4.3 ❯ Communication skills
Finding solutions

Warm-up
Ask students to imagine that they have been asked to come up with ideas for a new product for a company. What would they do to come up with as many ideas as possible? What way of generating ideas would they find most/least helpful? Why? Invite different students to share their ideas and answers with the class.

Lead-in

Students talk about brainstorming.

1A Look at the definition of *brainstorming* with students and go through the questions with them. Put them in small groups and give them time to discuss the questions, then invite students from different groups to share their ideas with the class. When discussing question 2, you may also wish to ask students if they can think of any disadvantages (e.g. students may say that brainstorming can be a time-consuming process; some team members may not contribute or try as hard as they would if they were trying to generate ideas on their own; some people may not feel comfortable speaking in front of / sharing ideas with a group).

> **2** Possible answer: boosts creativity and critical thinking, cultivates spaces where innovation can flourish, generates more ideas, it's safe as participants suspend judgement.

1B Refer students to page 127, go through the guidelines with them and check that they understand *defer*, *build on* (an *idea*), *introvert* and *remote worker*. If there is time, let students discuss the question in pairs or small groups first, then broaden this into a class discussion. Encourage students to give reasons for their answers.

Preparation: Finding ways to increase the magazine's revenue

Students read and think about the scenario for a roleplay.

2 If this is the first Communication skills lesson for your class, briefly tell students about *Lifestyle* magazine and the profile of its readership. Otherwise, elicit this information from students before you start. Go through the instructions with students and, if desired, teach this vocabulary from the text: *volume* (of *traffic*), *diversify*, *level off*, *revenue stream* and *-centric*. Give students time to read the text, then elicit the answer.

> Teo Doğan wants to change the business model for a number of reasons. Firstly, it's heavily dependent on advertising revenue. This small publisher does not have the resources to generate more business itself from advertisers. Its dependence on the PPC model via an advertising app means revenue is uncertain and unstable. There is also competition for advertising revenues from a growing number of online publications like *Lifestyle*. Readers use ad blockers which reduces the magazine's income. Finally, he mentions that competitors are already shifting and diversifying their business model to ensure multiple revenue streams.

Roleplay

Students roleplay a brainstorming meeting.

3A Put students in groups of three and explain the scenario: they all work for *Lifestyle* magazine and after reading Teo's report, they are having a brainstorming meeting to discuss the question 'How can we increase the magazine's revenue?' Assign roles (or let students choose), then refer students to their relevant pages and explain that they can use the ideas there and/or add ideas of their own. At this point, you may also wish to assign the role of meeting chair to one student per group (or let students decide). If useful for your class, spend a moment eliciting what the chair needs to do (e.g. keep an eye on how much time is spent on each stage of the meeting, encourage everyone to participate, make notes so the ideas can later be evaluated, confirm what has been agreed at the end of the meeting). Allow plenty of time for students to prepare for their meetings while you monitor and help them as necessary.

3B Students now roleplay their meetings. Set a time limit before they begin. During the activity, monitor and note down any points to highlight during feedback after Exercise 3C, but do not interrupt students' meetings.

3C In their groups, students now reflect on their roleplays. Go through the questions with them and give them time to discuss in their pairs, then invite different students to share their answers with the class. Finally, highlight any points you noted while monitoring.

Video

Students watch a video of a brainstorming meeting.

4 ▶ 4.3.1 Tell students that they are going to watch a video of a meeting similar to the one they have just had and go through the instructions with them. Give them a minute to read the questions, then play the video and check answers with the class.

1 James. He introduces the issue they will be brainstorming and reminds them of some brainstorming rules (e.g. generate lots of ideas, don't spend time on any one thing, focus on quantity, think 'outside the box').

2 Each person presents their idea briefly.

3 Yes, they sometimes build on each other's ideas (e.g. James mentions appeals for donations after Donna mentions crowdfunding and when James mentions events such as a book festival, Teo adds conferences with guest experts).

4 In Part 3 they are evaluating the paywall idea that Teo mentioned in Part 2.

5 In Part 4 they make decisions about what to do next. In this case, they need to do more market research to find out how viable a paywall would be.

6 It appeared to be very effective. They used good brainstorming techniques. The structure was very clear (generate ideas, defer judgement, etc.), they all contributed ideas, collaborated with the process and built on each other's ideas.

5 You could do this as a whole-class activity, checking answers as you go. Alternatively, you could let students complete the exercise individually, using their dictionaries and/or referring to the videoscript on page 147 to see the words used in context, then clarify meanings as necessary during feedback.

1 b **2** h **3** a **4** f **5** e **6** g **7** c **8** d

Extra activities 4.3

A This activity gives further practice of useful vocabulary from the video. Students saw the words and phrases in Exercise 5 of the Coursebook, so they should be able to complete this exercise without help. Ask them to work individually and then, if time allows, to compare answers in pairs before class feedback.

1 tiered **2** pitch in **3** tap into **4** ploughing money into **5** crunch time **6** throws up / has thrown up **7** a bit off the wall **8** lagging behind

Reflection

Students reflect on the conclusions from the video and their own approach to brainstorming.

6A Allow students to work individually first so that they can reflect on their own approach to generating ideas and finding solutions. Remind them to think about their answers to Exercises 1A and 3B and the guidelines from Exercise 1B on page 127. Encourage them to make notes so they can discuss their ideas with a partner in the next activity.

6B Put students in pairs or small groups to compare their reflections, then round up ideas in a class discussion.

MyEnglishLab: Teacher's Resources: extra activities; Interactive video activities

4.4 ❯ Business skills
Reporting and planning

GSE learning objectives

- Can infer meaning, opinion, attitude, etc. in fast-paced conversations between fluent speakers.
- Can present their ideas with precision and respond to complex lines of argument convincingly.
- Can exchange complex information on a wide range of matters related to their work.
- Can give a detailed account of a complex subject, ending with a clear conclusion.
- Can participate in extended, detailed professional discussions and meetings with confidence.

Warm-up

Write this quote on the board: *Life is a series of natural and spontaneous changes. Don't resist them; that only creates sorrow* (Lao Tzu, Chinese philosopher). Discuss the quote with the class. Do they agree that resisting change can only 'create sorrow'? Is it always a good idea to accept change as it comes? Do they themselves usually resist changes?

Lead-in

Students talk about different attitudes to change.

1A If there is time, get students to discuss the questions in pairs or small groups first, then invite different students to share their answers with the class.

1B Explain to students that the text and diagram explain the 'change curve concept': the different stages we usually go through following change. Give them time to read the text and look at the diagram and answer any vocabulary questions they may have. Put them in pairs and give them 2–3 minutes to discuss the questions, then broaden this into a class discussion.

1C You could discuss the question with the whole class or, if time allows, let students exchange ideas in the same pairs as for Exercise 1B and then get feedback from the class.

Possible answer

People often feel anxious about change when its purpose is unclear, when they feel insecure as to how they can support the change and particularly if they feel they will lose out during the change process. So management, in individual briefings and in larger group meetings, need to make clear the organisational purpose, the process and the individual benefits of change so that people can become engaged.

2 Students could do this in the same pairs as for Exercise 1B or in new pairs. Go through the instructions with them, check that they understand *transition* and give them a few minutes to discuss the questions in their pairs. As feedback, if you think your students would be comfortable doing so, invite them to share their experiences with the class.

Listening

Students listen to a meeting about change management.

3A 🔊 4.01 Go through the instructions with students and explain that they are going to hear the first part of the meeting between Jon, Maria and Paweł. Give them a minute to read the questions and ask them to make notes as they listen. Play the recording, twice if necessary, then check answers with the class.

> 1 The hotel visitor numbers across the group are down. The business is simply not there to warrant keeping all their service staff.
> 2 They could have: worked with lots of one-to-one meetings, dealt with emotions in private, talked things through, made clear what the changes meant in daily life.
> 3 A lot of anxiety and negativity was felt by staff in the Nordics about the investment in Poland, as it seemed to indicate a lack of focus/interest in the Nordic business, possibly a first step to closing down operations there.
> 4 Demotivation of internal staff because guests are starting to notice and feel it.

3B Do this activity with the whole class, eliciting answers from different students around the class.

> **Possible answer**
> People might not listen carefully to such management information. In any case, the logic of the change is very different to the impact of the change and people are naturally afraid of losing their jobs, etc.

4A 🔊 4.02 Explain to students that they are now going to hear the second part of the meeting and give them time to read the questions before they listen. Check that they understand *settle feelings* in question 2, then play the recording. Check answers with the class.

> 1 Organise some change workshops to explain what will be different and how roles might change.
> 2 The new webinar series to introduce the Polish expansion to staff.
> 3 60% of staff have signed up for a new homeworking option.
> 4 The negotiation with the lease holder to reduce the office space and get the cost advantage.

4B Again, do this as a whole-class activity, inviting different students to share their views with the class.

> **Possible answer**
> The statement is a generalisation but many managers do favour a focus on hard facts such as financial or technical data rather than a focus on emotions.

Useful language

Students look at useful language for discussing change.

5A Explain to students that they are going to look at useful language for discussing change and go through the headings in the table with them. Check that they understand *specifying*, *exploring* and *looking ahead*. Allow plenty of time for them to read the sentences which are already in the table and then to complete the gaps with sentences a–f (they could do this individually or in pairs). Check answers with the class, clarifying meanings as necessary. Make sure that students understand both the gapped sentences and the ones that are already in the table.

> 1 e 2 c 3 a 4 f 5 d 6 b

5B Refer students to audioscripts 4.01 and 4.02 on pages 153–154 and explain the activity. Give them 3–4 minutes to find and underline the expressions, then get them to compare answers in pairs. Get feedback from the class and list the phrases students identify on the board, for them to refer to when they do the roleplays in Exercise 6B.

> **Track 4.01**
> I think we're on track with the cost reduction programme, yes?
> We've let twenty people go now, right?
> It's not been at all easy to keep people motivated during the process.
> It's been very disruptive.
> You had your reservations at the beginning.
> We needed to make this change.
> The business is simply not there to justify keeping …
> We can't ignore the facts.
> It's our first time with this kind of change …
> I think we'll end up losing more people before the end of the year.
> There's a lot of uncertainty and confusion about what's happening.
> I did a big townhall speech and laid out the situation.
> I think our recent expansion into Poland has been part of the problem here.
> This generated a lot of anxiety and negativity.
> It's about diversification.
> We were doing this to safeguard jobs.
> We had to cut costs.
> **Track 4.02**
> Given people time to digest what happened.
> I think, basically, we should be looking at …
> By this time next week, we'll have started …
> Just make people aware that this is an opportunity, not a threat.

Extra activities 4.4

A This activity practises useful language for discussing change. Explain the writing task and remind students that they should try to use expressions from this lesson in their email. Also point out that they can add details of their own if they like. In weaker classes, you could let students plan their email in pairs. If time is short, they can then write their email as homework.

Model answer

Dear all,

I hope you are all well. Here is an update on my life in the States.

I am enjoying Chicago. It's been great. I've had lots of great experiences which have really triggered a lot of changes in me as a person. Mind you, I'm still struggling a little with English. I think if I'd made more time for English lessons before leaving for the USA, I wouldn't be having these problems.

I have a good apartment, but it's too far away from work. I should have found one closer to the office, to be honest. But it's OK. The project is going well and we're hoping to have all the software installed soon. By the end of this month, we'll have the project back on track.

The biggest insight for me is how important it is to be open and to accept people from different cultures. Developmentally, it's been great and I have really grown a lot. Emotionally, it's been tough, but that's also been good.

Look forward to seeing you guys in Paris again soon.

Regards,

Hugo

B This activity looks at some useful idioms from the listening. Students could do it individually or in pairs, using their dictionaries to help them if necessary. If there is time, encourage them to find the idioms in audioscripts 4.01 and 4.02 on page 153; seeing them used in context will help students work out their meanings. Check answers with the class, clarifying meanings as necessary and encourage students to record the idioms in their notebooks.

1 it sent mixed messages
2 we have really shot ourselves in the foot
3 people are over the moon
4 across the board
5 building a platform for the future
6 food for thought
7 it's been a steep learning curve

Task

Students roleplay an interview for a change management consultancy role.

6A Put students in groups of three, explain the scenario and activity and give them time to read the advertisement and ask you any questions they may have. Explain that they are going to roleplay the interview in the next activity and allow plenty of time for them to think about how they would answer the questions in the advertisement, while you monitor and offer help as necessary.

6B Tell students that they are going to take turns to be the interviewer, candidate and observer and explain that the observer should take notes of the candidate's answers during the interview. After each interview, the observer and interviewer should discuss the candidate's answers, give quick feedback on how well they think he/she handled the questions and decide whether he/she should be hired. Make sure students understand that they will be rotating roles and that they need to pause between interviews to give their feedback and make their decisions. Remind them to use language from Exercises 5A and 5B where possible. Before they begin, you may wish to give them 1–2 minutes to look at these exercises again and think about how they could use the different phrases during the interviews. Set a time limit for each interview and feedback session and ask students to begin. Monitor and note down any points to highlight during feedback, but do not interrupt the interviews or feedback sessions. As a follow-up, you could ask a few students to tell the class how well they think the interviews went and/or what they found easy or difficult about discussing change.

6C Give students 3–4 minutes to discuss the question in their groups, then broaden this into a class discussion. Finally, go over any points you noted while monitoring students' interviews in the previous activity.

MyEnglishLab: Teacher's Resources: extra activities; Useful language bank

4.5 ▶ Writing

Supply chain choices

GSE learning objectives

- Can write a detailed work-related report outlining issues and problems.
- Can write a detailed structured report on work-related topics.
- Can structure longer complex texts using a range of cohesive devices.
- Can write about complex subjects, underlining the key issues and in a style appropriate to the intended reader.
- Can systematically develop an argument giving the reasons for or against a point of view.
- Can correct structural errors in someone else's written report.

Warm-up

Elicit or give a brief definition of *supply chain*. Try to elicit the entities involved in a typical supply chain and write them on the board (e.g. *raw material supplier → manufacturer → wholesaler/distributor → retailer → customer*), then move on to the questions about supply chain management (SCM) in the Lead-in.

Lead-in

Students talk about supply chain management.

1 Do this as a whole-class activity, eliciting answers around the class.

Possible answer
SCM is the whole process upstream of assembly/production, starting with raw materials and passing through suppliers and their individual sub-suppliers. The main issues are that the assembly/manufacturing plant needs parts and components at their factory gate Just-In-Time, Just-In-Sequence, at a very good quality level, at a reasonable price and with suppliers able to respond flexibly to changes in demand and changes in product design. These days, all the companies in the supply chain will share production data on an integrated IT system, so that bottlenecks or shortages along the chain can be spotted and managed in time. A manufacturer might want to change a supplier if any of these issues are not working well.

2 Refer students to the report on page 139 and give them time to read it and choose the sentence which best summarises it, then elicit the correct answer. Point out the main structure of the report, but do not go into detail about the content yet, as students will look at this in the next activity.

> 2

Useful language

Students look at useful language for, and the main structure of, an internal recommendation report.

3 This is best done as a whole-class activity, checking and discussing answers as you go. After discussing the answers for each item, go through the relevant section of the Useful language box with the class.

Possible answers
1 In modern business writing, you don't just wait until the end to make a recommendation. You state it at the beginning as well, so that the reader has a context for the arguments that follow. It is also an example of the well-known discourse structure 'Say what you're going to say, say it, then say what you said'. This can be used in presentations and contributions to meetings as well as reports. It reinforces the message.
2 These words are often confused, with *proposal* being overused. The dictionary definitions are: *suggestion*: an idea, plan or possibility that someone mentions; *proposal*: a plan or suggestion which is made formally to an official person or group; *recommendation*: official advice given to someone, especially about what to do. So, a suggestion is something more open and informal and about something that is not necessarily important. Other people can easily agree, disagree or make other suggestions. It would not form the subject of a report. A proposal is more formal. It is about an important issue and will have details like cost and timescale and is often written down. A group of people will consider it, over a longer time period. A recommendation is advice rather than a plan. It can be spoken or written, formal or informal, important or not important.

3 Examples of linking expressions in the report:
Nevertheless, there have been … ; _In particular_, we have … ; **however**, it is difficult to … ; _In fact_, these points … ; _In summary_, I have strong doubts … ; _Of course_, we have a strong … ; … significant negatives, **above all**, the risk … ; I **therefore** recommend … .
They are not always at the beginning of a sentence. See examples in bold above.
4 a _On the plus side, … , but on the downside …_
b _Of course, / It's true that … ; however, … / On the one hand, … , but on the other hand…_
c _In general, … , although … / On the whole,… , but in this case …_
d _Firstly, … and secondly, …_
e _The main issue going forward is … / I therefore recommend that …_
5 Writers of business reports have to show that they have a balanced viewpoint that is based on careful consideration of the facts. Without measured language the writing might seem too dogmatic.

4 Get students to do this individually and then compare answers in pairs before class feedback. You could list the words/phrases students choose on the board for them to refer to during the writing task.

Extra activities 4.5

A–B These activities look at useful linking words and expressions. You could do Activity A as a quick whole-class activity, checking answers as you go. Students should then do Activity B individually. Explain that more than one word/phrase may seem possible for some items and point out that where that is the case, students should check if the word/phrase would fit better in another sentence. Check answers with the class.

A 1 d **2** a **3** b **4** c **5** h **6** g **7** e **8** f
B 1 However **2** In particular **3** In relation to **4** on the minus side **5** In general **6** Of course **7** In fact **8** Therefore

Optional grammar work
The recommendation report in Exercise 1 contains examples of inversion for emphasis, so you could use it for some optional grammar work. Refer students to the Grammar reference on page 119 and use the exercises in MyEnglishLab for extra grammar practice.

Task

Students write an internal recommendation report.

5A Put students in pairs, explain the activity and give them time to read the notes. Before they add their own ideas, let them ask you about any unknown words or anything else they may not understand. If you think your students will struggle to come up with ideas for the notes in their pairs, you could do this as a whole-class activity, inviting volunteers to contribute ideas and listing them on the board for students to refer to during the writing task. If students work in pairs, monitor during the activity and provide help as necessary.

5B Depending on the time available, students could plan their reports in class and write them for homework. Highlight the word limit and remind students to use the report in Exercise 1 to help them if necessary. Also remind them to use language from Exercises 3 and 4.

Model answer

Report: Review of suppliers for our SE Asia assembly plant

Background

We have just one supplier, QualTec, for our SE Asia assembly plant. The business relationship has been generally satisfactory up to now, with both positives (their quality levels are good) and negatives (their delivery times are not always reliable). But clearly, the fact that they are the sole supplier adds significant business risk for us. It should be noted that all our main competitors have several suppliers.

I believe that we need to review the whole situation in SE Asia in relation to our supply chain. Firstly, we should give a strong warning to QualTec that they need to be much more consistent with their delivery times. Secondly, we need to look for a second supplier anyway, regardless of the situation with QualTec.

Business case for reviewing SE Asia supply chain

QualTec have been our business partner for many years. There have been both positives and negatives in this relationship. Their quality levels are reasonable without being excellent. Other areas are very positive, in particular their input of accurate data into our IT system which helps us with our supply chain management. However, on the minus side there are serious issues with their late deliveries to our assembly plant. We operate a very tight Just-In-Time assembly line with just two days of stock at the factory gate. Any late delivery causes serious problems.

In fact, these points are secondary. The main issue is that QualTec have a monopoly position as our sole supplier. It's true that there are certain advantages in only having one supplier to deal with, but on the other hand it gives them too much bargaining power over price, etc.

In summary, I believe that the situation in SE Asia needs to be the subject of a management review. Of course, we could simply talk to QualTec about the delivery times and continue as we are; however, the situation with having just one supplier definitely leaves us in a weak position.

I therefore recommend that we give QualTec a formal warning that unless their delivery times become more reliable, we will end the business relationship. In addition, we should look for a second supplier so that we have greater control over the supply chain, for example, in areas such as responding to fluctuations in demand at our assembly plant and being able to negotiate on the price of supplied parts. A senior SCM team member needs time allocated to deal with all this and they can report back at the March meeting.

5C If students write their reports for homework, you could do this exercise in the next lesson. Put them in pairs and ask them to read each other's reports and discuss the questions. Encourage them to tell their partner what they think he/she has done well and suggest corrections/improvements where appropriate. During the activity, monitor and offer help as necessary. Students could then be asked to rewrite their reports based on their partner's feedback; they could do this in class or as homework.

MyEnglishLab: Teacher's Resources: extra activities; Writing bank; Interactive grammar practice

Grammar reference: Emphasis using inversion and fronting p.119

Workbook: p.23

Business workshop > 4
Disruption – planning ahead

GSE learning objectives

- Can understand the details in a linguistically complex audio recording.
- Can understand an extended hypothetical argumentation in a linguistically complex discussion.
- Can comment in detail on the content of a linguistically complex radio programme or podcast in which people describe reactions or opinions.
- Can talk about hypothetical events and actions and their possible consequences.
- Can present detailed, evidence-based arguments during work-related meetings.
- Can summarise orally information from different spoken sources, reconstructing arguments to present the overall result.
- Can give a detailed account of a complex subject, ending with a clear conclusion.
- Can summarise information from a linguistically complex presentation or lecture.

Background

Students read about a move towards autonomous transport in the UK and its impact on the economy and society.

1 Put students in pairs and give them time to read the background and discuss the questions. During class feedback, feed in information from the Note below as required. Finally, help students with any unknown vocabulary from the text.

Possible answers

1 Deciding the cost of using public roads, transportation for the public such as buses, trams and trains and keeping traffic moving in cities.
2 The consultants from the university in urban planning and city officials who work on congestion and parking issues.
3 The consultants from the automotive and petrol industries.
4 The increase in hybrid and electric cars has reduced the need for petrol.
5 Past disruptions and their impact on businesses and society.

Note

Many automotive manufacturers, think tanks, governments, insurance agencies and so on are beginning to look into the effects that autonomous vehicles will have on them, their businesses and on society as a whole. The car was one of the largest disruptors of the twentieth century and it is predicted that autonomous vehicles will be one of the largest of the twenty-first. The more technologically developed cars today already use technology for parking assistance, leading some people to feel that driverless technology is closer than we think. In order to test the ideas, a number of research laboratories and car manufacturers are beginning to experiment with the concept. These vehicles are not ready for urban driving, however, and governments need to look into ways to make testing them on city streets safe for all involved. Changes in the frequency and types of accidents have been predicted, giving insurance companies a number of options to consider. As this move towards autonomous transport grows more possible, planning for this major disruption is currently being taken seriously by all those involved.

Dealing with the problems

Students listen to a radio programme about the development and use of autonomous vehicles.

2 🔊 BW4.01 Put students in pairs, go through the instructions with them and give them 3–4 minutes to discuss their ideas. Alternatively, if time is short, students can brainstorm ideas as a whole class, then listen to the radio programme to check them; this option may be easier for weaker classes.

Some of the things mentioned include:
- technology: using a simulator, cameras, radar, sensors, making a virtual picture.
- car manufacturers: don't want to be left out, cars becoming more electronic, doing partly autonomous things like assisted parking.
- safety: cars are rehearsed so they know the route, adding data should make them more trustworthy, automated emergency braking reduces accidents by 15%; 93% of accidents caused by human error.
- urban driving: chaotic, cars, cyclists, people crossing the road, trucks unloading, reduce congestion and emissions.
- regulations including insurance: what laws will be made, how insurance companies are handling this, insurance companies may move away from small claims to bigger transport package, laws and standards need to be worked out.

3 🔊 BW4.01 Give students time to read the questions before they listen and check they understand *do away with* in question 4. Play the recording, twice if necessary, then check answers with the class.

1 They would free up time for commuters.
2 They use cameras, sensors and radar to create a picture around the vehicle.
3 assisted parking

4 He thinks that we have only really tested autonomous vehicles moving at very slow speeds in a predictable environment. He says there is a way to go, parking is much easier than actual driving, because the cars are moving slowly and the environment is known but that is far away from having no steering wheel.
5 The Department of Transport has devised a code of practice for tests with safety in mind.
6 As the large majority of accidents are caused by human error, insurance companies may need to expand beyond the simple accidents and provide other services in a bigger transport package.
7 Cars that take us everywhere will not exist in the near future, but cars will most likely get more and more automatic and better at being autonomous bit by bit.

4 Depending on the size of your class and the time available, you could ask students to discuss the questions in pairs or small groups first, then invite different students to share their ideas with the class. Alternatively, if time is short, discuss the questions with the whole class.

5 Put students in small groups, go through the instructions with them and check they understand the meanings of *parallel* and *advent*. Depending on the resources available to your students, you could let them do some online research to help them with ideas. Alternatively, you could prompt them with a few ideas from the answer key below if necessary. Give them plenty of time to discuss in their groups, then get feedback from the class. The activity can also be done as a whole-class discussion if you are short of time.

Possible answers
Businesses: fewer horse breeders; decrease in blacksmiths, stables, grooms, buggy makers; manufacturing grew and provided jobs; start of the assembly line which changed manufacturing; trucks used to transport goods.
Society: streets became cleaner; people moved to the suburbs, people became more independent, drive-through and fast-food restaurants opened, highways were built.
Advantages: cleaner streets; easier and quicker to travel; more jobs in urban areas; more independence in general and especially for women; people could move more easily to find work.
Disadvantages: air pollution caused by emissions leading to global warming; leading cause of death due to car crashes/accidents; people spending longer on the road than at home; fewer bicycles/trains, less public transport; people get less exercise.

Extra activities Business workshop 4

A This activity looks at useful vocabulary from the listening. If time is short, you could do it as a whole-class activity, checking answers and clarifying meanings as you go. Alternatively, get students to complete it individually. Encourage them to find the words in audioscript BW4.01 on page 159; this will help them work out their meanings. Check answers with the class, clarifying meanings as necessary.

1 e **2** c **3** i **4** f **5** j **6** a **7** h **8** b **9** g **10** d

An impact study

Students talk about the impact of a move towards autonomous vehicles on different sectors of the economy.

6A Put students in pairs and go through the instructions with them, pointing out that they need to think of as many specific examples as they can for each sector. Before they begin, look at the different sectors in the pie chart with them and check they understand each one. Encourage them to make notes and give them 3–5 minutes to discuss in their pairs. Get feedback from the class, listing the examples students come up with on the board.

Possible answers

Legal matters, insurance, traffic control: law suits regarding crashes, private car insurance, accident insurance, driving licences, driving schools, ownership of autonomous vehicles.
Transport of freight: lorries, trains, airplanes, ships, containers.
Land development, construction, infrastructure: urban planning, parking garages, no parking places on streets, streets can be wider, parking places for autonomous vehicles, need/no need for traffic lights, street signs, petrol stations changed to e-charging stations.
Automotive industry: car production plants, suppliers, sub-suppliers, raw materials, repair shops, car dealerships.
Commercial personal transport: taxis, limousines, rental cars and businesses.
Electronics, software, apps, digital media: development of new software for the autonomous cars, new apps needed, digital media necessary on streets for autonomous cars to find their way.
Petrol industry: cars becoming electric, oil fields and offshore rigs, petrol stations changing, no delivery lorries with petrol.
Medical services: ambulances need to change to autonomous, should be fewer accidents, insurance situation different, first-aid kits in cars, special lanes on highways for ambulances; program the autonomous vehicle.

6B Students should do this in the same pairs as for Exercise 6A. Refer them to the list of examples on the board and explain the activity. Allow 3–4 minutes for them to make their lists, then join pairs together into groups of four and ask them to compare their ideas and discuss the questions. As feedback, invite students from different pairs to share their ideas with the class.

Possible answers

Legal matters, insurance, traffic control: personal injury lawyers specialising in vehicle injuries, car insurance salespeople, traffic police, parking wardens, driving instructors, lawyers specialising in laws regarding autonomous vehicles.
Transport of freight: lorry drivers, train conductors and drivers, dock workers.
Land development, construction, infrastructure: urban planners, parking garage attendants, repair people for traffic lights, street painters, petrol station attendants.
Automotive industry: assembly line workers, engineers, people working for suppliers and sub-suppliers, miners, those who extract raw materials, people working in repair shops, car salespeople.

Commercial personal transport: taxi and limousine drivers, rental car businesses' employees.
Electronics, software, apps, digital media: software engineers, app developers.
Petrol industry: oil field or rig workers, petrol station attendants, lorry drivers.
Medical services: ambulance drivers, paramedics, emergency teams, first responders.

6C You could let students discuss the questions in the same pairs or groups of four as for Exercise 6B and then get feedback from the class or, if time is short, you could do this as a whole-class activity, inviting different students to contribute.

Extra activities Business workshop 4

B This activity practises useful collocations from the listening and is best done in two stages. Start by asking students to match the words in the first box with those in the second to make collocations. Check answers with the class, but do not focus on meaning yet. Ask students to complete the definitions with the collocations they have just formed; they could do this individually or, in weaker classes, in pairs. Check answers with the class, clarifying meanings as necessary.

1 drives on   2 virtual picture   3 code of practice
4 assisted parking   5 human error   6 fully fledged
7 steering wheel   8 free up time
9 leave someone out

Task: Present a choice

Students give a presentation on mitigating the impact of a major disruption and write a handout summarising their presentation.

7A Explain the scenario: students have been asked to come up with strategies and priorities focused on the impact of autonomous vehicles. Divide the class into four groups, assign roles A–D (or let students choose) and go through the instructions with them. Allow plenty of time for them to research ideas and point out that they need to be as specific as possible and think about everyone involved: the economy sector, everyday life and society as a whole. Encourage them to make notes so they can refer to them when they present their ideas in the next step. During the activity, monitor and provide help as necessary.

7B Put students in new groups with a representative from each of the original groups. If your class does not divide up into fours, some groups can have two representatives from some of the original groups. Explain that students should take turns to present the ideas from their original groups and then work out a set of recommendations for the think tank while considering the pros and cons of the suggestions made. Again, encourage them to make notes and go round offering help where necessary.

7c Tell groups that they will now take turns to present their recommendations from the previous step to the class. Explain that they will later need to put together a set of recommendations as a class, so while listening to the other groups, they should note down any key points they think are worth going over during the class discussion. Set a time limit for each presentation and allow groups a few minutes to prepare. When all the groups have given their presentations, hold a class discussion to put together a final set of recommendations for the think tank.

8 Students now write a handout summarising the key points of their presentations. Point out the word limit and remind students to keep it concise; encourage them to use bullet points. Also remind them that they can refer to their notes from the previous steps to help them. If time is short, they could plan their answers in class and then do the writing task for homework. For weaker classes, you may wish to let students plan their answers in pairs.

Model answer

Handout on recommendations to mitigate the impact of autonomous vehicles

Urban planning
- Rethink parking spaces on streets and in garages.
- Extra space could be used for pedestrians or cyclists.
- Garages could be used for storage areas.

Automotive industry
- Make sure that cars can handle chaotic city driving.
- Change to electric cars, retrain workers.
- Car dealerships will need help transforming their businesses, perhaps becoming owners of fleets of autonomous vehicles.

Governments and insurance companies – legal matters
- Determine/Regulate who will own vehicles and who will be responsible for maintenance.
- Reformulate insurance policies, offer wider range of services.
- Help retrain driving teachers, parking wardens, traffic police.

Petrol industry
- Invest in alternative methods of supplying fuel to vehicles.
- Convert petrol stations into charging stations.
- Retrain people in the industry to work with electric and autonomous vehicles.

MyEnglishLab: Teacher's Resources: extra activities

Review ◀ 4

1 1 repositioned **2** displacing **3** lifespan **4** agile
  5 venture **6** mainstay
2 1 established company **2** business landscape
  3 digital transformation **4** game changer
  5 novel approach
3 1 Had, seen **2** only, had **3** Given, has **4** wonder, has
  5 Suppose, should **6** Were, to
4 1 g **2** b **3** d **4** a **5** f **6** h **7** c **8** e
5 1 b **2** d **3** f **4** a **5** e **6** c

5 > Customer engagement

Unit overview

	CLASSWORK	FURTHER WORK
5.1 > **Marketing strategies**	**Lead-in** Students talk about factors that influence consumer behaviour. **Video** Students watch a video about influencing customer behaviour. **Vocabulary** Students look at vocabulary related to customer engagement. **Project** Students plan and present a marketing campaign.	**MyEnglishLab:** Teacher's resources: extra activities; Reading bank **Teacher's book:** Resource bank Photocopiable 5.1 p.150 **Workbook:** p.24
5.2 > **Persuasion**	**Lead-in** Students talk about using persuasion as a marketing strategy. **Reading** Students read an article about the importance of persuasion in the workplace. **Grammar** Students study and practise participle clauses. **Speaking** Students practise participle clauses by writing and talking about their financial decisions.	**MyEnglishLab:** Teacher's resources: extra activities **Grammar reference:** p.119 Participle clauses **Spoken English:** p.114 **Teacher's book:** Resource bank Photocopiable 5.2 p.151 **Workbook:** pp.25–27
5.3 > **Communication skills:** Presenting research data	**Lead-in** Students talk about data presentation. **Roleplay** Students present research data. **Video** Students watch a video of a presentation of research data. **Reflection** Students reflect on the conclusions from the video and their own approach to presenting and discussing research data.	**MyEnglishLab:** Teacher's resources: extra activities; Interactive video activities
5.4 > **Business skills:** Building relationships on trust	**Lead-in** Students talk about trust and trust-building strategies. **Listening** Students listen to a meeting about diagnosing problems among senior management. **Useful language** Students look at strategies and useful language for building trust in professional relationships. **Task** Students use trust-building strategies in a negotiation.	**MyEnglishLab:** Teacher's resources: extra activities; Useful language bank
5.5 > **Writing:** Advertising copy	**Lead-in** Students read and talk about the features of persuasive advertising copy. **Useful language** Students look at techniques and useful language for writing persuasive advertising copy. **Task** Students write advertising copy for a product.	**MyEnglishLab:** Teacher's resources: extra activities; Writing bank; Interactive grammar practice **Grammar reference:** p. 120 Groups of adjectives and gradable adjectives **Workbook:** p.28
Business workshop 5 > The art of persuasion	**Listening** Students listen to conversations about soft skills in the workplace. **Reading** Students read texts about skills needed in different companies. **Task** Students create and present a training course on teaching persuasive skills.	**MyEnglishLab:** Teacher's resources: extra activities

Business brief

The main aim of this unit is to explore how marketing influences customer behaviour. It looks at the techniques and strategies used by the marketing industry, the impact of digital technology and how consumers are targeted.

Since the digital revolution, marketing professionals have seen customer behaviour change radically. In the past, the **customer journey** started with a customer becoming aware of the product and brand, getting to know it, looking at and/or trying it, purchasing it and finally becoming a loyal customer. However, the enormous amount of information and decision-making power unlocked by the internet means that today's customer no longer behaves in this linear way. Instead, their decision-making process is much more **iterative**. Social media channels and search engines influence the customer journey by allowing customers to search for products using personalised criteria and search terms. This new model introduces the idea of a 'basket' of products being held for consideration. In order to get their products into the 'basket', marketers now have to make sure they provide as much information about them as possible.

Successful marketers need to understand what convinces customers to buy products and find better ways to persuade them. Many companies now look to the findings of researchers into '**persuasion science**', which draws on psychology, neuroscience and **behavioural economics** (the study of the effects that psychological factors have on how people make decisions about money), to help them develop and target their marketing strategies. Persuasion means changing the way people see the facts, so details can be very important in influencing them to make small changes in the attitudes and beliefs which affect their decisions. Researchers have identified a number of '**principles of persuasion**' which they believe can be **deployed** systematically to develop successful influencing strategies. For example, one of the principles of persuasion is scarcity, so a company may decide to make their product available only through certain designated suppliers to make it seem more exclusive and desirable. Another is **authority**, where a company may decide to commission a recognised expert, such as a scientist or doctor, to **endorse** their product to convince people of its validity. **Product placement**, also known as 'embedded marketing', is a marketing technique where references to specific brands or products, or images of the actual products, are incorporated into a film or television series. This taps into people's desire for **social proof**, looking to what others do to guide their own choices and behaviour.

This increased understanding of human psychology and behaviour has led to changes in the strategies used to influence the sale of physical products too. The **positioning of goods** in a store is a key example and the way supermarkets are laid out reflects this. Fruit and vegetables are deliberately positioned at the entrance to create an inviting impression of healthiness and freshness, while sweets, magazines and other small 'impulse buys' are positioned near the checkout for people to browse while waiting to pay. Using **buzzwords** and listing the benefits of a product on the advertising and packaging is another effective strategy; today's customers are tempted by products which claim to offer health benefits, so buzzwords such as 'superfood' can be extremely persuasive. **Pricing** is a fundamental marketing tactic which taps into a recognised human response: people's love of a bargain. The supermarket **multi-buy** strategy, for example, aims to persuade customers they're getting some of the goods for free, or at a reduced price. **Packaging** and **branding** goods to look like similar but more expensive market-leading products is another commonly used marketing strategy.

Customer engagement and your students

All students will have experienced marketing strategies in some way in their role as customers. They should all be able to think of examples of successful campaigns which have influenced them to buy products, whether online or in retail outlets. They will also have examples of targeted marketing based on analysis of their online purchase history and preferences, etc., which has reached them via social media. Students might like to compare which marketing channels are most successful in influencing them personally. In-work students may have participated in the creation of targeted marketing strategies in their organisations, including the analysis of customers' behaviour and the evaluation of the most appropriate techniques to influence them.

Unit lead-in

Draw students' attention to the unit title *Customer engagement* and elicit or give a definition of *engagement*. Ask students what they think *customer engagement* means in a business context and elicit or give a brief explanation. (There are many different ways to define the term, but a possible answer might be that customer engagement is about interaction between a customer and a brand, creating a relationship between an organisation and its customer base; this can be achieved via a series of methods, such as marketing campaigns, outreach via social media, personalised customer communications, etc.) Refer students to the quote and briefly discuss it with them. Do they think this is good advice? Why / Why not? Why would 'finding products for your customers' be a better option than 'finding customers for your products?'

5.1 ❯ Marketing strategies

GSE learning objectives

- Can recognise the use of persuasive language in a linguistically complex presentation or lecture.
- Can extract specific details from a TV programme on a work-related topic.
- Can recognise a wide range of idiomatic expressions and colloquialisms, appreciating register shifts.
- Can contribute to a group discussion using linguistically complex language.
- Can make a detailed, formal, evidence-based argument in a presentation or discussion.
- Can give a detailed account of a complex subject, ending with a clear conclusion.
- Can describe a business proposal in detail.

Warm-up

Put students in pairs and ask them to think about what influences them to buy a product and make a list. Give them 1–2 minutes to discuss in their pairs, then elicit answers around the class, listing students' ideas on the board. You could then ask them to decide, as a class, on the three most important factors.

Lead-in

Students talk about factors that influence consumer behaviour.

1 If time allows, let students discuss the questions in pairs or groups first and then as a class.

Video

Students watch a video about influencing customer behaviour.

2A Put students in pairs and go through the instructions with them. Give them 2–3 minutes to discuss the question in their pairs, then get feedback from the class. List students' ideas on the board.

2B Tell students that they are now going to watch the video and compare their ideas from Exercise 2A with the ones in the video. Play the video, then discuss the answers with the class. What other strategies does the video mention?

> Points mentioned are: location of items; choice of words to describe items; naming health benefits of certain foods; pricing (bargains, multi-buys, pricing above and below product price), packaging and design.

3 Give students time to read through the questions and options before playing the video and check they understand *buzzword, perceive, have an advantage over sb* and *multibuy*. With stronger classes, you could ask students to check if they can answer any of the questions before watching the video again and then watch to check/complete their answers.

> **1** a, b **2** b, c **3** a, c **4** b, c

4 Put students in pairs or small groups, give them 2–3 minutes to discuss the questions, then invite different students to share their answers with the class. Alternatively, if you are short of time, you could discuss the questions briefly with the whole class.

Extra activities 5.1

A ▶ 5.1.1 This activity provides students with extra listening practice. Explain the task and give students time to read the statements and ask you about any words they do not understand if necessary. Play the video, then check answers with the class. Students may need to watch the video twice for this activity: once to decide if the statements are true or false, then a second time to correct the false statements. In weaker classes, you may also need to pause the video to give them time to complete their answers.

> **1** F (Marketers need to understand what convinces customers to buy products because <u>customers have so many products to choose from</u>.)
> **2** T
> **3** F (Using the word 'superfood' to describe an item means that customers <u>may be persuaded</u> to buy it.)
> **4** F (Customers are often tempted to buy items which <u>they think will be the answer to all their ills</u>.)
> **5** T
> **6** F (Eating healthy foods <u>cannot cancel out</u> the bad effects of foods which are less healthy.)
> **7** T
> **8** T
> **9** F (Some product lines owned by stores have packaging that <u>makes them look similar</u> to well-known brands.)
> **10** T

Vocabulary: Influencing customer behaviour

Students look at vocabulary related to customer engagement.

5 Put students in pairs. Go through the instructions with them and make sure they understand how the 'magic table' works: they should match definitions 1–9 with the words and phrases in the table, writing the question number in the box for each word/phrase. If their answers are correct, the total for each row and column will be 15 (for example, the top row would be definition 6 for *mislead*, definition 7 for *paradox* and definition 2 for *well-being*: 6 + 7 + 2 = 15). Check answers with the class, clarifying meanings as necessary.

> Top row: **6** mislead **7** paradox **2** well-being
> Middle row: **1** be positioned **5** be inclined
> **9** persuasive
> Bottom row: **8** tactic **3** inviting **4** reasonable

6 Explain to students that when the sentence halves are matched, each sentence is a definition of the phrase in bold in it. Get them to complete the exercise individually, then check answers with the class, checking understanding of the phrases in bold.

> **1** d **2** a **3** i **4** e **5** h **6** f **7** c **8** g **9** b

7 Ask students to do this individually. Encourage them to read the whole text quickly before attempting to complete the gaps and also to look carefully at the words around each gap – this will help them work out what type of word is missing each time. Check answers with the class.

> **1** buy into **2** persuasive **3** well-being **4** tap into
> **5** cancel out **6** tactic **7** inclined **8** mislead

8 Put students in pairs, explain the activity and go through the list with them before they begin. Give them 2–3 minutes to discuss in their pairs, then get brief feedback from the class.

Extra activities 5.1

B This activity practises vocabulary related to customer engagement. Students saw it in Exercises 5 and 6 in the Coursebook so they should be able to complete the exercise individually. Before they begin, point out that the sentences are definitions of the words/phrases in bold and students need to choose the correct options in italics in order to complete the definitions. If there is time, get them to compare answers in pairs before class feedback.

> **1** are different from each other
> **2** have a plan they haven't yet shown **3** pricing
> **4** want to buy the product **5** can claim
> **6** a fair price **7** you **8** easy

Project: Planning a marketing campaign

Students plan and present a marketing campaign.

9A Put students in pairs and explain that they are going to plan a marketing campaign for a new food product. Go through the instructions and questions with them and encourage them to make notes in answer to each question. If time is short or if you think your students will struggle to come up with ideas, you could make a few suggestions, e.g. some ideas for superfoods could include:

- aquafaba – the water that legumes are cooked in: can be used to replace egg whites in vegan food; is composed of carbohydrates, proteins and soluble plant solids
- kombu – edible kelp: contains glutamic acid, which is responsible for the basic taste umami and iodine; is also a source of dietary fibre
- sea asparagus: grown in water, not soil; has folic acid and vitamin B9; can help detoxify the liver.

'Food from another country that is not well-known internationally' can be food students have experienced while travelling, food their family ate if they came from another country, food students have learnt about from other students, etc.
If you have a large class, you could put students in small groups instead, so that there are fewer campaigns to be presented to the class the next stage.

9B Pairs now prepare a short presentation of their campaign and then video it on their smartphones. Alternatively, you could ask them to give the presentations to the class instead. You may also wish to ask them to write up a plan for their campaign, outlining the different steps and what they think will be important to make their product successful. Again, allow plenty of time for students to prepare their presentations, while you monitor and offer help as necessary.

> **Model answer / outline for presentation**
> <u>Our plan for a lunch packet that can be purchased at a university or company cafeteria</u>
> **1** Find interesting name for product – stress that it is convenient, healthy and tasty.
> **2** Come up with buzzwords referring to healthy qualities, e.g. supplies energy, but is low in calories.
> **3** Position in separate area of the cafeteria – it is take-away. Perhaps near the door but separate from the normal line for food with special display.
> **4** Priced a bit higher than the cheapest items, but less expensive than the highest-priced ones.
> **5** Sold in a reusable see-through container which people could fill again when buying this special lunch.
> **6** Adverts on social media, especially in groups students/ employees use.
> **7** Selling points include reusable container, health benefits, reasonably priced, tasty, convenience, advertised where students/employees will see it.

9C Pairs now take turns to show (or give) their presentations to the class. Explain that each presentation will be followed by a Q&A session, so during the presentations the class should note down any questions they want to ask the presenters afterwards. You could round off the activity with a class vote on which campaign students think would be the most successful.

MyEnglishLab: Teacher's Resources: extra activities; Reading bank
Teacher's book: Resource bank Photocopiable 5.1 p.150
Workbook: p.24

5.2 ❯ Persuasion

GSE learning objectives

- Can understand complex arguments in newspaper articles.
- Can evaluate evidence presented in a linguistically complex argumentative text.
- Can identify inferred meaning in a linguistically complex text.
- Can add information using appended clauses with 'being' and/or passive participles.
- Can participate in linguistically complex discussions about attitudes and opinions.

Warm-up

Write this quote on the board: *Persuasion is often more powerful than force.* (Aesop, Greek fabulist and storyteller) Discuss the quote with the class. Do they agree? What do they think makes a person persuasive? Do they think they are good at persuading others?

Lead-in

Students talk about using persuasion as a marketing strategy.

1 Put students in pairs, explain the activity and go through the diagram with them before they begin. Check that they understand *empathetic*. Give them a few minutes to discuss in their pairs, then invite different students to share their answers with the class.

> Students' own answers, but other ideas might include: convenience, necessity, basic needs, health, price, rewarding or treating oneself, spontaneous purchase.

❯ Spoken English
p.114: I'd buy things at the drop of a hat

1 🔊 SE5   Tell students that they are going to hear people discussing the question they discussed in the Lead-in: *What persuades you to buy something?* Explain the activity and give them time to read the questions first, so they know what to listen for. Encourage them to make notes in answer to the questions while listening, then play the recording, twice if necessary and check answers with the class.

Possible answers
1 He now buys out of necessity; he thinks more carefully about his purchases than before.
2 her (girl)friend
3 He's gullible and will buy things when people recommend them; he doesn't think about whether he needs them.
4 It's a hormonal release. It's a sense of euphoria.
5 He prefers to have things that are of a higher quality.

Yes/No questions and answers

2 🔊 SE5   Explain the activity and give students time to quickly read through the extracts before they listen. Play the recording again, then check answers with the class. Note that in weaker classes, students may need to listen twice or you may need to pause the recording for them to complete their answers.

1 Really?
2 Oh, absolutely.
3 So do you think it depends
4 Certainly does with me.
5 It works?
6 So you feel you're
7 I think so
8 Oh, definitely. Definitely.
9 people talk about that, don't they?
10 They do, yeah.
11 I don't think you'd allow yourself to feel that, would you
12 Absolutely not

3 Tell students that the phrases they wrote in Exercise 2 were different types of *yes/no* questions and answers. Explain the activity and go through the question and answer types with the class. If necessary, elicit or give an example for each question type. Give students time to complete the exercise, individually or in pairs, then check answers with the class.

> **a** 3 **b** 5,6 **c** 9,11 **d** 1 **e** 2,4,7,8,10 **f** 12

Reading

Students read an article about the importance of persuasion in the workplace.

2 Tell students that they are going to read an article about the importance of persuasion in the workplace. Explain that they do not need to worry about detailed information in the text for now; they should read it quickly to understand the main ideas and choose the best summary. Give them time to read the text, then elicit the answer. After class feedback, you may wish to check understanding of this key vocabulary from the text, before students read it in more detail in the activity that follows: *behavioural economics, deploy, spark* (v.), *concede, debunk, shift* (n.), *reciprocity, scarcity*.

> 1

3 Give students time to read the questions and check that they understand *cut through* in question 3. Explain that they need to read the text in more detail this time and ask them to underline the parts of the text that give them the answers. If there is time, get them to compare answers in pairs before checking with the class. During feedback, elicit both the answers and the parts of the text where students found each one.

Possible answers

1 The decision-making is done by marketers and Human Resources executives.
2 Professor Cialdini feels that when businesses, those making policies or other individuals make small changes, these can have a large impact.
3 Executives who are pressed for time generally want to find a simple solution and are glad to be given specific, practical ideas.
4 He says that we cannot change someone's behaviour in this way; they are more influenced by the context something is presented in than by having all the information about it.
5 This is possible when the message is linked to human motivating factors.
6 They feel that the method changes the way people see the facts they are presented with.

Extra activities 5.2

A This activity provides students with extra reading practice and is best done in two stages. Start by asking students to complete the matching task individually; explain that the sentences paraphrase ideas from the reading text. Check answers with the class, then ask students to find the ideas in the article which the sentences paraphrase. Get them to compare answers in pairs before class feedback.

1 d **2** j **3** h **4** i **5** b **6** g **7** c **8** f **9** a **10** e

Original sentences:

1 *Robert Cialdini ... create a new role of Chief Persuasion Officer (CPO).*
2 *Targeting the business audience, it follows the success of other behavioural economics books.*
3 *The book's mission is to ... by providing customers and employees with all the information available.*
4 *reciprocity – having favours performed for them, people feel obliged to return them*
5 *authority – people look to experts to show them the way*
6 *scarcity – the less available the resource, the more people want it*
7 *liking – the more people like others, the more they want to say yes to them*
8 *consistency – people want to act in a way that is consistent with their values*
9 *social proof – people look to what others do in order to guide their own behaviour*
10 *You cannot underestimate the importance of small things in influencing people.*

Grammar: Participle clauses

Students study and practise participle clauses.

4 Go through the information about participle clauses with students and look at the example with them. Highlight that *persuasion science* is the subject of the sentence and that the

underlined clause refers to it, it gives extra information about it and the sentence would make sense and be grammatically correct without the participle clause. You could then do the rest of the exercise with the whole class, checking answers as you go, or ask students to do it individually. After checking answers, refer students to the Grammar reference on page 119 and go through it with them, clarifying any points as necessary.

1 *Drawing on psychology, neuroscience and behavioural economics, 'persuasion science' should be deployed in a systematic way.* (The underlined phrase refers to 'persuasion science'.)
2 *Targeting the business audience, it follows the enormous success of other behavioural economics books.* (The underlined phrase refers to the pronoun 'it' which, in turn, refers back to the book *The Small Big: Small Changes That Spark Big Influence.* This item is the one that students are most likely to find difficult and you may have to encourage them to give the longer answer, rather than just saying that the clause refers to 'it'.)
3 *Linking their message to human motivations, readers are able to increase their ability to persuade others by making tiny shifts in their approach.* (The underlined phrase refers to 'readers'.)
4 *reciprocity – having favours performed for them, people feel obliged to return them.* (The underlined phrase refers to 'people'.)

5–6 Get students to complete both exercises individually. Remind them that they can refer to the Grammar reference on page 119 if they need help. Check answers with the class.

5 1 c **2** f **3** a **4** d **5** b **6** e
6 1 sitting **2** Having launched **3** decreasing
    4 Having been repaired **5** put out **6** Living

7 Explain the activity and look at the example with students. Remind them to look carefully at the verb forms in the underlined parts of the sentences in order to decide if they need an *-ing* or an *-ed* participle. Check answers with the class, clarifying any errors as necessary.

1 The method <u>created to help influence customers</u> has been successful.
2 <u>Having suggested an idea for the campaign,</u> / <u>After suggesting an idea for the campaign,</u> she got a promotion.
3 <u>Travelling to work by train,</u> he was able to spend time working on his presentation.
4 <u>Being fluent in Italian,</u> I had no problem taking part in the meeting in Rome.
5 The economy did not do well last quarter, <u>forcing several companies to go out of business</u>.
6 <u>Having offices all over the world,</u> we have employees who speak many languages.
7 <u>Handled carefully,</u> it will work well for years.
8 <u>The product going on sale tomorrow</u> should be a real cash cow. / <u>Going on sale tomorrow, the product</u> should be a real cash cow.

Extra activities 5.2

B This activity gives further practice of participle clauses. Ask students to complete it individually and remind them that they can refer to the Grammar reference on page 119 if they need help. Check answers with the class, clarifying any points/errors as necessary.

1 Using **2** Having read **3** Approaching
4 appointed **5** Being **6** damaging **7** addressing
8 Not having concentrated

Speaking

Students practise participle clauses by writing and talking about their financial decisions.

8A Ask students to work individually. Explain the activity and, if necessary, give them an example for the first item, e.g. *Being naturally cautious, I always look for the least risky investment option.* While students are writing their sentences, monitor and check that they are using participle clauses correctly; correct any errors as necessary.

8B Put students in pairs or small groups and give them 3–4 minutes to compare and discuss their priorities. As feedback, elicit sentences from different students, inviting the class to say to what extent they agree or disagree with each student's priorities.

MyEnglishLab: Teacher's Resources: extra activities
Grammar reference: Participle clauses p.119
Spoken English: p.114
Teacher's book: Resource bank Photocopiable 5.2 p.151
Workbook: pp.25–27

5.3 ⟩ Communication skills

Presenting research data

GSE learning objectives

- Can extract key details from quantitative data in complex business documents.
- Can effectively discuss the meaning and implications of research data.
- Can critically evaluate the effectiveness and appropriateness of a presentation.
- Can understand nuances of meaning in a linguistically complex presentation or lecture.
- Can recognise a wide range of idiomatic expressions and colloquialisms, appreciating register shifts.
- Can compare and evaluate different ideas using a range of linguistic devices.

Warm-up

Ask students if they have ever attended a presentation, or read material, in which data was being presented (e.g. research or statistical data). In what ways was this data presented? Did they find it easy to understand/remember? Why / Why not?

Lead-in

Students talk about data presentation.

1 If time is short, you could do this as a whole-class activity, eliciting the answers to question 1 from the class, then inviting a few students to share their answer to question 2 with the class. Alternatively, put students in pairs and give them 2–3 minutes to discuss the questions, then discuss the answers with the class. Tell students that they do not need to read the information in each graph/chart in detail at this point; they only need to identify the type of visual used each time. During feedback, you may wish to share some of the information from the Note below with the class.

1 a bar chart, two doughnut charts, a table

Note

- Data analytics is also known as 'data analysis'. It can refer to both qualitative and quantitative techniques used by businesses to gather data, so the formats can vary widely. An organisation collects and analyses data about its customers/users and business processes. User data is categorised and analysed to study consumer behaviour, preferences, trends and patterns. The data is then used to inform fact-based decision-making and improve performance. For example, an online publisher such as a magazine or newspaper might collect data related to user behaviour, preferences, interests and demographics to help it adapt the content, layout, design and overall strategy in order to increase reader engagement.
- Analytics assumes that the best predictor of future behaviour is past and present behaviour. It tries to use this assumption to predict future outcomes.
- The internet, digital technology and smart devices make it easier than ever for businesses to track consumer activity and gather vast amounts of data these days.
- Metrics are not to be confused with key performance indicators (KPIs). A KPI is used to measure performance and success. A metric is simply a number within a KPI that helps track performance and progress.

Preparation: Presenting and discussing research findings

Students read and think about the scenario for a roleplay.

2 If this is the first Communication skills lesson for your class, briefly tell students about *Lifestyle* magazine and the profile of its readership. Otherwise, elicit this information from students before you start. Go through the instructions with students and point out that they will need to read the information on the dashboard in more detail this time. Students could complete the activity individually or, in weaker classes, in pairs. If they work individually, you could get them to compare their answers in pairs before class feedback.

Possible answers
- Visits by section: three most popular are society, entertainment and technology; sport is less popular
- Attention time: society, travel and technology get the most, people spend less time on entertainment; sports and style fare less well here and in 'unique visitors'
- Geographical data: 46% of readers outside UK, so need for global/international content
- How users access content: 75% through phones, but only 20% use the magazine's app
- Reader poll: 8 out of 10 readers would recommend to a friend and a quarter would pay for content

Roleplay

Students present research data.

3A Put students in pairs and explain the activity. Point out that they are each going to present two sections of the dashboard and remind them of the key findings they identified in Exercise 2. Allow plenty of time for students to prepare for their roleplays, while you monitor and help as necessary. Encourage them to make notes.

3B In their pairs, students now take turns to present their data. Explain that after each presentation, they should discuss the implications and conclusions that can be drawn from the data. During the roleplays, monitor and note down any points to highlight during feedback after Exercise 3D, but do not interrupt students.

3C Students should do this in the same pairs as for Exercises 3A and 3B. They should now look at the Reader poll in Exercise 1 and discuss the conclusions they can draw from the data. Again, monitor during the activity and note down points to highlight in feedback.

Possible answer
It seems that readers are highly satisfied with the online magazine but only about a quarter of them would be willing to pay for content. One implication is that it could be worth introducing a subscription service. Another implication is that the magazine could use these metrics to adapt the content more closely to readers' interests.

3D Let students discuss the questions in their pairs. Encourage them to give reasons for their answer to question 1. Have a brief feedback session, inviting a few students to share their answers with the class and then highlighting any points you noted while monitoring.

Video

Students watch a video of a presentation of research data.

4 ▶ 5.3.1 Go through the instructions with students and give them time to read the questions before they watch the video. Check they understand *in a favourable/unfavourable light* and *go into (sufficient) depth*. Play the video, then check answers with the class.

Possible answers
1 Teo highlighted the most important points rather than going into great detail in order to adapt the presentation to his audience.
2 He reported the low use of the magazine app in an unfavourable light and the fact that over a quarter (26%) of readers in the poll said they would be willing to pay for content in a favourable light.
3 He showed the different graphs, charts and tables he wanted to refer to on separate slides. He also indicated by pointing where exactly he wanted to direct his audience's attention.
4 It seems he did, as when he offered to show them more detailed information, James felt it wasn't necessary.
5 In data-driven decision-making, data is at the centre of the process. It's the primary (and sometimes the only) input. You rely on data alone to decide the best path forward. In data-informed decision-making, data is a key input among many other variables. You use the data to build a deeper understanding of what value you are providing to your users. For example, if the magazine finds that clickbait headlines get a lot of page views, which is a key metric for the publisher, they might be tempted to have more such headlines. This would be a data-driven decision. However, if the 'bounce rate' is high, that means users are not engaged and therefore clickbait titles could be alienating users and damaging the brand's reputation. So a data-informed decision would be to find out what topics readers are curious about but want to know more about and improve the user experience.
6 To use analytics more consistently and regularly to help shape content. There are pros and cons of using metrics, but they can show useful patterns and help the magazine shape its content strategy.
7 Students' own answers, but at this stage it can reasonably be argued that they may not have enough data to make that decision.

5 Tell students that they are going to look at some useful vocabulary from the video and explain the activity. Get them to complete it individually and encourage them to refer to videoscript 5.3.1 on page 148 to see the words in context; this will help them work out their meanings. Check answers with the class, clarifying meanings as necessary.

1 c 2 e 3 f 4 a 5 b 6 g 7 j 8 i 9 d 10 h

Extra activities 5.3

A This activity practises key vocabulary from the video. Ask students to complete it individually and point out that they can refer to the definitions in Exercise 5 of the Coursebook if they need help. Check answers with the class, clarifying meanings as necessary.

1 cater to 2 tempting 3 fare badly 4 churn out
5 bland 6 resonate 7 boost 8 the bottom line
9 improve the odds 10 rate

Reflection

Students reflect on the conclusions from the video and their own approach to presenting and discussing research data.

6A Allow students to work individually first so that they can reflect on their own approach. Go through the sentences with them before they begin and remind them to think about their ideas from Exercise 3D and the answers to Exercise 4.

6B Put students in pairs or small groups to compare their reflections, then broaden this into a class discussion. You could round off the task by sharing some of the tips from the Note below with the class.

> ### Note
>
> **Tips for presenting and discussing research data**
> - Focus only on the main findings/key data in your presentation. In some departments or companies, a greater level of in-depth presentation may be required.
> - Adapt the presentation to the audience, i.e. simplify if necessary.
> - Round figures up or down to help the audience understand them more easily.
> - Repeat key figures in different ways to aid the audience's memory.
> - Decide if you want to present/discuss the findings in a favourable or unfavourable light.
> - Make sure your data and visuals are clear and easy to read, both on slides and on handouts.
> - Explain what the figures mean if necessary. Analyse the data, draw logical conclusions and make recommendations.

MyEnglishLab: Teacher's Resources: extra activities; Interactive video activities

5.4 ❯ Business skills

Building relationships on trust

> ### GSE learning objectives
> - Can infer meaning, opinion, attitude, etc. in fast-paced conversations between fluent speakers.
> - Can follow a work-related discussion between fluent speakers.
> - Can propose a range of different options in a complex negotiation.
> - Can switch between formal and informal language during a work-related discussion to build rapport.

Warm-up

Depending on the time available and the size of your class, write some or all of the questions below on the board and get students to discuss them in pairs or small groups for 2–4 minutes. When the time is up, invite students from different pairs/groups to share their answers with the class.

- *Who is the most trustworthy person you know?*
- *Why do you trust them?*
- *What makes a person trustworthy?*
- *What kind of people don't you trust? Why?*
- *Have you ever lost trust in someone? Why? How did that make you feel?*
- *How would you feel if someone betrayed your trust?*
- *Do you think trust is an easy thing to earn?*

Lead-in

Students talk about trust and trust-building strategies.

1A Put students in pairs and give them time to discuss the questions, then get brief feedback from the class. Encourage them to elaborate. List students' ideas for question 3 on the board.

1B Tell students that they are going to read an extract from an article about trust and if you wish, teach these words from the text: *empathy, ethical (principles), competence*. Ask them to read the text quickly and make notes in answer to the questions – remind them to refer to the list on the board for question 1 – then discuss the answers with the class.

Listening

Students listen to a meeting about diagnosing problems among senior management.

2A ◀ 5.01 Go through the instructions with the class and explain the scenario and task. Give students time to read the questions before they listen, then play the recording. Note that students, especially in weaker classes, may need to listen twice in order to check/complete their answers.

> 1 Ellie asks Frank if he is getting over the flu. It's a personal question, showing empathy and care. Frank is positively surprised that Ellie remembers, which helps to build the relationship.
> 2 Ellie shows empathy: showing her knowledge that Dieter travels a lot and empathy that it must be exhausting.
> 3 Simone points out the similarity of their professional lives with lots of travel and a common experience of Budapest and Poland. For Dieter, this helps to build trust because he senses that Simone understands his organisation.
> 4 Ellie confirms that Simone's experience means she is familiar with the challenges Frank is facing. Frank's own confirmation means that he may be beginning to believe that Konnect is the right partner.

2B Discuss the question with the whole class. Encourage students to give reasons.

> **Possible answer**
>
> Overall, it was a productive start to the meeting. The communication strategies used by Ellie and Simone create a positive atmosphere and a good sense of rapport and begin to build trust effectively.

3A ◀)5.02 Tell students that they are going to hear the next part of the meeting and give them a minute to read the questions. Play the recording, twice if necessary, then check answers with the class.

> 1 Frank says staff lack project- and people-management skills.
> 2 Ellie does not like to propose training solutions before consulting to find out the real needs of a client's organisation, which she doesn't understand at the moment in the case of DeutschTek.
> 3 A lack of resources, a lack of time and lack of skills. Processes may be an issue.
> 4 Simone is suggesting that the Berlin office's expectations may be unrealistic for the newly created offices and this is creating pressure on the new local offices to perform too quickly.
> 5 To conduct, over a period of two weeks, a meeting with the local management involved, conduct one-to-one interviews, observe meetings locally and write a diagnostic report with recommendations.

3B Again, discuss this question with the whole class. If necessary, let students refer to audioscript 5.02 on page 154.

> **Possible answer**
> Ellie and Simone demonstrate their understanding of the company's context and offer solutions based on real experience. When Dieter says addressing the issue is urgent, Simone reassures him and offers an important observation about why rushing won't help. Ellie says that they will work transparently and highlights that her and Simone's aim is to help DeutschTek reach their goals, which is another good way to build trust. Frank concludes by saying it's 'a win–win situation', suggesting he has full confidence in Konnect.

Useful language

Students look at strategies and useful language for building trust in professional relationships.

4A Before students complete the exercise, go through the headings and sentences in the table with them and check that they understand each one. Then get them to complete the exercise individually and check answers with the class, clarifying meanings as necessary. In weaker classes or if time is short, you could also do this as a whole-class activity, eliciting the correct category for each phrase as you go.

> 1 e 2 g 3 a 4 c 5 h 6 d 7 f 8 b

4B Students could do this individually or in pairs. Check answers with the class, eliciting both the expressions students have identified and the categories from Exercise 4A they belong in. Encourage students to record the expressions by adding them to the correct category in the table.

> **Possible answers**
> **Show empathy:** *I imagine it's quite demanding to be on the road so much.*
> **Establish similarity:** *Seems we have a lot in common when it comes to travel.*
> **Establish similarity:** *We probably know the same hotels.*
> **Trust first:** *We knew you had everything under control.*
> **Prove competence:** *Obviously you know your business better than we do, but I'm sure we can help.*
> **Be open:** *To be transparent, at this point in time we don't have the data to make any recommendations.*
> **Demonstrate integrity:** *Well, I'm very happy to stick to the promise I made when we met …*
> **Establish similarity:** *Our main objective in business is to help you reach your goals …*

> **Extra activities 5.4**
>
> **A** This activity looks at the trust-building strategies students saw in the Lead-in. Explain the task and if necessary, let students read the extract in Exercise 1B of the Coursebook again. Ask them to complete the exercise individually, then check answers with the class.
>
> > 1 f 2 b 3 a 4 h 5 c 6 g
>
> **B** Students now practise more expressions for establishing similarity or common ground. Go through the instructions with them before they begin and encourage them to read the whole dialogue first, before they complete the gaps. Check answers with the class. If there is time, you could get them to practise their dialogue in pairs.
>
> > 1 No, really 2 What a coincidence
> > 3 Same here / Me, too 4 Me, too / Same here
> > 5 So have I 6 Neither do I 7 We should
> > 8 That's good for me, too

Task

Students use trust-building strategies in a negotiation.

5A Tell students that they are going to roleplay a meeting in which, in addition to negotiating a good deal, they will have to use trust-building strategies to help them develop a long-term relationship with their counterpart. Put them in pairs, assign roles A and B and go through the instructions with them. Give them time to read the professional context and ask you about anything they do not understand. Then refer students to their respective information on pages 126 and 130 and give them time to read it, while you monitor, answering any questions they may have and offering help as necessary. Make sure they are clear about their objective and point out that in addition to achieving the best result for their company, they should aim to build trust and a long-term relationship with the other speaker.

Remind them to refer to the trust-building strategies in Exercise 1B (and 4A) and try to use a good range of language from Exercise 4. Allow plenty of time for students to prepare for the meeting, while you go round and provide help as necessary. In weaker classes, you may wish to group Student As and Student Bs together first, to briefly discuss their roles and prepare for their meetings together before returning to their original pairs. Set a time limit for both the preparation stage and the negotiation meeting.

5B Students now hold their meetings. Remind them of the time limit before they begin and during the roleplay, monitor and note down any points to highlight during feedback. When they have finished, get them to discuss what went well and what could be improved: Did they use trust-building strategies from Exercise 1B? Did they use them effectively? If not, what could they do differently next time? Give them a few minutes to discuss in their pairs, then broaden this into a class discussion. Finally, highlight any points you noted while monitoring.

MyEnglishLab: Teacher's Resources: extra activities; Useful language bank

5.5 ⟫ Writing
Advertising copy

GSE learning objectives

- Can write promotional materials using descriptive language to advertise a product or service.
- Can write work-related materials using persuasive language.
- Can give detailed written feedback on the effectiveness of a piece of work-related correspondence.

Warm-up

Draw students' attention to the lesson title *Advertising copy* and check that they understand the meaning of *copy* in this context (written material that is for publication). You may also wish to tell them that a *copywriter* is someone who writes copy as a profession. Ask them if they read advertising copy in magazines, newspapers, etc. or on websites. Does it influence their decision to buy a product? Do they ever buy products because of it? If so, when?

Lead-in

Students read and talk about the features of persuasive advertising copy.

1 Go through the instructions with students and make sure they understand the difference between the features and benefits of a product. If there is time, let them brainstorm ideas in pairs or small groups first. Elicit ideas around the class and list them on the board for students to refer to when they do Exercise 2.

Possible answers
Features: blanket with sleeves, choice of colours, washable fabric
Benefits: keeps you warm, helps you feel relaxed, makes an ideal gift

2 Explain the activity and give students time to read the four alternative versions of the advertising copy for the Slanket on page 140. If there is time, get them to discuss the questions in pairs or small groups first; remind them to refer to the list on the board. Discuss the answers with the class.

Possible answers
How the style and language differs
Version 1: very short, but still covers all the important points; use of bullets; could be highly effective if the reader doesn't have time to read a lot of text.
Version 2: begins with a strong opening hook and then the benefit concisely stated in a single sentence; uses bullets for factual information that the customer might be looking for; ends with a short paragraph that restates the benefits and then has a call to action in the form of a rhetorical question.
Version 3: short full sentences rather than bullets; begins with a hook to get interest, then uses *Seriously* to establish a tone of a friend talking to a friend; continues with emotional language based on the feelings of cosiness and family; factual information is at the end.
Version 4: short full sentences; the hook is the whole first paragraph that builds a scenario familiar to the reader; the scenario contains a known problem and the product solves this problem; very informal style with exclamation mark, repetition of *nice and ...* (typical of informal speech), use of *super-, bestie;* direct call to action at the end with the imperative *Give others ...* ; closes with a very small amount of factual information.

Useful language

Students look at techniques and useful language for writing persuasive advertising copy.

3 This is best done as a whole-class activity, checking and discussing answers as you go. After discussing the answers for each item, go through the relevant section of the Useful language box with the class. For question 1, make sure students understand what a *hook* is in writing (an opening sentence which 'hooks' the reader's attention and makes them want to keep reading; examples of different types of hooks include: a question, a quotation, a strong statement, an interesting fact or statistic). For question 4, point out that these techniques can be combined; for example, *bright and bold or calm and classic* in version 3 is a) repetition of sounds, b) repetition of a structure and c) a contrast. For question 5, elicit or tell students who Aristotle was (a Greek philosopher). Also note that versions 1 and 2 are more factual and use reason (*logos*), while 3 and 4 have more feelings and emotion (*pathos*). On the internet, credibility (*ethos*) is established by things like star ratings for products, endorsements by brand ambassadors and, of course, the reputation of the brand name itself.

1 The structure in the table is followed loosely and flexibly in the four texts:
  • Opening hook: not present in 1, present in the other versions.
  • Product info: the features are prominent in 1 and 2, but left to the end with less detail in 3 and 4; the benefits are kept short and simple in 1 and 2, but are described in more detail in 3 and 4.
  • Persuasive techniques: in general 1 and 2 have a balance of facts and emotional persuasion, while 3 and 4 try to persuade more openly.
  • Call to action: not present in 1, rhetorical question at end of 2, rhetorical question just after halfway in 3, imperative at end of 4.

2 They are included to capture the style of informal speech and so establish an informal, friend–friend relation of confidence between reader and writer. By being so informal, they also surprise the reader and make them pay attention.

3 A rhetorical question is a question that does not necessarily expect an answer, although sometimes the speaker answers it themselves. People use rhetorical questions to create interest; they open an issue in the listener's mind and the listener is waiting for closure. Examples: 1 none; 2 *Why not buy one for yourself and another for a loved one?*; 3 *And why buy only one? Who wouldn't love a Slanket?*; 4 *Wouldn't it be great if there was a solution?*

4 Version 1: *watching TV, relaxing on the sofa or reading a book*
Version 2: *thick and warm, but lightweight and practical; passionate purple, ruby red, winter white and basic beige*
Version 3: *bright and bold, or calm and classic*
Version 4: *nice and large, nice and warm, nice and lightweight*

5 To some extent, but they are more relevant when persuading people in a context such as a meeting, a presentation or a report.

6 1: *ideal for keeping you warm and cosy* + all four bullets; 2: *keep you cosy and warm all winter long; thick and warm but lightweight and practical; suitable for a wide range of activities from watching TV on the sofa to wrapping up on an overnight flight*; 3: *As soon as you feel the softness of this luxury fleece, you'll know what we mean; You'll feel so cosy as you snuggle down on the sofa to watch a series or chat with a family member;* 4: whole first para; *nice and large, nice and warm, nice and lightweight; It will keep you super-cosy all evening long as you watch a series or chat with your bestie; A Slanket also solves your gift-giving problems;* last para.

4 Get students to do this individually and then compare answers in pairs before class feedback. You could list the words/phrases students choose on the board for them to refer to during the writing task.

Extra activities 5.5

A This activity looks at the persuasive techniques students discussed in Exercise 3 of the Coursebook. Explain the activity and if necessary, refer students back to questions 3–6 in Exercise 3. Go through them with the class and elicit examples from the website copy and/or the Useful language box in the Coursebook. Ask students to complete the exercise individually, then check answers with the class.

1 **d** *Combining/Offering/Bringing*; **e** *tradition/modern, sophisticated/minimalistic, complex/simplicity*; **f** the three sentences of the text
2 **b** *perfect pecans, crunchy cashews, wonderful walnuts*; **c/d** *no added …*
3 **a** *Does it sound like Paradise?*; **f** *on the beach, at lunchtime, after lunch*; **g** *Imagine yourself…*
4 **a** *Where would you like to go today?*; **b** *relaxed/refreshed/ready*; **d** *longest meeting agenda/toughest presentation questions*; **g** *Where would you like to go today? / you'll arrive relaxed*

Optional grammar work

The website copy in Exercise 1 contains examples of groups of adjectives and gradable adjectives, so you could use it for some optional grammar work. Refer students to the Grammar reference on page 120 and use the exercises in MyEnglishLab for extra grammar practice.

Task

Students write advertising copy for a product.

5A Put students in pairs and give them time to read the information on page 141. Explain that in their pairs, they need to identify any unknown words in the text, divide them equally between them, look up their words in their dictionaries and then explain their meaning to their partner. For a shorter activity, you could ask students to work in groups of three instead, or you could do this with the whole class, explaining the meanings of any unknown words yourself.

5B Students now write their own advertising copy for the Snuggle Slippers. Allow some time for them to plan their work while you monitor and offer help as necessary. Remind them to refer to the model texts on page 140 and the Useful language box. Also encourage them to look again at the questions in Exercise 3 before they write their advertising copy. Set a time limit for both the planning stage and the writing task. If time is short, the writing task can be set for homework.

Model answer

Version 1 (more conventional, mix of facts and feelings)

Our microwavable Snuggle Slippers are the ultimate in comfort while you watch TV, work at your computer or get ready for bed. Your nice, warm feet will feel so good.

- Choice of colours: cream, blue or black.
- Available in three sizes.
- Non-slip soles.
- Easy to use – just place in the microwave for 90 seconds and the heat lasts for an hour.
- Uses natural seeds inside the slippers to retain heat.
- Not machine washable. Can be wiped with a damp sponge.

Snuggle Slippers are the treat you deserve. The warmth of the luxurious soft fabric will aid relaxation and relieve tired and aching feet. Also makes an ideal gift – your daughter would love these just as much as your granny.

Version 2 (more openly persuasive, more feelings and emotions)

You are at home. Maybe watching TV, or chatting to loved ones, or using your tablet. There is just one problem – your feet are cold. Wouldn't it be nice to have the comfort of some new slippers? But not just any slippers. You need the warmth and luxury of our super-soft microwavable Snuggle Slippers.

Yes, that's right – microwavable slippers! Our slippers contain natural flax seeds – just pop them in the microwave for 90 seconds and the slippers retain their heat for a whole hour. The luxurious soft fabric and non-slip soles will add to your comfort. Get ready to relax. Bedtimes and weekends are going to feel soooooo good.

And why not give them as a gift as well? Imagine their smiling faces as they see what you bought them. Best. Present. Ever.

5c If students do the writing task as homework, you could do this activity in the next lesson. Put them in pairs and ask them to read each other's copy and then explain to their partner how their copy will persuade readers to buy the Snuggle Slippers. They should then give their partner feedback, thinking about the techniques and language from Exercise 3. Which techniques did their partner use? Did he/she use them effectively? What has he/she done well? What could be improved? Students could then rewrite their copy based on their partner's feedback; they could do this in class or as homework.

MyEnglishLab: Teacher's Resources: extra activities; Writing bank; Interactive grammar practice

Grammar reference: Groups of adjectives and gradable adjectives p.120

Workbook: p.28

Business workshop > 5
The art of persuasion

GSE learning objectives

- Can follow a group discussion on complex, unfamiliar topics.
- Can infer meaning, opinion, attitude, etc. in fast-paced conversations between fluent speakers.
- Can take full notes on points made during meetings on a wide range of work-related topics.
- Can get the gist of specialised articles and technical texts outside their field.
- Can contribute to a group discussion using linguistically complex language.
- Can contribute fluently and naturally to a conversation about a complex or abstract topic.
- Can participate in extended, detailed professional discussions and meetings with confidence.
- Can give a detailed account of a complex subject, ending with a clear conclusion.
- Can summarise group discussions on a wide range of linguistically complex topics.

Background

Students read about a university considering implementing the skill of persuasion in its curriculum.

1 Put students in pairs and give them time to read the background and discuss the questions. Check answers with the class. During class feedback, feed in information from the Note below as required. Do not go into detail about soft skills and hard skills yet, as students will discuss these in the next activity.

Possible answers

1. They understand business theory, but they are not good at communication in the workplace.
2. Skills that help people communicate more effectively with each other.
3. Hard skills help them to run a business.
4. They were reluctant to add it because of its negative image.
5. They see it as a way to communicate effectively, understand the viewpoint of another person and get one's message across.

Note

Many universities and companies feel that it is essential to teach soft skills as well as business skills to students and employees. These skills are also transferable, meaning that they can be used in different jobs throughout a company or one's career. An important skill which is not often explicitly taught is the art of persuasion, as it is seen as being manipulative. However, the importance of the skill has become clearer in recent years and is one of the skills most in demand in the workplace. It is necessary in a wide range of fields including sales, social media, PR, financial services, human resources and medicine. For this reason, both companies and universities are taking another look at this skill and discussing how to best implement it in training courses and curricula.

Listening to different views

Students listen to conversations about soft skills in the workplace.

2A Students could do this in the same pairs as for Exercise 1 or in new pairs. Give them 2–3 minutes to discuss the questions, then discuss them with the whole class. Make sure they understand the difference between hard skills and soft skills (hard skills are concrete, quantifiable abilities or skill sets such as proficiency in a foreign language, computer programming, web design, typing speed; soft skills, also known as 'people skills', are more subjective skills which are harder to quantify. They are self-developed attributes and qualities and are related to one's personality. For examples of soft skills, see the answer key below). List students' ideas for question 2 on the board.

> **Possible answers**
> 2 good listening skills, time management, resilience, creativity, teamwork, flexibility, problem-solving, empathy

2B 🔊 BW5.01 Explain the scenario and activity and refer students to the list on the board. Encourage them to make notes while listening, play the recording, then discuss the answers with the class.

> The soft skills mentioned are: persuasion, adaptability, communication, time management, leadership, critical thinking, teamwork, flexibility, collaboration, integrity, problem solving, perseverance, creativity, dealing with criticism, working under pressure, active listening

3 🔊 BW5.01 Give students time to read the questions, then play the recording, twice if necessary. Check answers with the class.

> 1 The graduates have very good knowledge of business but need more training in soft skills.
> 2 Magda says that retail staff find it difficult to sell products, not everyone wants to take part in team work and employees request such training sessions.
> 3 Magda undertook some research using the latest literature and spoke to employees. Werner spoke to managers in his company.
> 4 His employees mentioned integrity as a skill and perhaps they didn't feel that integrity and persuasion could be combined.
> 5 They say that persuasion can be combined with communicating with emotion and logic, integrity, active listening, teamwork, collaboration, creativity and adaptability.

4 If time is short, briefly discuss the questions with the whole class, nominating a few different students to answer. Alternatively, let students discuss in pairs or small groups first, then get feedback from the class.

> **Possible answer**
> The leaders of the business school team could look at curricula of other universities and business schools/programmes, talk to a variety of HR personnel, put together a database of job advertisements pinpointing the soft skills required, interview alumni who are in work, talk to people in companies in different sectors.

5 🔊 BW5.02 Go through the instructions with students and point out that they need to make notes about which skills each of the speakers feels should be included in the new course. Play the recording, then check answers with the class.

> **Joelle Dubois:** persuasion, empathy (understanding someone else's point of view and acknowledging their feelings and situation), critical thinking, flexibility, teamwork, integrity, clarifying and summarising
> **Lucio Ricci:** persuasion, perspective-taking (imagining the situation from someone else's point of view), active listening, goal-setting
> **Noah Klein:** persuasion, rapport, good communication (skills), asking friendly questions, adaptability, teamwork, talking about mutual benefits, clarifying and summarising (summing up the discussion)

Extra activities Business workshop 5

A–B These activities practise vocabulary from the listening. Ask students to complete them individually, using their dictionaries if necessary. If there is time, encourage them to find the words/phrases in audioscripts BW5.01 and BW5.02 on pages 159–160; seeing the words used in context will help them work out their meanings. Check answers with the class, clarifying meanings as necessary.

> **A** 1 Collaboration 2 emphasise 3 perseverance
> 4 pushback 5 adaptability 6 values 7 proceed
> 8 convey 9 rapport 10 integrity
> (Not needed: capable)
> **B** 1 promote in-house 2 critical thinking
> 3 top-notch 4 training audit 5 goal-setting
> 6 solid knowledge 7 working under pressure
> 8 interpersonal skills 9 active listening
> 10 top of the list

Analysing alumni needs

Students read texts about skills needed in different companies.

6 Draw students' attention to the section title *Analysing alumni needs* and teach or elicit the meaning of *alumnus*. Go through the instructions with students, making sure they understand the scenario and what they need to do. Before they read, you may wish to look at the skills at the bottom of each chart with them and check understanding of each one. Give students time to read the statements and match them to the bar charts. Check the answers to the matching task, then get students to discuss why they think each set of skills is important in each alumnus' job; they could do this in pairs or small groups. Allow 2–3 minutes for this, then get feedback from the class.

> Alumnus 1 C Alumnus 2 A Alumnus 3 B

Task: Create a course

Students create and present a training course on teaching persuasive skills.

7A Put students in small groups and explain the scenario: they all work for the same company / study at the same university and have been asked to design a new course on teaching persuasive skills. Go through the instructions and the points to discuss with them and point out that they should consider all the information they have so far from the discussions from Exercises 3 and 5 and the alumni needs analysis from Exercise 6. Also point out that they should make notes on each of the points in the two boxes; they will need to refer to them later. There is a lot for students to discuss and decide on, so allow plenty of time for this stage. During the activity, monitor and provide help as necessary.

7B Join groups together into larger groups and explain that students now have to compare and discuss their choices from Exercise 7A and agree on one final idea for each of the points discussed. Remind them to refer to their notes from Exercise 7A and to update them as necessary during their discussion. Again, monitor and help students as necessary during the activity.

7C Students should do this in the same groups as for Exercise 7B. Explain that they are now going to put together a detailed course outline to describe their course. Encourage them to separate it into clear sections (e.g. *Skills to include, Length, Format, Number of participants*) and to use bullet points. To help them, you could give them an empty 'template' on the board (following the format of the model answer below), which they can then complete for their course; this would be particularly helpful for weaker classes. Set a time limit before students begin and while they are writing, monitor and offer help where needed.

Model answer

Course to teach persuasion skills
Skills to include:
- rapport (1 lesson)
- active listening, perspective-taking and empathy (1 lesson)
- goal-setting (1 lesson)
- learning to compromise, finding mutual benefits (1 lesson)
- practising adaptability and flexibility (1 lesson)
- small talk and communication skills (1 lesson)
- working on teamwork and collaboration (1 lesson)
- critical thinking (1 lesson)
- clarification and summarising (1 lesson)
- practising the individual skills (1 lesson)
- persuasion with integrity (1 lesson)
- assessment by other course participants, while watching simulation or roleplay, general discussion, takeaways (1 lesson)

Length: 1 semester
Format: once a week for one semester (12 meetings), 90 minutes long
Number of participants: up to 20

7D Groups now take turns to present their course outlines to the class. Make sure that all students in each group have a chance to present, e.g. two students could present the Skills to include section, another student the Length and Format section, a fourth student could talk about the goal of the course and its expected outcome and so on. If necessary, give groups 3–4 minutes to prepare for their presentations and think about what they are going to say. When all groups have given their presentations, the class votes for the best course, i.e. the one which they think is most feasible and will bring the best results for students.

MyEnglishLab: Teacher's Resources: extra activities

Review ◀ 5

1 1 b 2 a 3 b 4 a 5 b 6 b 7 b 8 a 9 a 10 a
2 1 tactic 2 persuasive 3 paradox
  4 cancel each other out
  5 there is more to it than meets the eye
  6 has no legal standing
3 1 Published 2 Combining 3 Aimed / Being aimed
  4 guided / being guided 5 Having had
4 1 best if you decide 2 honest with you
  3 to see you again 4 share the view
  5 go with your idea 6 needs and expectations
5 1 three points; contrast (*high – reasonable, global – local, tomorrow – today*)
  2 repetition of sounds ('h' and 'r' in first sentence); rhetorical question
  3 rhetorical questions; three points; repetition of words (*our airline*); repetition of structures (*Which airline … ?*)

The business of tourism

6

Unit overview

	CLASSWORK	FURTHER WORK
6.1 **The impact of tourism**	**Lead-in** Students talk about tourism in their country/region. **Video** Students watch a video about tourism in Iceland. **Vocabulary** Students look at vocabulary related to tourism and hospitality. **Project** Students hold a class debate on tourist accommodation.	**MyEnglishLab:** Teacher's resources: extra activities; Reading bank **Teacher's book:** Resource bank Photocopiable 6.1 p.152 **Workbook:** p.29
6.2 **Evolving tourism**	**Lead-in** Students talk about tourism trends in their country/region. **Reading** Students read an article about tourism in Spain. **Grammar** Students review and practise past tenses and discourse markers. **Writing** Students write an article about tourism in their country.	**MyEnglishLab:** Teacher's resources: extra activities **Grammar reference:** p.120 Review of past tenses and discourse markers **Teacher's book:** Resource bank Photocopiable 6.2 p.153 **Workbook:** pp.30–32
6.3 **Communication skills:** Business networking	**Lead-in** Students talk about networking in a business setting. **Roleplay** Students roleplay a networking conversation. **Video** Students watch a video of people networking. **Reflection** Students reflect on the conclusions from the video and their own approach to networking.	**MyEnglishLab:** Teacher's resources: extra activities; Interactive video activities
6.4 **Business skills:** Exploring options	**Lead-in** Students discuss a quote on storytelling. **Listening** Students listen to a conference presentation about environmental responsibility. **Useful language** Students look at techniques and useful language for storytelling in presentations. **Task** Students tell a story about a life lesson.	**MyEnglishLab:** Teacher's resources: extra activities; Useful language bank **Spoken English:** p.114
6.5 **Writing:** Email to a business partner	**Lead-in** Students read and discuss an email exchange about travel arrangements. **Useful language** Students look at useful language for, and the structure and content of, emails enquiring about and confirming arrangements. **Task** Students write an email and a reply as part of the planning of an international sales conference	**MyEnglishLab:** Teacher's resources: extra activities; Writing bank; Interactive grammar practice **Grammar reference:** p. 121 *If* and alternatives **Workbook:** p.33
Business workshop 6 Sustainable business travel	**Listening** Students listen to a meeting about a company's response to corporate image damage. **Listening and Reading** Students listen to a phone call and read a text about developing a sustainable corporate travel policy. **Speaking** Students create and present a corporate travel policy and write an email summarising recommendations.	**MyEnglishLab:** Teacher's resources: extra activities

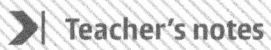

Business brief

The main aim of this unit is to look at the business of tourism and how the hospitality industry can impact on local economies.

The **hospitality and tourism** industry is one of the fastest growing sectors in many of the world's economies, but it is one that poses many challenges. **Tourism development**, **tourism policy** and **planning and destination development** all need to be carefully evaluated by national and local governments alike and innovative solutions found if they are to maximise the benefits of the industry for their economies, while minimising the impact on the natural and social environment. One of the most common problems is lack of adequate **infrastructure**, the transport systems, services and facilities available for people to use in and around tourist areas. In many places, infrastructure development has failed to keep pace with the speed of the growth in tourism and simply cannot cope with the number of people now wanting to use it. Governments need to invest in things like the road systems and shopping and medical facilities to support businesses in the hospitality sector. At the same time, they need to be sensitive to **environmental issues** caused by **overdevelopment**, ensuring that plans are put in place to protect the natural landscapes and sites of national importance that have attracted **mass tourism** in the first place.

In some places, like the Lake District in northern England, radical policies have been put in place to control the number of tourists allowed into a **tourist hotspot** at the height of the season, to try and mitigate the impact on the local area. While this works as a solution to protect the natural environment and normalise life for the local citizens who live there, it is seen as controversial by some, as it effectively limits free enterprise and the number of potential customers available to businesses. However, the growth of the tourist industry can also provide new income streams and new business opportunities for the local population in areas which may be remote and otherwise underdeveloped. For example, in many areas in Europe and America the inadequate provision of accommodation to meet the increasing demand has led to the exponential growth of **new business models** like Airbnb to make up the shortfall in the new **tourist traps**. This allows more people from the local economy to participate in the **tourist boom** by renting out rooms in their homes in places where there aren't enough hotels and guesthouses.

Seasonal tourism also vastly increases the market for local goods and **employment prospects** for local residents. Hotels, restaurants and other catering outlets need to be supplied and many of them prefer to use local ingredients to capitalise on their connection to the area. Other **local enterprises** producing gift items like textiles, artwork and ceramics can also benefit directly from the increase in customer numbers, but even local tradesmen are likely to see an increase in business as the demand for building and renovation work to supply facilities grows. Tourism is also important for the contact it gives people with nationals from other countries and how they see their own area. It can help people view their home town with new eyes, re-evaluate its merits and renew their sense of identity. Awareness of, and comparison with, other tourist destinations can also help local businesses to develop and offer better facilities and a wider range of services to their customers. Growth in the hospitality industry affords many more opportunities for employment too, with **new career paths** opening up in areas such as hotel management, events planning, catering and human resources.

International hospitality and your students

Most students will have been on holiday somewhere, whether in their own country or abroad and so will have experience of the hospitality industry as customers. Pre-work students may also have had experience of seasonal work within the hospitality industry themselves. Most students should also be aware of the tourist attractions and facilities within their own areas and they may know of issues around the impact of increasing tourist numbers on the local environment. Many students will also have travelled abroad and will be able to compare impressions of hospitality services and attitudes to tourism in various destinations around the world.

Unit lead-in

Elicit a brief description of the photo and ask students if they recognise the landmark shown (Badaling, the most visited section of the Great Wall of China). Then draw their attention to the quote and briefly discuss it with the class. You may wish to share with them an extended version of this excerpt of Andersen's autobiography: *To move, to breathe, to fly, to float, To gain all while you give, To roam the roads of lands remote, To travel is to live.* What are students' thoughts on the quote? Do they themselves enjoy travelling? Do they ever travel on business?

6.1 ❯ The impact of tourism

GSE descriptors

- Can extract specific details from a TV programme on a work-related topic.
- Can understand most TV news and current affairs programmes.
- Can contribute fluently and naturally to a conversation about a complex or abstract topic.
- Can participate in extended, detailed professional discussions and meetings with confidence.
- Can clarify points they are trying to make in an academic discussion, using linguistically complex language.

Warm-up

Discuss these questions with the class: *What do you think are the benefits of tourism for a country? Are there any drawbacks? Should governments try to improve domestic tourism or try to attract international tourists instead? Why? What do you think is the best way for a country to attract more tourists?*

Lead-in

Students talk about tourism in their country/region.

1 Discuss the questions with the whole class; check that students understand *sustainable* in question 2. For questions 2 and 3, encourage them to elaborate.

Video

Students watch a video about tourism in Iceland.

2 ▶ 6.1.1 Tell students that they are going to watch a video about tourism in Iceland and if desired, teach this vocabulary from the video: *underbelly, unspoilt, solitude, photobomb, interior, be overtaken by (tourism), infrastructure, boom, cut-price.* Give students a minute to read the questions, then play the video and check answers with the class.

Possible answers

1 It's seen as a destination with beautiful nature, where you can enjoy *solitude and spectacular, wide, open spaces.* (Ólöf)
2 Some Icelanders think it's grown too fast and visitors say some places/hotspots are crowded. Although it's successful as a tourist destination, some residents are unhappy that it has got out of control, e.g. rent has gone up in favour of Airbnb lets at the expense of local people.

3 ▶ 6.1.1 Explain the activity and ask students to complete the exercise individually. Note that the speakers are listed in the order they appear in the video. However, as they are seen more than once in different soundbites, make sure students understand before they watch a second time that they will need to move backwards and forwards through the extracts as they check their answers. Play the video for students to check their answers, then go through them with the class, clarifying meanings as necessary. Check that they understand the meanings of both options in italics each time, i.e. both the correct and the incorrect one.

1 putting a **2** overrun **3** tourists **4** decline
5 infrastructure **6** banned **7** citizens **8** rentals
9 tourism **10** Airbnb rentals

4 Put students in pairs or small groups and give them 3–4 minutes to discuss the statements. As feedback, invite different students to share their views with the class.

Extra activities 6.1

A ▶ 6.1.1 This activity provides students with extra listening practice. Ask them to answer as many of the questions as they can before watching again, then play the video for them to check/complete their answers. Check answers with the class.

1 b **2** a **3** c **4** a **5** c **6** b **7** b **8** a

Alternative video worksheet: Hospitality trends

1 If time is short, discuss the questions with the whole class. Otherwise, let students discuss in pairs or small groups first, then get feedback from the class.

> **1–2** Students' own answers
> **3** Possible answers:
> A porter is someone in charge of the entrance to a hotel. They may offer to carry guests' bags.
> A concierge is someone in a hotel whose job is to help guests by telling them about places to visit, where to eat out, etc.
> A housekeeper is someone who is employed to manage the cleaning in a hotel (similar to 'chambermaid' which is now an old-fashioned term).

2 ▶ ALT6.1.1 Tell students that they are going to watch a video about trends in the hospitality sector and give them time to read the questions. Ask them to make notes in answer to the questions while watching and play the first part of the video (0:00–3:15). Check answers with the class.

> **Possible answers**
> **1** Alexa is a digital concierge that is used for requesting housekeeping items or asking for information about the hotel. It has been introduced at some Marriott hotels.
> **2** By using a translation app.
> **3** Customising the settings in the room to suit the guest's needs, e.g. the lighting or room temperature, or if they want certain amenities.
> **4** Visitors can stay at Collective Retreats, which are luxury tents opposite Manhattan (on Governors Island) and wake up to views of the river and the Statue of Liberty. The emphasis is on the experience rather than material goods. (But the tents all have electricity, high-quality sheets and towels and there are luxury bathrooms.)
> **5** The luxury tents cost 75 US dollars a night.

3 ▶ ALT6.1.1 Tell students that they are going to watch the next part of the video and explain the activity. Give them time to read the questions and options first and check that they understand the meanings of *initiative*, *cater to* and *insulate*. Play the video (3:16–5:00), then check answers with the class.

> **1** b **2** a **3** b **4** c

4 This activity looks at useful collocations from the video. You could do it as a whole-class activity, checking answers and clarifying meanings as you go. Alternatively, let students attempt the exercise individually, then check answers with the class, clarifying meanings as necessary.

> **1** digital **2** luxury **3** sustainable
> **4** environmentally
> (Not used: experiential, season)

5 ▶ ALT6.1.1 Explain the activity and point out that the first letter of each missing word is given. Encourage students to read the extracts quickly before attempting to complete the gaps. Also remind them to look carefully at the words around each gap and think about what type of word is needed in each case. Play the video again for students to check their answers, then go over the answers with the class, clarifying meanings as necessary. You may wish to write (or invite students to write) the answers on the board, so students can check their spelling.

> **1** concierges **2** housekeeping **3** sustainability
> **4** hospitality **5** initiative **6** sustainable **7** conscious
> **8** Solar **9** quirky **10** feature

6 Put students in pairs or small groups and give them 3–4 minutes to discuss the questions. Then invite students from different pairs/groups to share their answers with the class.

7 Tell students that they are going to write an article about hotels in the future and go through the instructions and the ideas/expressions in the box with them. Remind them to think carefully about a) their readers: who they are writing the article for, b) how they can organise their ideas into clear paragraphs, c) an interesting title they can give their article and d) how they can use the ideas and expressions in the box. Point out the word limit and in weaker classes let students plan their articles in pairs. If time is short, this writing task can also be assigned as homework.

> **Model answer**
> Hotels in the future
> Many people often imagine future hotels as having tiny, pod-like rooms and robot concierges, like in some business hotels in Japan. However, future hotels will not only have technological features, such as online check-in and personalised lighting or heating, but they will also be designed to be environmentally friendly to accommodate the increasing demands of the twenty-first century traveller.
>
> One of these demands is the growing trend for experiential travel. This involves the visitor experiencing a destination actively from a social, cultural, historic, culinary or environmental perspective. One example could be going to a music festival in the UK with local people and staying in a luxury tent, or 'glamping', in order to have a more memorable travel experience.
>
> Another trend in hospitality is sustainability. Travellers today are far more environmentally conscious than they used to be. They might expect a jug of filtered water in their room and soap in the bathroom rather than plastic bottles. They don't expect their towels to be washed every day and are aware of recycling and saving water and energy.
>
> The idea of robot receptionists or concierges is still far away. In reality, the majority of guests would prefer to deal with a human rather than a robot. It is unlikely that Alexa will replace all human concierges. Interpreting and translating different languages is a very difficult task to programme. Any standard international hotel will still require staff with excellent communication skills who can speak several languages and deal with guests' queries in a friendly and polite manner.

Vocabulary: Tourism and hospitality

Students look at vocabulary related to tourism and hospitality.

5 Go through the words in the box with students before they begin and get them to complete the sentences individually. Alternatively, you could let them complete the sentences using their dictionaries, then clarify meanings during feedback.

> **1** Photobombing **2** World Heritage Site, national importance **3** record season **4** guesthouse
> **5** Destination development **6** Infrastructure **7** hotspot

6 Explain the activity and point out that by matching the words, students will create two different types of item: compound nouns (e.g. *guidebook*) and collocations (e.g. *wildlife park*). Highlight that some of the compound nouns are written as one word and others as two words and that some are hyphenated. Ask them to do the exercise individually and then to compare answers in pairs. During class feedback, clarify meanings as necessary.

> bed and breakfast, eco-resort, hotel chain, mainland, marine life, mass tourism, national park, overdevelop, package tour, tour guide, tourist trap, whale-watching, white-sand beach

7 Students should do this individually. Encourage them to read the text quickly first to get the gist before attempting the exercise and point out that they may need to change the form of some words. Check answers with the class.

> **1** national parks **2** eco-resorts **3** marine life **4** hotspot
> **5** white-sand **6** Infrastructure **7** mainland **8** package
> **9** guesthouses / bed and breakfast **10** hotel chains
> **11** tour guides **12** overdeveloped

8 First, get students to complete the questions individually; again, remind them that they may need to change the form of some words. Check answers with the class. Then put students in pairs to discuss the questions. Encourage them to give reasons for their opinions. Give them 3–5 minutes to discuss in their pairs, then elicit ideas around the class.

> **1** guesthouse / bed and breakfast
> **2** hotspots / World Heritage Sites / mass tourism
> **3** destination **4** infrastructure **5** chains

Extra activities 6.1

B This activity practises vocabulary from the lesson. Ask students to do it individually and point out that they may need to change the form of some words. Also explain that they will need to write more than one word in some gaps, but the first letter of only the first word is given each time. If time allows, get them to compare answers in pairs before class feedback.

> **1** national parks, hotspots **2** eco-resorts
> **3** guesthouses **4** World Heritage Sites **5** bed and breakfast(s), hotel chains **6** national importance
> **7** destination development **8** record season
> **9** infrastructure **10** tourist traps, mass tourism

Project: The tourist accommodation debate

Students hold a class debate on tourist accommodation.

9A Divide the class into two groups, A and B and tell them that they are going to hold a class debate. Go through the instructions with students, then refer each group to their information and give them time to read it. Point out that they have to prepare persuasive arguments irrespective of their own opinion; this task encourages critical thinking. Set a time limit before they begin and during the activity, monitor and help students with prompts/ideas for their arguments if necessary (see answer key below).

> **Suggested arguments for the proposal**
> - Hotel developments will create more jobs for local people and attract more visitors, which will boost the local economy.
> - Visitors will not depend so much on Airbnb and guesthouses.
> - It will lead to improvements in infrastructure, depending on demand.
> - Providing hotel accommodation will help to reduce rent for local residents because the cost of renting is rising due to demand from holiday lets.
>
> **Suggested arguments against the proposal**
> - Hotel development will attract more visitors and spoil the local area, e.g. with littering.
> - International hoteliers only care about making a profit and not sustainable growth.
> - Infrastructure (roads, public toilets, service stations, cafés and restaurants, etc.) is already limited.
> - Hotels will compete with residents who offer accommodation in guesthouses and want to make money out of holiday rentals.
> - Local people won't be able to enjoy the countryside peacefully because it will be overrun by tourists and coach tours.

9B Before starting the debate, encourage students to use vocabulary from the lesson as well as persuasive language; you may wish to write a few expressions on the board for them to refer to during the activity, e.g. *This will not only … , but also … ; I am certain/sure this will … ; This/It will definitely … ; That's why we strongly recommend … ; We therefore recommend/suggest …*

In large classes or if short of time, you could appoint two or three speakers who will present the arguments prepared by the whole group. To extend the activity, you could get students to vote (individually) for or against the proposal at the end of the debate. Tell them that they may change their minds depending on the strength of the speakers' arguments. In the event of a tie, your vote can be considered in order to reach a final decision. If there is time, you could also hold brief Q&A sessions after each group has presented their arguments; students could write down questions as they listen to the different speakers and then ask their questions before voting.

MyEnglishLab: Teacher's Resources: extra activities; Reading bank
Teacher's book: Resource bank Photocopiable 6.1 p.152
Workbook: p.29

6.2 > Evolving tourism

GSE descriptors

- Can understand complex arguments in newspaper articles.
- Can use the past perfect with adverbial clauses of time.
- Can use the past perfect continuous in a range of common situations.
- Can structure longer complex texts using a range of cohesive devices.
- Can use a range of verb tenses to convey nuances of meaning in an academic text.
- Can write a research report including detailed analysis and evaluation of own and others' work on the topic of investigation.

Warm-up

Put students in pairs and ask them to think about how tourism has changed in the last few years. Encourage them to think about their own country/region as well as global trends they may have heard of. Give them a few minutes to discuss in their pairs, then elicit ideas around the class.

Lead-in

Students talk about tourism trends in their country/region.

1 Put students in pairs and give them 2–3 minutes to discuss the question. Check that they understand the meaning of *high-end* before they begin. As feedback, invite students from different pairs to share their answers with the class; encourage them to elaborate.

Reading

Students read an article about tourism in Spain.

2 Before students do this exercise, you may wish to teach some vocabulary from the text, e.g. *upmarket, blend, homage, entity, cachet, shift, prompt, infancy.* Ask them to complete the exercise individually and before they begin, point out that they only need to *scan* the text in order to find the answer to the question; they should not read it in detail at this point. Check the answer with the class.

Possible answer

Spain has traditionally been a destination for mass tourism: *It more or less invented mass tourism in the 1960s.* But things are changing: *the guests in the Heritage Madrid Hotel are wealthy Europeans and Latin Americans, … there is a drive to upgrade Spain's image.*

3 Ask students to work individually, explain the activity and tell them to ignore the words and phrases in bold and italics for now. Encourage them to look carefully at the words around the gap each time, paying particular attention to linking words, time phrases, pronouns, etc., which may refer to the previous sentence or clause and therefore help them choose the correct extract. If there is time, get them to compare answers in pairs before checking with the class.

1 b **2** j **3** h **4** i **5** c **6** g **7** f **8** e **9** a **10** d

4 Put students in pairs and give them 2–3 minutes to discuss the question. If there is time, you could join pairs together into groups of four to exchange ideas before getting feedback from the class.

Extra activities 6.2

A This activity provides students with extra reading practice. Ask them to complete it individually and encourage them to underline the parts of the text that give them the answers. Check answers with the class.

1 Wealthy Europeans and Latin Americans looking for a mix of gastronomy and high-end shopping.
2 There had been a significant increase: 52.7 m tourists had visited Spain in 2010 and spent €48.9 bn, while 2017 saw 81.9 m visitors who spent €87 bn.
3 France
4 There is a drive to upgrade Spain's image and it's now more fully developed; in the 1960s its main exports were *oranges, shoes and sunshine.* There is a shift toward luxury.
5 Luxury destinations such as Aubocassa, an olive farm in Mallorca that offers a luxury oil tourism experience, as well as a boom in top restaurants in the Basque country and Catalonia.
6 Tourist numbers have fallen but total tourist spending has increased.
7 Rowdy tourist behaviour had led to calls for limits on short-term rentals (e.g. restrictions on holiday lets like Airbnb).
8 Ms García Castelo says that luxury tourism in Madrid is still in its infancy compared to cities such as Paris, Rome or London.

Grammar: Review of past tenses and discourse markers

Students review and practise past tenses and discourse markers.

5A Let students attempt the exercise individually first and then get them to compare answers in pairs if time allows. Go through the answers with the class, eliciting the form for each tense and pointing out its use in each example. You could then refer students to the Grammar reference on page 120 and clarify any points as necessary.

Past Simple: e existed; **g** had, wasn't; **j** seemed, had, was needed
Past Continuous: i was booming
Past Perfect Simple: c had visited
Past Perfect Continuous: e hadn't been growing;
Present Perfect Simple: d has led; **h** has (long) been
Present Perfect Continuous: a have been falling

5B Do this as a whole-class activity, checking answers as you go. Then refer students to the section on discourse markers on page 121 of the Grammar reference and go through it with them, again clarifying any points as necessary.

Points in time: **c** in 2010; **f** now; **g** in the 1960s; **i** when (also in the text: *while, the previous year*)
Duration: **d** in recent years (i.e. over the last few years, not a specific point in time); **h** long (= for a long time)
Sequence: **b** first

6A Students should do this individually. Ask them to ignore the underlined words for now and point out that more than one answer may be possible for some items. Also remind them that they can refer to the Grammar reference on page 120 if they need help. Check answers with the class, clarifying any errors in the use of past tenses as necessary.

1 were running / ran, (had) founded, took over, launched
2 have shrunk, have been demanding
3 brought / had brought
4 was, was, were feeling / felt

6B Students do this individually. Get them to compare answers in pairs before class feedback.

1 in the 1990s 2 the previous year 3 in 2018
4 more recently 5 during 6 Then (also, first, next, after, after that, finally, previously, etc.)

7A Do this as a quick whole-class activity. Note that in order to find what *it* refers to, students will need to find the sentence in the text (paragraph 3, gap 3 / sentence h).

it = Spain *when* = the 1960s

7B Ask students to complete the exercise individually and if necessary, do the first item as an example with the class. Get them to compare answers in pairs if time allows, then check answers with the class.

1 *that* = the history; *which* = the three-star hotel
2 *This* = the idea of technology and easy air travel shrinking distances
3 *That* = the €9.2 bn revenues from the luxury market
4 *This* = the idea that hotel revenue for five-star hotels in Barcelona went down; *they* and *their trips* refer to luxury clients and to the trips of these luxury clients

Extra activities 6.2

B This activity gives further practice of past tenses. It is a consolidation exercise, so it would be better for students to do it individually. Before they begin, point out that more than one answer may be possible for some items. During class feedback, elicit all the possible answers for each item.

1 gave 2 walked 3 were moving 4 were milling
5 taught / were teaching 6 had fallen
7 had knocked 8 told 9 happened
10 made / were making 11 hadn't resonated
12 had lived

C Students now practise discourse markers. Again, ask them to work individually and point out that there may be more than one possible answer for some items. If there is time, get them to compare answers in pairs before class feedback.

1 it/this 2 what 3 as/while/when 4 Our
5 while/when 6 We 7 That 8 what

Writing

Students write an article about tourism in their country.

8 This writing task can be done in class or as homework. You could also let students plan their article in pairs in class, making notes and then write it at home. Point out that they should a) think about their audience (the article is for an in-flight magazine), b) give their article a suitable title, c) organise their ideas into clear paragraphs and d) write between 220 and 260 words. Encourage them to use past tenses and discourse markers in their writing. If you think your students will struggle, you could hand out the article in the answer key below for them to use as a model text.

Model answer

A history of tourism development in Cuba

Cuba boasts seven UNESCO World Heritage Sites and has long been an attractive destination for tourists. In 2017, tourism in Cuba generated over 4.5 million arrivals and it is one of the main sources of revenue for the island.

Until 1898, Cuba was Spain's closest colony to the United States. During the first part of the twentieth century, Cuba continued to take advantage of investments and created industries to support mostly U.S. interests. Its proximity to the USA also helped Cuba's economy grow quickly.

However, relations between Cuba and the USA deteriorated rapidly after the Cuban Revolution. The island was then cut off from its traditional market by an embargo. This led to a decline in tourism to record low levels because a travel ban had been imposed on U.S. citizens visiting Cuba.

Unlike the USA, Canada normalised relations with Cuba in the 1910s and Canadians have been visiting Cuba on vacation for many years. In fact, about one third of its visitors are Canadians.

Since 1980, the Cuban government has changed its policies and has allowed for small private businesses. It has also been creating programmes aimed at boosting international tourism. In recent years, the USA has re-established diplomatic ties with Cuba and the island's tourism industry is expected to benefit from these relations in the near future.

MyEnglishLab: Teacher's Resources: extra activities
Grammar reference: Review of past tenses and discourse markers p.120
Teacher's book: Resource bank Photocopiable 6.2 p.153
Workbook: pp.30–32

6.3 › Communication skills
Business networking

Warm-up

Put students in pairs or small groups and ask them to discuss the questions below. After they have discussed in their pairs, elicit answers around the class.
- *How would you start a conversation with someone sitting next to you at a conference?*
- *How would you keep the conversation going?*
- *Do you find it easy to make small talk with people you meet for the first time? What about people at your place of work/study? What (if anything) do you find difficult?*

Lead-in

Students talk about networking in a business setting.

1A Go through the definitions in the box with students, then draw their attention to the quotation. If there is time, get them to discuss it in pairs or small groups first, then elicit answers around the class.

> The quotation suggests that 'farming' refers to building and maintaining good long-term relationships – it's a slow process like sowing seeds – whereas 'hunting' is going out to sell quickly and aggressively 'going in for the kill' like a hunter.

1B Explain the activity, give students time to read the statements and check they understand *extrovert*, *introvert* and *strike up a conversation*. Let them think about their answers individually first, then put them in pairs to discuss them. As feedback, invite a few students to share their views with the class. Note that students' answers will vary, depending on personal attitudes, behaviour, experiences and reflections on their own culture and the culture of others they have met for business, if they work in international settings. The perception of small talk in business differs across cultures.

1C Depending on the time available and the size of your class, you could discuss the question with the whole class or let students discuss in pairs or small groups first.

> **Possible answers**
> It might include preparing what you're going to say in advance (e.g. have your opening pitch), thinking of some open questions, interesting anecdotes and appearing calm and confident even if you're not feeling it – 'fake it 'til you make it' – which is Charlie's suggestion in the next exercise.

Preparation: Networking at a trade fair
Students read and think about the scenario for a roleplay.

2A If this is the first Communication skills lesson for your class, briefly tell students about *Lifestyle* magazine and the profile of its readership. Otherwise, elicit this information from students before you start. Explain the activity and give students time to read the text and email, then elicit the answer. Also check students understand *insight* and *drum up* from the texts.

> Donna Johnson expects Susan to cover the travel trade fair event, network and spread the word about *Lifestyle* magazine and try to drum up some advertising from exhibitors.

2B Explain to students that they are now going to read a message Susan sent to a colleague, Charlie, after reading Donna's email and Charlie's reply. Get them to complete the activity individually, then check answers with the class.

> Susan feels anxious as she says networking does not feel natural to her. Students make their own comments on Charlie's advice, but it seems good.

Roleplay

Students roleplay a networking conversation.

3A Put students in pairs, explain the scenario and activity and assign roles (or let students choose). Refer them to their relevant information and give them time to read it while you monitor and help them with any questions they may have. Point out that their main aim is to *network* and encourage them to think about how they can achieve this without being too 'direct'; they should try to be polite and friendly. In weaker classes, you may wish to let students prepare in A–A and B–B pairs, then ask them to go back to their original pairs for their roleplays.

3B Students now roleplay their conversations in their pairs. During the roleplays, monitor and note down any points to highlight during feedback, but do not interrupt students.

3C Students should do this activity in the same pairs as for Exercises 3A and 3B. Go through the questions with them before they begin and encourage them to give reasons for their answers to questions 3 and 4. Give them 3–4 minutes to discuss in their pairs, then invite different students to share their answers with the class.

Video

Students watch a video of people networking.

4 ▶ 6.3.1 Tell students that they are now going to watch a video of a conversation similar to the one they have just had and go through the instructions with them. Give them time to read the questions before they watch and check they understand *transmit* (*confidence*) and *steer* (*the conversation*). Play the video, then check answers with the class. The video presents some useful networking techniques, but for extra support, you may wish to share some of the information in the Note below with your students; you could do this now or after Exercise 5.

1 She initiates the conversation by saying her name and her company name. She smiles confidently and shakes his hand.
2 Very well. She talked about the readership figures and profile of the readers and their interest. She showed him the website and also mentioned that readers were interested in independent holidays and would be a suitable target audience for Martin's boating holiday business. It seems that she did a good job promoting the magazine and avoided direct selling while networking.
3 When Martin mentioned the high cost of living in Munich, she said, 'Believe me, it's the same in London'.
4 She seemed to steer the conversation successfully. She asked Martin open questions to get him talking about himself and his business. She responded positively to what he said. She showed she's a good listener. She also tried to connect with him on a level other than business when she mentioned they had had similar experiences living in Munich and London (high cost of living). She didn't dominate the conversation or remain silent so there's a clear sense of give-and-take in the conversation.
5 It's quite indirect and polite, not 'hard sell'.
6 She did very well (see the points made for 4).

Note

Business networking – a model for a good proactive approach
- Before walking up to a stranger, have your (elevator) pitch prepared: yourself, your company and your goals with the listener. Avoid direct selling. Build your credibility.
- Use their name immediately.
- Make good conversation. Prepare in advance so you have an idea where you want to steer the conversation. Find the mutual connection. Try connecting with them on a level other than business; people often bond through overlapping areas of interest, no matter what they are.
- Be authentic. Talk about things you are passionate about as well; it doesn't have to be a shared interest.
- Become a good listener. Sincere interest is a form of generosity, even flattery and encouragement. Decide which comments are worth following up on.

5 Tell students that they are going to look at some useful vocabulary from the video and ask them to complete the exercise individually or, in weaker classes, in pairs, using their dictionaries if necessary. Check answers with the class, clarifying meanings as necessary. If time is short, this can also be done as a whole-class activity, checking answers and clarifying meanings as you go.

1 e 2 d 3 a 4 b 5 c 6 f 7 g 8 i 9 h 10 j

Extra activities 6.3

A This activity practises the vocabulary from Exercise 5 in the Coursebook. Ask students to complete it individually and remind them that they can refer to the definitions in the Coursebook if they need help. Check answers with the class.

1 f 2 j 3 d 4 a 5 h 6 b 7 i 8 g 9 e 10 c

Reflection

Students reflect on the conclusions from the video and their own approach to networking.

6A Allow students to work individually first so that they can reflect on their own approach from the roleplay. Go through the statements with them before they begin and remind them to think about their answers to Exercise 3C and the techniques Susan used in the video. You could also remind them of the information from the Note above, if you have shared it with them. If not, this would also be a good point to do so, to help students reflect on their approach to networking.

6B Put students in pairs or small groups to compare their reflections, then broaden this into a class discussion. Encourage students to elaborate.

MyEnglishLab: Teacher's Resources: extra activities; Interactive video activities

6.4 ➤ Business skills
Storytelling in presentations

GSE descriptors

- Can understand the details of extended and linguistically complex talks on a range of political, environmental and social issues.
- Can understand stories being told by a fluent speaker using colloquial language.
- Can understand the use of irony to emphasise a speaker's meaning.
- Can identify analogies and metaphors used to support a position in a linguistically complex presentation or lecture.
- Can narrate a story in detail, giving relevant information about feelings and reactions.
- Can tell a detailed anecdote using linguistically complex language.

Warm-up

Draw students' attention to the lesson title and ask them if they have ever attended a presentation where the speaker told their audience a personal story. If they have, invite them to tell the class if they found it interesting and why / why not. If not, ask if they think a personal story might make a presentation more interesting; encourage them to give reasons.

Lead-in

Students discuss a quote on storytelling.

1A Look at the quote with students and check that they understand *constitute* and *arsenal*. Put them in pairs and give them 2–3 minutes to discuss the question, then invite students from different pairs to share their views with the class.

1B If time is short, discuss the questions briefly with the whole class, nominating a few different students to answer. Alternatively, let students discuss in pairs or small groups first, then get feedback from the class.

Possible answers

1 Many stories include a protagonist facing a tough challenge, villains, emotions and dramatic twists in the plot, a happy ending in which the main protagonist triumphs and a moral message.
2 Stories can be useful in many more formal situations, such as in an opening conference speech or during a presentation, but also during meetings to make a point clearer, or even during a job interview to put across an insight in an engaging and innovative way. Stories are also part of much social conversation, although these are often to entertain rather than to make a point.
3 Great storytellers generally speak with confidence and fluency; they are able to use their voice and body language to maximum effect to add drama to their narrative; and they have a good sense of timing – to slow down and speed up, to use pauses – to add impact to what they say. They are able also to shift tone and make a serious point at the end of the story.

Spoken English
p.114: Mate, stop it – 'cause it's annoying

1 ◀) SE6 Explain to students that they are going to hear people discussing question 2 from Exercise 1B and refer them to the question. Explain the activity and give students time to read the questions so they know what to listen for. Encourage them to make notes in answer to the questions while listening, then play the recording. Note that students, especially in weaker classes, may need to listen twice and/or you may need to pause the recording to give them time to note down their answers.

Possible answers
1 She describes them as anecdotes; telling real stories about real people.
2 Storytelling can make something that might be unpleasant or uninteresting seem (subconsciously) more engaging; it can encourage you to buy a product.
3 When having a difficult conversation (such as reprimanding someone) or an honest one, the use of storytelling can help to give an (indirect) example or a vision of the future.
4 One of the woman is very enthusiastic about the power/effect of storytelling; the other woman is sceptical of its use and prefers to tell other people her opinion in a direct way.
5 PowerPoint might be seen as boring; people might switch off.

Vague language

2 ◀) SE6 Explain the activity and give students time to quickly read through the extracts before they listen. Play the recording again, twice if necessary, then check answers with the class, clarifying meanings as necessary.

1 I suppose   2 and so on   3 there you go, see
4 if you like   5 might sound strange   6 maybe
7 in some ways   8 or something   9 Apparently
10 here we go again

3 You could do this as a whole-class activity, checking answers as you go. In feedback, point out the function of vague language – to make what we say sound non-specific, less direct or less factual – and explain that it is very common, especially in spoken English. Tell students that using vague language will help make their English sound more natural and also make it easier for them to speak fluently as it can help them communicate using less precise vocabulary and also buy them some thinking time in a conversation.

Vague language (used to sound less direct or non-specific): I suppose, so on, if you like, might sound strange, maybe, in some ways, or something, apparently
Phrases not used to sound less direct: there you go, see (used when you think you have proved to someone that what you are telling them is right); here we go again (used when something is starting to happen again)

Listening

Students listen to a conference presentation about environmental responsibility.

2 ◀) 6.01 Go through the instructions with students and give them time to read the questions. Ask them to make notes in answer to the questions as they listen, then play the recording, twice if necessary. Check answers with the class.

1 Ángel wants to propose ways to make the construction industry's environmental impact as positive as possible; to show greater leadership as an industry.
2 He is/was a member of Ángel's team; a top sales performer. He decided to leave the company.
3 A mixture of shock and disappointment.
4 Don't forget your good people. However capable they are, they also need appreciation.
5 If we forget to care about what is important to us, eventually we lose it.
6 The relationship with the environment, because too much hotel building is damaging the environment, killing wildlife and generating huge amounts of pollution.

3 ◀) 6.01 Before you play the recording again, give students 2 minutes to go through the questions and options and ask you about any they do not understand. To check answers, you could play the recording again and tell students to ask you to pause when an answer is heard.

1 a   2 a   3 a   4 a   5 a   6 b

4 Discuss the question with the whole class. Encourage students to give reasons.

Useful language

Students look at techniques and useful language for storytelling in presentations.

5A Explain to students that the diagram outlines the structure of the story in Ángel's presentation and go through it with them. Check that they understand *turning point*, *irony* and *moral*. You may wish to point out that irony is often used as a rhetorical device, to add 'texture' to a story – in this particular case, humour. Ask students to complete the exercise individually, then check answers with the class.

1 b **2** i **3** a **4** e **5** c **6** f **7** h **8** j **9** g **10** d

5B Get students to do this individually and then compare answers in pairs before class feedback. List the strategies and language students need to identify on the board and encourage them to record them in their notebooks.

Humour (irony): *Not that I'm offering myself as a model. Far from it.; That will become abundantly clear very shortly.*

Repetition of short sentences/chunks: *A great guy. Top salesman. Great with customers.; … lots of nice travel, no hassles with team conflict … leadership was easy.; He felt totally ignored. He felt very undervalued.; … we're not caring enough.; We're not managing …*

Rhetorical questions: *What was going on?; How could this happen?; Was it all a bad dream?*

Powerful contrasts: *I was appreciating Tom because he was very good for me. I wasn't really appreciating Tom for being Tom.*

Simple logic: *If you forget to care, you lose the relationship.; And if we want to prevent a relationship disaster with our environment … , then we have to do three things.*

Added detail to enrich the story: *I just wanted to relax in my room; Really nice place. Amazing pasta. Anyway, I was eating mussels, he was eating bruschetta.*

Emphasise own mistakes: *And I suddenly realised this was my big mistake.*

Introducing highlights of the story: *In the end, … ; At the end of the day, … ; So …*

Strong demands: *We have to … ; We must … ; And we need to …*

Extra activities 6.4

A This activity looks at useful expressions for storytelling in presentations. Let students attempt the exercise individually, using their dictionaries if necessary, then check answers with the class, clarifying meanings as necessary.

1 No sooner had I **2** Without thinking
3 To my relief **4** By an amazing coincidence
5 In the meantime / To cut a long story short
6 that very morning **7** to cut a long story short / in the meantime **8** It just shows

B This activity introduces adverbs of opinion, which are often used to add impact to a story. Look at the instructions with students and before they begin, go through the adverbs in the box with them. Alternatively, let them complete the exercise on their own and clarify meanings as necessary during class feedback.

1 Strangely **2** Fortunately **3** Sadly **4** Funnily
5 Amazingly **6** Obviously **7** Frankly **8** stupidly

Task

Students tell a story about a life lesson.

6A Explain the activity and encourage students to change or add a few details to their version of the story. Remind them to use techniques and language from Exercises 5A and 5B. In stronger classes, you could ask students to prepare their own personal story about a lesson learnt after an interesting and/ or challenging travel/holiday experience. Encourage them to build stories which have drama and interest for the audience and which lead to a conclusion with a clear lesson or moral.

6B Put students in pairs and explain that they are now going to tell each other their stories and then give their partner feedback. Explain that while their partner is telling his/her story, they should listen for techniques and expressions from Exercises 5A and 5B and make notes. They should then tell their partner what they liked about the way he/she told the story, what they think could be improved and whether they think he/she used a good range of language and techniques from the lesson. During the activity, monitor and note down any interesting points that arise; highlight these in a brief feedback session at the end.

MyEnglishLab: Teacher's Resources: extra activities; Useful language bank
Spoken English: p.114

6.5 ❯ Writing
Email to a business partner

GSE descriptors

- Can write clear and precise emails intended to create rapport and put the addressee at ease.
- Can write about complex subjects, underlining the key issues and in a style appropriate to the intended reader.
- Can respond in writing to other people's arguments in an appropriate style.
- Can switch between formal and informal styles in email as needed during negotiations.
- Can give detailed written feedback on the effectiveness of a piece of work-related correspondence.

Warm-up

Discuss these questions with the class: *If you ran a tourist agency in your country, what services would you offer? Would you have different offers for tourists from different countries? Why / Why not?*

Lead-in

Students read and discuss an email exchange about travel arrangements.

1 Explain the scenario and activity and give students time to read the emails and brainstorm ideas for both questions; they could do this in pairs or small groups. Allow 3–4 minutes for them to discuss in their pairs/groups, then elicit ideas around the class.

Possible answers

1 A request for more free time on the itinerary for individuals to go exploring/shopping on their own; a list of recommended restaurants with different types of cuisine; a visit to a whisky distillery as an alternative to the Loch Ness trip.
2 Trying to get an idea from Yongli how many other groups she might send to Edinburgh this year; finding out if there are other local trips that might interest the groups that Yongli sends; asking if there are other services he can provide, such as a list of recommended restaurants.

Useful language

Students look at useful language for, and the structure and content of, emails enquiring about and confirming arrangements.

2 This is best done as a whole-class activity, checking and discussing answers as you go. After discussing the answers for each item, go through the relevant sections of the Useful language boxes with the class.

1 No, it wouldn't. It is expected – even in business to business communications like these – to make some sort of polite remarks to help build a constructive working relationship.
2 James answers all the questions appropriately, although this will always be subjective. His response to the question of price is short and to the point, but this is fine in business-to-business communications when you are giving the answer you are assuming the reader wants to hear. His response to the question about making the trip shorter is longer, as he needs to – politely – say this is *not* possible. His answer to the question about the mini-coach is also slightly longer, even though he is telling Yongli what she wants to hear. This is because of a previous mistake in this area and James wants to rebuild trust on this point.
3 Yongli mentions confusion over the coach on a previous occasion and does so appropriately. James deals with it professionally and politely, though again this will always be subjective.
4 See the notes for 1 above. A suitable conclusion to a business-to-business email is also important for building good working relationships, though some nationalities will appreciate it more than others.

3 Get students to do this individually and then compare answers in pairs before class feedback. You could list the words/phrases students choose on the board for them to refer to during the writing task.

Extra activities 6.5

A–B Explain to students that when put in the correct order, the items in Activities A and B will form two emails: the first one is an email to a recruitment agency and the second one is the reply. Ask them to complete both exercises individually and then, if there is time, to compare answers in pairs. Remind them to think carefully about the structure of each email; encourage them to refer to the Useful language boxes on page 66 and the model emails on page 141 of the Coursebook if they need help. Check answers with the class.

A 1c 2f 3a 4e 5h 6d 7g 8b
B 1d 2b 3c 4a 5g 6h 7f 8e

Optional grammar work

The emails in Exercise 1 contain examples of *If* and alternatives, so you could use them for some optional grammar work. Refer students to the Grammar reference on page 121 and use the exercises in MyEnglishLab for extra grammar practice.

Task

Students write an email and a reply as part of the planning of an international sales conference.

4A Put students in pairs and go through the instructions with them. Give them time to read the background information on the two companies on page 142 and ask you about anything they do not understand. Explain that they are going to write two emails discussing the sales conference – an initial one from Anna at Greenleaves and then a reply from Yumei at City Events – and draw their attention to the notes. Give students time to read them and again, answer any questions they may have before moving on to the next activity. You may also wish to point out that Chinese style of names is surname first, followed by given name. For example, Jia Yumei would be Ms Yumei Jia in Western style.

4B Students now write their emails. Depending on the time available, they could plan them in class and then write them for homework. Remind them to look again at the questions in Exercise 3 and think carefully about the structure of their emails: they should organise them into clear paragraphs and try to use numbering, bullet points, etc. where appropriate. Also remind them to use language from the Useful language boxes and refer to the model emails on page 141 if they need help. If they write their emails in class, set a time limit for the planning and writing stages.

Model answers

Initial email

Hi Yumei,

This is Anna Meier from Greenleaves. Do you remember me? We were in contact when City Events helped us to organise a very successful sales conference in Singapore last year.

The good news is that we want to organise another similar conference next year, with the delegates staying four nights, June 4–7. However, we don't want to use the same hotel. We had issues with the size of the conference hall and the food supplied for the buffet lunches was bland and unimaginative. We would like City Events to find a suitable hotel for this event next year, using your local knowledge.

I have a few questions:

1 How much extra will you charge to find a hotel? We will make the booking from our end; we simply need you to research the alternatives and check on the facilities and suitability. To give an idea of the high service level we expect from the hotel, last year we had an issue with the flower display on the main stage. We asked for a large display of fresh flowers next to the speaker's podium, but the flowers were never changed and looked tired by the end of the conference.

2 Last year, the gala dinner on the final night was held at the hotel. This was not the best solution because the delegates already knew the restaurant and the event consequently did not feel particularly special. Next year, we would like to go to an outside restaurant, so can you also research this? The restaurant needs to seat 200 people in a separate private area and there must be a varied menu suitable for all tastes (including special dietary requirements) and, of course, good food.

If you need any more details please don't hesitate to get back to me. It will be great to work with you again and I am sure that together we can organise another fantastic event.

Best wishes,

Anna Meier

Executive Assistant to the VP (Sales)

Reply

Hi Anna,

Yes, of course I remember you! I was at the conference in my role as Event Coordinator and I met many of your colleagues. I recall the very impressive opening talk given by the Vice-President.

We would be delighted to work with you again as your local event management partner. Of course we can do all the research you require – we have very good local knowledge of hotels and restaurants based on over ten years of organising conferences and promotions here in Singapore. And I understand completely your point about the quality of the buffet lunches last year. If you wish, we could look into the possibility of external caterers providing the lunches, regardless of which hotel you eventually choose.

With regard to your specific questions:

1 For next year, I'm afraid we will have to increase our price to you by 15%. There will be extra work doing the research you have asked for (on-site visits, etc.) and in addition our staff costs are now higher than before. This is in line with what is happening in the labour market in Singapore more generally.

2 Yes, we have in mind a restaurant that will be an excellent setting for your last night gala dinner. It is a venue that specialises in corporate events and can hold 200 people. After dinner, delegates can move upstairs where you will have exclusive use of their rooftop bar. The venue is quite expensive, but we have used it before and the food is excellent.

Finally, let me apologise about the flowers on the stage of the conference hall. As you know, this was covered by the contract you had with the hotel and was not under our immediate control. But as the event organiser, we should have been more proactive and talked to the hotel as soon as the issue was raised by you. I will make sure that next year we monitor all aspects of the event and respond quickly if you raise any concerns.

We look forward to welcoming you here in Singapore next June. I am sure your sales conference will be a great success. We find that events such as these act as team-building opportunities and contribute greatly to staff motivation, employee loyalty and talent development in the client company. They can be seen as an investment in human resources rather than a cost.

Jia Yumei

4c If students write their emails as homework, this activity can be done in the next lesson. Put them in pairs and ask them to read each other's emails and discuss the questions. As a follow-up, you could ask them to rewrite their emails using their partner's suggestions; they could do this in class or as homework.

MyEnglishLab: Teacher's Resources: extra activities; Writing bank; Interactive grammar practice
Grammar reference: *If* and alternatives p.121
Workbook: p.33

Business workshop ❯6
Sustainable business travel

GSE descriptors

- Can interpret the main message from complex diagrams and visual information.
- Can critically evaluate the effectiveness of slides or other visual materials that accompany a linguistically complex presentation or lecture.
- Can contribute fluently and naturally to a conversation about a complex or abstract topic.
- Can follow a work-related discussion between fluent speakers.
- Can confidently argue a case in writing, specifying needs and objectives precisely and justifying them as necessary.

Background

Students read about a furniture manufacturer looking to develop eco-friendly travel practices.

1 Before students read, you may wish to teach/check understanding of these words from the background: *ergonomic*, *undergo*, *credentials*, *hard-hitting*. Put students in pairs and ask them to read the text and discuss the questions, then check answers with the class.

1 It manufactures and supplies ergonomic office furniture made from sustainable wood.
2 Green, sustainability, very environmentally focused.
3 A recent television documentary has criticised the company's contribution to air pollution from air and car transport, of failing to use video conference technology as an alternative and of its use of expensive international hotel chains which contribute little to local economies in developing countries.
4 She has scheduled a meeting with her management team and organised a telephone conference with a consulting company as she is deeply concerned about the situation.

Initial response

Students listen to a meeting about a company's response to corporate image damage.

2 🔊 BW6.01   Explain that Anja, the CEO of Reformula, is having a meeting with the management team to discuss the situation students read about in the Background. They should note down what each participant thinks of the documentary and how he/she thinks the company should respond. To help students, you could write the names of the participants on the board (*1 Anja, 2 Jan, 3 Agata, 4 Łukasz*) and encourage them to make notes in two columns for each speaker: one headed *Opinion about the programme* and one *Ideas on how to respond*. This might be particularly helpful for weaker students, as it will help them follow the recording more easily. Play the recording, twice if necessary, then check answers with the class.

> **Anja**
> Opinion: Agrees with Łukasz that a balanced approach is necessary.
> Ideas: There is a need to respond constructively and robustly, but also quickly. Wants Łukasz to send an email which specifies which data he plans to research, so the management team can agree on the scope.
>
> **Jan**
> Opinion: Accepts the general criticism made by the programme.
> Ideas: The company needs to change and find ways to reduce business travel; it needs to use technology more and create effective travel policies that align to its values.
>
> **Agata**
> Opinion: Programme was sensationalist, unbalanced and unfair. The company has a reasonable travel policy related to business needs.
> Ideas: The company needs to be practical and not overreact and reduce business travel. Relying on technology for client meetings might have a negative impact on sales.
>
> **Łukasz**
> Opinion: Agrees partly with Jan and Agata; the programme was sensationalist and there is a need to remain practical, but the company should have a clear sustainable travel policy. Believes more data is necessary to respond.
> Ideas: Investigate the situation and come up with a balanced and measured response. Ask people in his team to collect more data for the management meeting next week.

3 Tell students that after the meeting, Łukasz prepared some visuals for the team. Draw their attention to the visuals and give them time to look at them and ask you about any unknown vocabulary. Then explain the activity and give them time to answer the questions; they could do this individually or in pairs. When they are ready, invite different students to share their ideas with the class.

Extra activities Business workshop 6

A–B These activities look at idioms from the listening. You could do Activity A with the whole class, checking answers and clarifying meanings as you go. Students can then complete Activity B individually, referring to the definitions in A if they need help. Check answers with the class.

> **A 1** b **2** g **3** c **4** a **5** d **6** f **7** e
> **B 1** paying lip service **2** a knee-jerk reaction
>   **3** Off the top of my head **4** see eye to eye
>   **5** hold our nerve **6** face up to **7** play it by ear

C–D These activities practise opinion and commenting adverbs from the listening. Again, you could do Activity C with the whole class, checking answers as you go. Point out or elicit that these adverbs are often used to indicate an opinion or to comment on something and make sure students understand their meanings before moving on to the next exercise. Get students to complete Activity D individually, then check answers with the class.

> **C 1** d **2** c **3** e **4** a **5** f **6** b
> **D 1** Presumably **2** Technically **3** Undoubtedly
>   **4** Frankly **5** Strangely **6** Sadly

External support

Students listen to a phone call and read a text about developing a sustainable corporate travel policy.

4 🔊 BW6.02   Remind students that Anja has scheduled a telephone conference with Lighten Up, a consulting company which helps organisations develop eco-friendly travel practices. Go through the instructions with students and give them time to read the questions so they know what to listen for, then play the recording and check answers with the class.

> 1 She wants to get a few insights to prepare for her management team meeting next Monday and then plan how to use Lighten Up as consultants.
> 2 One: to just pay lip service to the issue programme and in reality do nothing; two: to create an effective, sustainable travel policy and approach for years to come.
> 3 The lack of structure and random data of the initial fact-finding by Łukasz on current company practices.
> 4 Financial, e.g. getting travel costs under control, responsible business tourism using local hotels and restaurants; environmental, e.g. reducing pollution by reducing travel; health/lifestyle, e.g. reducing stress and risk.
> 5 She's worried because it is a bigger issue than she realised. He assures her that once the upfront work is done to set up the policy, it's easier to manage the system.
> 6 Effectively. He is direct but supportive, making clear the full challenge of creating a proper sustainable travel policy in order that Anja is able to do it right.

5 Explain that students are going to read Björn's preliminary findings for Reformula. Look at the questions with them and give them time to read the text. Depending on the time available and the size of your class, you could get students to discuss the questions in pairs or small groups first, then invite different students to share their views with the class. Alternatively, if time is short, discuss the questions with the whole class.

Task: Create a travel policy

Students create a corporate travel policy and write an email summarising recommendations.

6A Put students in small groups and explain the scenario: they are all consultants from Lighten up and they have been asked to help management at Reformula create a sustainable travel policy. They should review all the information they have on Reformula, as well as the options Björn presented to Anja and try to add a few ideas of their own if they can. Depending on the resources available to your students, you could let them do some quick online research, to help them with ideas. Encourage them to make clear notes of the key points; explain that they will need them for the next stage, when they will prepare a presentation for the management team at Reformula.

6B Groups now create their proposed policy and prepare to present it to the management team. Go through the instructions with them and point out that they need to outline their recommendations in detail; highlight the different aspects they need to consider: *scope, targets, roles and responsibilities, reporting, documentation, policy communication*. Also explain that they should be prepared to answer questions about their policy from the management team. Allow plenty of time for this preparation stage and while students are working, monitor and provide help as necessary.

6C Join groups together into larger groups and explain that they are going to take turns to present their proposed policy. The students listening will take on the role of the Reformula management team. Explain that there will be a Q&A session at the end, so while listening to each presentation, the 'management team' should note down questions they want to ask the 'consultants'. Set a time limit for each presentation and Q&A session and ask groups to begin. As an optional follow-up, you could ask representatives from different groups if they thought the policy that was presented to them would be an effective one and why / why not.

7 Depending on the time available, students could do this writing task in class or as homework. Point out the word limit and encourage students to refer to their notes from Exercise 6A. If you think this would be helpful for your students, you could let them plan their emails in their original groups (or in pairs) and then write them individually.

Model answer

Dear Anja,

Following our meeting, I would like to outline our recommendations for Reformula's new sustainable travel policy, to include initial short-term goals as well as longer-term objectives.

Firstly, in the short term, we would recommend focusing on car and air travel and setting new internal travel policy guidelines, as these are relatively manageable. For example, aim to reduce air and car travel by 10% within a two-year period. Additionally, aim to increase the number of electric cars in your corporate fleet to 20% within the same period. It would also make sense to set a new target for the use of video conference technology, for example 25%, as this would align with the above.

To achieve some quick cost benefits, we believe it is advisable to cut the number of business class flights back to the level of five years ago, within a two-year period. You could also set a realistic limit on the cost of hotel accommodation at €125 per night and encourage stays in local hotels to maximise money flow into local economies. We estimate the total savings to be in the region of €1 million.

For the longer term, I suggest we support you to develop a strategic approach to travel policy with the creation of a Sustainable Travel Officer, who would define and communicate policy both internally and externally. In order to ensure that this role has the necessary authority, we recommend a direct reporting line to the Management Board. We would be delighted to assist you in developing a job description for this role and to support you in the recruitment process.

I will talk you through these recommendations in more detail when we meet next week.

With kind regards,

Garry Griffin

MyEnglishLab: Teacher's Resources: extra activities

Review ◀ 6

1 1 World Heritage Site **2** infrastructure
  3 overdeveloped **4** package tour **5** mainland
2 1 national importance **2** record season **3** hotel chains
  4 mass tourism **5** marine life
3 1 had already sold **2** had been selling **3** worked
  4 was working **5** has called **6** has been calling
4 1 I'd like to tell you about something
  2 after a lot of thinking
  3 then all of a sudden
  4 problem with all of this was
  5 has taught me an important lesson
  6 that's what I want to talk about today
5 1 confirm, remain **2** Apparently, confusion
  3 free, directly **4** long-term, relationship
  5 hear, again **6** know, re **7** feedback, unfortunately
  8 assure, happen

7 ›› Managing conflict

Unit overview

		CLASSWORK	FURTHER WORK
7.1 › **Workplace clashes**	**Lead-in**	Students talk about conflict in the workplace.	**MyEnglishLab:** Teacher's resources: extra activities; Reading bank
	Video	Students watch a video about conflict in the workplace.	**Teacher's book:** Resource bank Photocopiable 7.1 p.154
	Vocabulary	Students look at vocabulary related to talking about and managing conflict.	**Spoken English:** p.115
	Project	Students conduct and then discuss a class survey on managing conflict in the workplace.	**Workbook:** p.34
7.2 › **The road to reconciliation**	**Lead-in**	Students talk about resolving conflict in the workplace.	**MyEnglishLab:** Teacher's resources: extra activities
	Listening	Students listen to conversations on resolving staff conflict.	**Grammar reference:** p.121 Hedging and tentative language
	Grammar	Students study and practise structures for hedging and tentative language to express doubt.	**Teacher's book:** Resource bank Photocopiable 7.2 p.155
	Speaking	Students talk about defusing conflict in the workplace.	**Workbook:** pp.35–37
7.3 › **Communication skills:** Giving support and guidance	**Lead-in**	Students discuss different aspects of their working style and behaviour.	**MyEnglishLab:** Teacher's resources: extra activities; Interactive video activities
	Roleplay	Students roleplay a conversation in which they offer support and guidance to colleague.	
	Video	Students watch a video in which a colleague is given advice and guidance about a difficult situation at work.	
	Reflection	Students reflect on the conclusions from the video and their own approach to giving support and guidance.	
7.4 › **Business skills:** Mediating conflict	**Lead-in**	Students talk about mediating conflict in the workplace.	**MyEnglishLab:** Teacher's resources: extra activities; Useful language bank
	Listening	Students listen to conversations that deal with conflict at work.	
	Useful language	Students look at strategies and useful language for mediating conflict between others.	
	Task	Students roleplay mediations between colleagues.	
7.5 › **Writing:** Report on workplace conflict	**Lead-in**	Students read and discuss an HR report on a conflict in a sales department.	**MyEnglishLab:** Teacher's resources: extra activities; Writing bank; Interactive grammar practice
	Useful language	Students look at useful language for, and the main structure of, a report on workplace conflict.	**Grammar reference:** p.122 Prepositions and prepositional phrases
	Task	Students write a report to tactfully explain a workplace decision.	**Workbook:** p.38
Business workshop 7 › International team conflict	**Reading**	Students read survey results on team collaboration and a blog post about culture and conflict.	**MyEnglishLab:** Teacher's resources: extra activities
	Listening	Students listen to an interview with international team members.	
	Task	Students create and present a training programme for new international teams and write a blog post summarising a presentation.	

Business brief

The main aim of this unit is to examine how **conflict** can arise in different situations in the workplace and look at the possible factors which can lead to misunderstanding and strategies to deal with conflict.

Working closely with teams of colleagues can be rewarding, but there can also be challenges and conflicts. People bring different personalities, different styles, opinions and expectations to the workplace and some conflict is inevitable, not just between people (management and employees) but also relating to everyday issues (between individual colleagues), to ideas, strategies or policies and in some cases, to relationships with clients. In order for the conflict not to **escalate** and become destructive, it's important that it is channelled, or managed, so that it leads to outcomes which are beneficial and not detrimental.

The most common causes of conflicts at work are differences in working styles, poor performance and bad communication. One thing new employees find particularly challenging is balancing personal and professional commitments. This can lead to problems in their time management and performance. They may also need time to learn about business etiquette, what it's OK to do, or not to do and how to relate to people in different roles and at different levels within the organisation.

Managers being overly involved in team members' work and **micromanaging** them is also a common cause of conflict. This leads to a lack of trust and can adversely affect the employees' **morale**; they feel they are constantly being monitored and not trusted to work autonomously. It can be hard to achieve the right balance between seeing the bigger picture and worrying about the detail, but managers need to be aware of the fine line between guiding someone and continually look over their shoulder. It's important to learn when to step back and let staff work independently.

Conflict resolution is a way for two or more parties to find a peaceful solution to a disagreement. When trying to resolve a serious conflict, some companies prefer to use a third-party – a different manager or an outside agent – to act as a **mediator**. In business management theory, conflict resolution is often explained as a five-step process.

Step 1: Identify the source of the conflict.
The more information available about the cause of the conflict, the easier it is to help resolve it. The manager or mediator needs to give both parties the chance to share their side of the story in order to get a sound understanding of the situation and show **impartiality**.

Step 2: Look beyond the incident.
The source of the disagreement might be a minor problem that occurred months ago, but over time the two parties' attitudes have become **entrenched** and they have become generally **hostile** towards each other instead of addressing the root cause. The mediator must help them to look beyond the **triggering incident(s)** to see the real cause.

Step 3: Request solutions.
The mediator solicits the parties' ideas on how to make things better between them. It's important that they are an active listener, aware of every verbal nuance, as well as a good reader of body language. The aim is to get the disputants to stop fighting and start cooperating to resolve their issues.

Step 4: Identify solutions both disputants can support.
The mediator identifies the most acceptable course of action from the suggestions, not only from the perspective of the disputants, but in terms of the benefits to the organisation.

Step 5: Agreement.
The mediator needs to get the two parties to shake hands and agree to one of the alternatives identified in Step 4. Some mediators may even write up a contract in which actions and time frames are specified.

Conflict management and your students

In-work students may be able to think of examples of conflict in their place of work and talk about the steps taken to resolve it. Some may have experienced conflict themselves with management or colleagues and be willing to discuss the causes and results. Pre-work students may have experienced conflicts during their studies, for example between students and a member of staff, or during temporary summer employment and be able to reflect on how that arose and how it was handled.

Unit lead-in

Draw students' attention to the unit title and check they understand *conflict*. Then refer them to the quote and demonstrate or ask them to show you what a *clenched fist* is. Ask them what a clenched fist might represent (anger, aggressiveness) and how it contrasts with shaking hands (e.g. a handshake might convey trust or a peaceful intention). Briefly discuss the quote with the class. Can students explain it in their own words? How important is it to be willing to 'shake hands' in the workplace? Why?

7.1 ❯ Workplace clashes

GSE descriptors

- Can understand most TV news and current affairs programmes.
- Can extract specific details from a TV programme on a work-related topic.
- Can recognise a wide range of idiomatic expressions and colloquialisms, appreciating register shifts.
- Can describe the results and consequences of a specific action taken by an employee.
- Can answer questions in a survey using linguistically complex language.
- Can discuss findings from a research study.
- Can summarise group discussions on a wide range of linguistically complex topics.
- Can make proposals to resolve conflicts in complex negotiations.

Warm-up

Draw students' attention to the lesson title *Workplace clashes* and teach or elicit the meaning of *clash*. Ask: *Have you ever experienced 'clashes' (or conflict) at your place of work/study?* Ask for a show of hands, then invite different students to share their experiences with the class. What was the situation? How did it make them feel? How did they handle it? Note that some students may be uncomfortable describing what happened exactly; if so, they can just talk about how the situation made them feel and how they handled it.

Lead-in

Students talk about conflict in the workplace.

1 If there is time, put students in pairs or small groups to brainstorm ideas for both questions, then get feedback from the class. Alternatively, do this as a quick whole-class activity, eliciting a few ideas around the class.

Possible answers

1 someone wanting to dominate, people not sharing information, someone trying to micromanage, someone not pulling their weight
2 different expectations in levels of performance, different personalities, different working styles, an employee who is negative/demotivated

---

❯ Spoken English
p.115: Some people just don't like change

1 ◀) SE7 Explain to students that they are going to hear people discussing question 2 from the Lead-in and refer them to the question. Remind them of the meaning of *conflict* – elicit a brief explanation – then explain that they should make notes in answer to questions 1–5 while listening. Give them a minute to look at the questions so they know what to listen for, then play the recording, twice if necessary. Check answers with the class.

Possible answers

1 He couldn't concentrate because other colleagues were gossiping/chatting. This led to conflict and frustration.
2 The boss listened to the man's problem and offered to find him a quiet space.
3 Some people block other people's ideas; some people take credit for other people's ideas; some people just don't like change.
4 Companies contain (different) individuals with a variety of personality types; it's difficult for different types of people to work well together.
5 More senior staff might pick on more junior staff.

Informal language

2 ◀) SE7 Explain the activity and give students time to quickly read through the extracts before they listen. Play the recording again, then check answers with the class. Do not focus on the meanings or functions of the informal phrases yet, as students will look at them in more detail in the activity that follows.

1 sure enough 2 shot up 3 'cause
4 it's a pain 5 full stop 6 'cause 7 've gotta
8 's not gonna

3 This is best done as a whole-class activity, checking answers and clarifying meanings as you go. Alternatively, you could let students attempt the exercise individually and then compare answers in pairs and clarify meanings during class feedback.

a 3/6, 7, 8 b 5 c 4 d 1 e 2

Video

Students watch a video about conflict in the workplace.

2 ▶ 7.1.1 Put students in pairs, tell them that they are going to watch a video about managing conflict in the workplace and explain the activity. Go through situations a–e with them and check they understand the meanings of *demotivated*, *pull your weight* and *praise* (v.). Give them 2–3 minutes to discuss in their pairs, then play the video and check answers with the class.

a, b, c, e

3 ▶ 7.1.1 Allow students to read through the statements before they watch again, then play the video. In weaker classes, students may need to watch twice for this activity: once to decide whether the statements are true or false and then a second time to correct the false statements. Get students to compare answers in pairs before class feedback.

> 1 T
> 2 F Rowena says employees may get upset if ~~the boss~~ **a team member** is not working as hard as them.
> 3 T
> 4 F There can be challenges getting on with ~~very experienced~~ **new** co-workers.
> 5 T
> 6 F James has ~~never~~ had **some** problems working in an open-plan environment.
> 7 F James ~~admits to being a micromanager~~ **has experienced being micromanaged** by a boss, managing every detail of his work.
> 8 T
> 9 F It's **not a good idea** ~~best~~ to implement a new policy such as hot-desking without asking staff for their opinion.
> 10 F In case of conflict, Rowena's advice is to speak to your ~~board of managers~~ **line manager or HR**, who are always willing to talk.

4 If time is short, discuss the questions briefly with the whole class, nominating a few different students to answer. Otherwise, let students discuss in pairs first, then get feedback from the class.

> **Possible answers**
> 1 People might get demotivated and become less productive; people might leave the job, or ask to be moved to a different department; people might argue; when people are new to the company, they might take a while to adapt, etc.
> 2 Companies should issue clearly defined steps to employees and conduct training about what to do if someone has a problem with other people at work.

Extra activities 7.1

A ▶ 7.1.1 This activity provides students with extra listening practice. Explain the task and point out that they need to use three or four words in each gap. Give them time to read the extracts first, then play the second part of the video (from 2:47), twice if necessary and check answers with the class.

> 1 trusted to do   2 affect my morale
> 3 really affects people   4 connected or close to
> 5 raised the issue with   6 a big thing happening
> 7 take on board   8 conflict does arise
> 9 not dealt with quickly   10 is having some issues

Vocabulary: Conflict in the workplace

Students look at vocabulary related to talking about and managing conflict.

5 Depending on the level of your class, you could go through the words in the box with students before they complete the sentences or let them use their dictionaries during the activity and then clarify meanings during class feedback.

> 1 morale   2 on top of your game   3 raise the issue with
> 4 line manager   5 micromanaging

6 Give students a minute to read the sentences and elicit the word which completes the sentences. Then check that students understand the meanings of the phrases in bold.

> line

7 You could let students attempt the exercise individually or in pairs, using their dictionaries to look up any unknown words and then clarify meanings during feedback. Alternatively, teach or elicit the meanings of the words in the box before they begin. Encourage them to read the text quickly first to get the gist before completing the gaps. Round off the task by asking students how helpful they think the advice in the article is for dealing with 'high-conflict people'. Would they add to or change any of it?

> 1 mediator   2 clashes   3 irrational   4 escalate
> 5 resolution   6 blame   7 struggle   8 confrontational
> 9 provocation   10 criticism   11 empathy

8 Explain to students that the words/phrases in bold are from Exercises 5–7 and get them to do the exercise individually. Check answers with the class.

> 1 c   2 d   3 a   4 b   5 e

9 Put students in pairs and give them 3–4 minutes to discuss the question. Encourage them to use vocabulary from the lesson. Then invite students from different pairs to share their experiences with the class.

Extra activities 7.1

B This activity practises key vocabulary from the lesson. It is a consolidation exercise, so it would be better for students to do it individually. Go through the instructions with them and point out that although the same word completes both sentences in each pair, the *form* of the word may be different in each sentence (e.g. an infinitive in the first sentence and an *-ing* form in the second one). If necessary, do the first item as an example with the class.

> 1 resolution   2 issue   3 a escalates   b escalating
> 4 morale   5 a blame   b blaming   6 clash
> 7 confrontational   8 line

Project: HR survey for managing conflict

Students conduct and then discuss a class survey on managing conflict in the workplace.

10A Put students in small groups, go through the instructions and list of topics with them and explain the activity. Explain that their questions could be about minor conflicts between colleagues and managers, between departments or between employees and management, e.g. company policy they are unhappy about or have not been consulted on, such as hot-desking, ways of working, working conditions, work schedules or distribution of work. It is best to avoid questions about conflict involving personality clashes and salaries if your students are all from the same organisation. During the activity, monitor and offer help as necessary.

Possible answers

1 What are the most common causes of conflict in your place of work/study?
2 Which of these things cause the most conflict in your place of work/study: hot-desking, working in an open-plan office, working in teams, distribution of work, sharing resources, the air-conditioning/heating? Anything else? (Please specify.)
3 How well is conflict handled in your organisation, e.g. on a scale of 1 to 5 (where 1 = it is handled badly or not dealt with and 5 = it is handled very effectively)?
4 How could conflict be dealt with more effectively in your place of work/study?
5 How satisfied are you with the way your line manager deals with conflict at work?
6 How satisfied are you with the way the HR department helps to mediate conflicts?
7 When a conflict arises, do you feel you can approach your line manager or HR and ask for mediation?
8 Think of a recent conflict situation in your place of work/study. Were you prepared to reach a compromise with the other party? Why / Why not?
9 What suggestions do you have for dealing with conflict at work more effectively?
10 If management could do one thing in the workplace to help conflict from escalating, what should it be?

10B Students now carry out their survey. Explain that they should join another group and take turns to ask and answer their questions. Encourage them to make notes, as they will need to present their findings to the class later. When all students have had a chance to answer the other group's questions, they should join another group and ask the new group their questions. They can continue to do this for as long as time allows.

11A–B Ask students to return to their original groups and explain that they are now going to present their findings to the class and make recommendations for dealing with conflict based on the respondents' answers. The presentations should be quite short, with 4–5 recommendations from each group. In larger classes, you could join two or three smaller groups together and ask them to share their questions and findings and decide on the most important points to present to the class. Allow groups plenty of time to prepare and encourage them to use language for making recommendations, e.g. *We recommend/suggest … ; Our findings show it is best to … ; It would be a good idea to … ; To sum up, we strongly recommend …* When they are ready, ask them to begin their presentations and explain that as they listen to other groups, they should note down points they have in common.

11C As a class, students now agree on a five-point plan for dealing with conflict. In larger classes, students can work in sub-groups and then share their plan with the rest of the class.

MyEnglishLab: Teacher's Resources: extra activities; Reading bank
Teacher's book: Resource bank Photocopiable 7.1 p.154
Spoken English: p.115
Workbook: p.34

7.2 > The road to reconciliation

GSE descriptors

- Can follow a work-related discussion between fluent speakers.
- Can infer meaning, opinion, attitude, etc. in fast-paced conversations between fluent speakers.
- Can recognise a wide range of idiomatic expressions and colloquialisms, appreciating register shifts.
- Can contribute to a group discussion using linguistically complex language.
- Can participate in extended, detailed professional discussions and meetings with confidence.
- Can convey finer shades of meaning precisely by accurately using a wide range of modification devices.

Lead-in

Students talk about resolving conflict in the workplace.

1 Put students in pairs or small groups and give them 3–4 minutes to discuss the options and the pros and cons of each one. Before they begin, check that they understand the meanings of *get off to a good start*, *mediate* and *exacerbate*. Get feedback from the class. Alternatively, if time is short, do this as a quick whole-class activity: ask for a show of hands for each option, then elicit ideas around the class.

Possible answers

a Might make you feel better in the short term, but will not resolve the conflict and may escalate it.
b Might avoid any further conflict in the short term, but will not resolve it and risks creating other conflict elsewhere.
c Shows a willingness to address the issue and be constructive, but the other party may not engage, or it may create further tension.
d Probably means some kind of resolution will result, but makes the conflict formal and risks an outcome that you are not happy with.

Listening

Students listen to conversations on resolving staff conflict.

2A 🔊 7.01 Explain the context and activity and give students time to read the questions before they listen. Ask them to make notes in answer to the questions while listening. In weaker classes, you may need to pause after each answer is heard to give students time to write their answers and/or play the recording a second time.

Possible answers

1 Claire's line manager, Zoe, has said she isn't working well in the team / isn't pulling her weight and that she hasn't been meeting her targets.
2 Claire admits she's been leaving early on some days to take her brother to medical appointments. She says that Zoe sets unrealistic goals and isn't understanding enough. She tried to talk to Zoe but without success: *She always says 'not now'.*
3 Claire would probably like the HR manager, Ruth, to talk to Zoe and for Zoe to be more flexible regarding her working hours and personal situation.

2B Discuss the questions with the whole class. Encourage students to give reasons.

Possible answers

I've been talking to Claire and she seems quite stressed at the moment. I understand her brother has been ill. How is she doing in the team? Would you like to tell me about it?

Claire might need more flexibility with her working hours for a while. What do you think? Would that work for you?

Could we possibly sit down together and come up with a solution that works for everyone?

It could be a difficult conversation because Claire has spoken to Ruth in HR rather than to her line manager, Zoe, directly.

3 🔊 7.02 Explain the activity and give students time to read the questions and options. Check that they understand *micromanage* and *delegate*, then play the recording, twice if necessary. Check answers with the class.

> **1** a **2** b **3** a **4** b **5** b **6** a **7** b **8** a

4A 🔊 7.03 Tell students that they are going to hear the next part of the meeting and explain the activity. Give them time to look at the notes before they listen and play the recording. Do not go over the answers yet, as students will discuss them in the next activity.

1 Treat the other person with respect; be polite and discuss constructively.
2 Recognise that the other person is not 'being difficult', but has a different position. This could relate to personality, values or ways of working.
3 Try to understand their point of view and use active listening (make eye contact, listen carefully, nod and allow the person to finish before you start talking).
4 Listen actively. Don't jump to conclusions.
5 Establish the facts that might affect your decision together.
6 Be open to the idea that a third position may exist. What seems fair to you may seem unfair to a colleague.

4B Students could discuss the questions in pairs or small groups and then share their views with the class or, if time is short, you could do this as a whole-class discussion.

Extra activities 7.2

A 🔊 7.02 Explain the activity and ask students to complete it individually. Play the recording for students to check their answers, then go through them with the class.

> **1** b **2** a **3** b **4** c **5** b **6** a **7** b **8** c **9** c **10** a

Grammar: Hedging and tentative language

Students study and practise structures for hedging and tentative language to express doubt.

5A Draw students' attention to the Grammar heading and explain that hedging devices and tentative language are very important in conversation; they can 'soften' what we say and make it sound less direct and more polite. Give students a minute to underline the tentative language in Ruth's sentence and check answers with the class.

> But <u>perhaps</u> if you explained your family situation to her <u>a little</u>, you <u>might</u> find she <u>would</u> be more understanding.

5B Explain the activity, draw students' attention to the box on the left and elicit one or two examples for each item (for item c, explain that these are verbs that can soften what we say, e.g. *This appears to be …* instead of *This is …*). You could refer students to the Grammar reference on page 121 and go through it with them now or after they do the exercise. During feedback, clarify any points about the words/phrases in bold as necessary.

> a **2** should be able to **3** might **5** it would **8** Could **10** might
> b **2** some kind of **4** a few minor, certain **6** quite a few **10** quite
> c **5** I thought **6** It appears **7** was wondering **10** seems, seem
> d **1** a bit earlier **6** to a certain extent **8** possibly
> e **3** It's conceivable **9** It is possible

6 Ask students to work individually and look at the example with them before they begin. If there is time, get them to compare answers in pairs before checking with the class. If you would like to offer students more practice, you could get them to roleplay Ruth and Claire's conversation from Exercises 2A and 2B using hedging and tentative language.

> 2 Francine <u>seems to be</u> under <u>quite</u> a lot of pressure to meet <u>some of</u> her production targets.
> 3 <u>It's possible that</u> Luigi was being <u>slightly</u> unfair when he told the department to cut costs.
> 4 Don <u>sometimes doesn't think</u> about how his goals <u>may</u> affect others.
> 5 Jan's team <u>doesn't appear to be</u> working together <u>very</u> effectively.
> 6 If we buy new equipment, we'll <u>possibly</u> improve output <u>to a certain extent</u>.
> 7 <u>It might not be</u> a good idea to buy new equipment because we <u>don't have a lot of money</u> now.
> 8 <u>I was wondering if we could</u> have a face-to-face meeting to identify the problems and resolve <u>certain</u> issues.

Extra activities 7.2

B This activity gives further practice of hedging and tentative language. Tell students that there may be more than one correct answer for some items and if necessary, do the first item as an example with the class. Get students to compare answers in pairs before class feedback.

Possible answers

1 I was wondering if / whether he could be slightly better at managing his time.
2 (Actually,) Zoe is a bit / a little / quite demanding and often / sometimes likes to check my work.
3 Ignoring the issue could possibly / would probably make matters worse (, actually).
4 Pascal seems (is) a bit / a little / quite serious – he doesn't tend to chat with colleagues at the water fountain.
5 I was wondering if you talked to him, whether you might (possibly) / would get to the root of the problem.
6 It's conceivable / possible this new strategy might exacerbate quite a lot of tensions.
7 It may be possible to mediate as long as the other party is willing to listen a little / to some degree.
8 It seems / appears (that) we need to build up a certain degree of trust in this organisation.
9 It's possible / conceivable / likely that listening a little more carefully will (probably) prevent a conflict from getting out of hand.
10 Be a bit / a little more positive, keep the conversation polite and avoid blaming the other person if possible.
11 You might (possibly) be able to reach an agreement if you find some common ground.
12 It might be a good idea to ask your team members to make a little more (of an) effort to understand one another's motivations sometimes.

Speaking

Students talk about defusing conflict in the workplace.

7 Put students in groups, explain the activity and give them time to read the scenario. Encourage them to refer to the principles in Exercise 4A. Allow 3–5 minutes for them to discuss in their groups, then invite students from different groups to share their ideas with the class.

MyEnglishLab: Teacher's Resources: extra activities
Grammar reference: Hedging and tentative language p.121
Teacher's book: Resource bank Photocopiable 7.2 p.155
Workbook: pp.35–27

7.3 › Communication skills
Giving support and guidance

GSE descriptors

- Can contribute fluently and naturally to a conversation about a complex or abstract topic.
- Can ask questions politely in difficult situations or on emotional or controversial topics.
- Can give detailed advice on a wide range of subjects using linguistically complex language.
- Can recognise a wide range of idiomatic expressions and colloquialisms, appreciating register shifts.
- Can compare and evaluate different ideas using a range of linguistic devices.

Warm-up

Discuss these questions with the class: *What would you say is your strongest quality? If you could change one thing about the way you work, what would it be?*

Lead-in

Students discuss different aspects of their working style and behaviour.

1A Explain the activity and give students time to read the statements. Check they understand *pessimist*, *moveable* and *prioritise*, then give them 1–2 minutes to tick the statements which are true for them.

1B Put students in pairs to compare and discuss their answers to Exercise 1A and the questions in this activity. Check they understand *(cause) friction* before they begin. Allow 3–4 minutes for students to discuss in their pairs, then elicit answers around the class. During feedback, you may wish to feed in some of the information from the note below.

Note

People behave in different ways and have different working styles. Differences can sometimes make working together difficult. We may not always get on with other people and this can even lead to personality clashes and conflict. Here are some examples of types of differences.

1 Optimists and pessimists: a naturally upbeat person might find it difficult to work with someone who has a more negative attitude and vice versa. The 'glass-half-full' person may feel the pessimist is constantly complaining and looking for flaws instead of focusing on finding solutions. The pessimist might be a more cautious person and more aware of the risks of optimism bias.
2 'Ideas' people like to challenge the status quo and 'action' people, the 'doers', get impatient with discussion and want to make things happen.
3 Extroverts often dominate conversations; introverts may find it hard to contribute to meetings with extroverts present.
4 Some people are more organised and structured in how they work; some are more spontaneous and flexible.

5 Teamwork vs. individual work: some people work well alone while others thrive in teams.
6 Attitudes to cooperation and competition.
7 Attitudes to time and deadlines.
8 Direct and indirect communication styles.

Many people feel that having a harmonious office or team dynamic is the ideal scenario. But it can also be argued that with more diverse personalities and working styles, there's a little more friction, which can make for a more innovative and creative team. For instance, when a person has a direct communication style and holds strong opinions, other team members may explain themselves more clearly and share ideas more openly too. However, there is always a fine line between helpful and harmful friction.

There are many free online personality tests students can do to find out more about their working style.

Preparation: Giving guidance to a team member

Students read and think about the scenario for a roleplay.

2 If this is the first Communication skills lesson for your class, briefly tell students about *Lifestyle* magazine and the profile of its readership. Otherwise, elicit this information from students before you start. Explain the scenario and activity and go through the questions with students before they read. Ask them to complete the activity individually and, if there is time, to compare answers in pairs before class feedback.

> **Possible answer**
>
> Yvonne did the interview with the Cambridge professor, which had been assigned to Charlie and for which he had prepared the questions. Charlie might be feeling upset because Yvonne appears to have done a number of things without including or consulting him. She finished a task without him and sent it to their editor without copying him into the email. She appears to be taking the credit for the interview that he prepared.

Roleplay

Students roleplay a conversation in which they offer support and guidance to colleague.

3A Put students in pairs and go through the instructions with them. Explain that the conversation is between the Assistant Editor at *Lifestyle* and a journalist who is finding it difficult to work with Yvonne Williams. Assign roles, refer students to pages 129 and 133 and give them time to read their information while you monitor and help them with any questions they may have. Set a time limit before students begin and monitor and offer help as necessary during the activity.

3B Set a time limit for the roleplays and ask students to begin their conversations. During the activity, monitor and note down any points to highlight during feedback after Exercise 3C, but do not interrupt students' conversations.

3C In their pairs, students now reflect on their roleplays. Go through the questions with them before they begin and check they understand *reassure* and *take sides*. Again, monitor and note down any points worth highlighting; go through these and any points you noted during the roleplays in a brief feedback session at the end of the activity.

Video

Students watch a video in which a colleague is given advice and guidance about a difficult situation at work.

4 ▶ 7.3 1 Explain that students are going to watch a conversation similar to the one they have just had and answer some questions. Give them time to read the questions, check they understand *reframe* and ask them to make notes in answer to the questions while watching. Play the video, then check answers with the class.

> 1 Charlie uses the opportunity to respond to Donna's comment about different styles. He's also keen to show that he realises his manager is a very busy woman and that he's not complaining and doesn't want to make a big issue about it.
> 2 He's right to some extent because his approach has been to avoid the situation and say nothing, which he admits has not had the effect he was hoping for, so he needs to find another solution.
> 3 She says he brings a lot to the magazine.
> 4 When Charlie asks her for her advice directly: *What would you do?* Before this point she's guiding Charlie by asking him questions to find out more about the situation and how he reacted. This helped him to see that he should have said something and then brought out his concerns and why he didn't.
> 5 Donna doesn't want to take sides, as this is not a serious conflict that needs management involvement. She tries to coach Charlie informally to deal with the situation himself.
> 6 Donna expresses the view that they make a good team despite – or perhaps because of – their different styles. This has the effect that Charlie expresses the value in his collaboration with Yvonne.
> 7 Possible answer: When staff feel they are capable of coming up with the solutions themselves they may be empowered by their manager's support. Donna does give Charlie direct advice when he asks for it, but she also tries to guide him to his own conclusions and solutions by asking questions about what actions he took and what he could do differently in future.

5 Tell students that they are going to look at some useful vocabulary from the video and explain the activity. Get them to complete it individually or in pairs and encourage them to look carefully at the words around each word/phrase in bold to help them choose the correct definition. Let them refer to the videoscript on page 149 if necessary and find the phrases there. Check answers with the class, clarifying the meanings of the words/phrases in bold as necessary.

> **1** d **2** e **3** a **4** j **5** i **6** c **7** h **8** g **9** f **10** b

> **Extra activities 7.3**
>
> **A** This activity practises the vocabulary from Exercise 5 of the Coursebook. Ask students to complete it individually and let them refer to Exercise 5 if they need help. Check answers with the class.
>
> > **1** upfront **2** word **3** daunting **4** gripe
> > **5** get a word in edgeways **6** have a lot on my plate
> > **7** niggle **8** outraged **9** hindsight **10** sour

Reflection

Students reflect on the conclusions from the video and their own approach to giving support and guidance.

6A Allow students to work individually first so that they can reflect on their own approach. Go through the questions with them before they begin and remind them to think about their ideas from Exercise 3C.

6B Put students in pairs or small groups to compare their reflections, then get feedback from the class.

MyEnglishLab: Teacher's Resources: extra activities; Interactive video activities

7.4 > Business skills
Mediating conflict

GSE descriptors

- Can follow a work-related discussion between fluent speakers.
- Can recognise a speaker's feelings or attitude in linguistically complex speech.
- Can rephrase controversial statements into more neutral language.
- Can negotiate a solution to a dispute (e.g. an undeserved traffic ticket, blame for an accident).
- Can make proposals to resolve conflicts in complex negotiations.

Warm-up

Draw students' attention to the lesson title *Mediating conflict* and teach or elicit the meaning of *mediate* and *mediation*. Ask them if they have ever been in a situation where conflict in their place of work (or study, e.g. when working in a team) was mediated by a third party. Ask those who answer 'yes' to share their experiences with the class. Who mediated? Did it help resolve the conflict?

Lead-in

Students talk about mediating conflict in the workplace.

1A Tell students that they are going to read a blog post about mediation skills and explain the activity. You may wish to teach these words from the text before they read: *cross-cultural, seek (to do something), assumption, constructively*. Give students time to read the text and answer the questions – they could do this individually or in pairs – then check answers with the class.

1 Workloads have increased. Companies have embarked on frequent rounds of change. Cross-cultural and language differences at work frequently result in misunderstanding.
2 In the one-to-one meeting, to seek to understand the issue and to challenge any negative assumptions. In a later joint meeting, the mediator enables individuals to understand each other and to find solutions.
3 Good listening and summarising skills.

1B If time allows, let students discuss the question in pairs or small groups first, then elicit answers around the class. Encourage students to give reasons.

Listening

Students listen to conversations that deal with conflict at work.

2A 🔊 7.04 Explain the scenario and activity and give students time to read the questions before they listen. Ask them to make notes in answer to the questions, then play the recording, twice if necessary and check answers with the class.

1 Paul is asking for data about Carmen's clients from Europe.
2 Carmen doesn't have the data or the time to put the data together. She feels Paul should focus on his own job and not try to do hers. She's worried that he may contact her clients' offices in the USA.
3 Carmen feels Paul is disrespectful in telling her and her team how to do their jobs.

2B Do this as a quick whole-class activity, inviting different students to share their views with the class.

Possible answer

Well. Barry listens to clarify why Carmen is frustrated. He challenges her rather negative assumptions about Paul, insists on the need to stay positive and solve the problem and gains her agreement that they hold a meeting with Paul in Paris that weekend.

2C If there is time, get students to discuss the question in pairs or small groups, then get feedback from the class. Otherwise, discuss the question with the whole class. Encourage students to give reasons.

3A 🔊 7.05 Explain that students are now going to hear the call that Barry makes to Paul in his role as a mediator, to help resolve the conflict between Paul and Carmen. Before they listen, you may wish to teach this vocabulary from the recording: *battle of wills, perspective, come across (as), talk something through (with someone), common ground*. Play the recording – twice if necessary – then check answers with the class.

1 Paul's following Susan's suggestion that the different sales regions cooperate and share data more. He feels the USA can learn a lot from Carmen's success in Europe.
2 Carmen doesn't want to cooperate.
3 Paul's lack of explanation of why he is emailing and his direct communication style. Also, the fact that Carmen is not a native English speaker may be causing a misunderstanding.

3B Discuss the question with the whole class.

Possible answer

Effectively. Barry clarifies Paul's positive motivations for communicating and identifies possible reasons for the misunderstanding that has caused the conflict. He is also adept at challenging Paul's assumptions and building the awareness and commitment necessary to have a constructive joint meeting in Paris.

4A 🔊 7.06 Tell students that they are going to hear the meeting between all three parties and give them time to read the questions. Again, note that students may need to listen a second time in order to check/complete their answers.

1 To help Carmen and Paul both reset their collaboration in the sales team.
2 A lack of engagement with the need for more collaboration from Carmen and her team.
3 The emails from Paul and their style, which she feels are telling her how to do her job.
4 Not to make assumptions; to be fair and understand others' positive motivations and not judge.
5 He asks Paul to explain and then forces Carmen to acknowledge and respond using short questions: *Can you respond to what Paul said?, You hear what he says about his motivations?, But you can see where he's coming from?*
6 Paul should visit Carmen in Europe on a learning visit.

4B Discuss the question with the whole class.

Possible answer

It's very important to make people in a conflict situation respect the fact that others may hold very different views to them but which are valid from their point of view. It's important not to contradict or undermine them (especially if feelings are involved) by claiming to 'own' the truth and suggesting that others' views are false.

Useful language

Students look at strategies and useful language for mediating conflict between others.

5A Explain to students that the model shows ten key steps to mediating conflict between others in the workplace. Give them some time to look at the model and steps a–e and check that they understand *stick to, ground rules, challenge* (v.) and *perceived*. Get them to complete the exercise individually or in pairs, then check answers with the class.

1 e **4** c **7** a **8** b **10** d

5B Ask students to work individually and, if time allows, get them to compare answers in pairs before checking with the class. During feedback, clarify meanings as necessary.

1 i **2** g **3** a **4** e **5** d **6** c **7** f **8** b **9** j **10** h

5C Refer students to audioscripts 7.04, 7.05 and 7.06 on pages 155–156 and explain the activity. Give them 4–5 minutes to find and underline the expressions and then match them to the appropriate skills from the model in Exercise 5A, then get them to compare answers in pairs.

Possible answers

1 **Clarifying the mediator role:** *I've been asked to step in so I'm having a chat to you both separately, and then I thought we could meet …*
2 **Stressing common objectives:** *Everybody wants to resolve this.; Listening to you both, I think we can find common ground and a way to move forward.; I honestly think that there's a lot of common ground.*
3 **Identifying and sticking to ground rules:** *Let's share views openly, stay positive and find some solutions to move forward.*
4 **Discovering individual views:** *Can you tell me from your point of view what's been happening?*
5 **Forcing people to listen to each other:** *Let me summarise what I'm hearing.; But you can see where he is coming from?*
6 **Challenging perceptions:** *Have you actually discussed this with him?; Be careful of assumptions.; I think this communication style may be creating a misunderstanding.*
7 **Confirming the perceived point of conflict:** *So, for you, a lack of collaboration is the real issue?*
8 **Exploring solutions:** *Any other suggestions from your side?*
9 **Proposing a way forward:** *OK, so given this, can we decide on a way forward?*
10 **Summarising the final agreement:** *So … we've agreed to meet in Paris, and plan to come up with a solution there.*

Extra activities 7.4

A This activity provides further practice and consolidation of the strategies and functional language from this lesson. Ask students to complete it individually and get them to compare answers in pairs before class feedback.

a 3 **b** 9 **c** 8 **d** 1 **e** 2 **f** 5 **g** 10 **h** 7 **i** 6 **j** 4

B This activity introduces some useful idioms from the listening and can be done in two parts. Start by asking students to find the idioms in audioscripts 7.04, 7.05 and 7.06 and complete them. Check answers with the class, then move on to the matching task. If time is short, you could do it quickly with the whole class, checking answers as you go.

1 hatchet, g **2** at, d **3** point, e **4** air, f **5** wills, b
6 eye, a **7** foot, h **8** injury, c

C Students now practise the idioms in Activity B. Ask them to complete the exercise individually and, before they begin, point out that they may need to change the form of some words. Check answers with the class.

1 at loggerheads **2** clear the air **3** see eye to eye
4 battle of wills **5** bury the hatchet **6** add insult to injury **7** sore point **8** got off on the wrong foot

Task

Students roleplay mediations between colleagues.

6A–B Put students in groups of three and explain that they are going to roleplay mediations between colleagues. Point out that they are going to hold three different conversations and that they are each going to take on the role of mediator once. Start with Mediation 1. Assign roles A–C, refer students to their relevant pages and explain that Student A is the mediator. Give them time to read their information and prepare for the conversations; they should do this individually. Remind them to use the steps from the mediation model in Exercise 5A and the useful language in Exercises 5B and 5C. Set a time limit for the preparation stage and then for the roleplay and when students are ready, ask them to begin. At the end of the mediation, give them 2–3 minutes to reflect on their roleplay: Did they follow the steps from the mediation model? How effective were they? Did they use language from Exercises 5B and 5C? What went well? What could be improved? During the activity, monitor and note down any points worth highlighting, for a brief feedback session at the end. Follow the same steps for Mediations 2 and 3. When students have completed all three roleplays and follow-up discussions, conduct a brief feedback session, going over any points you noted while monitoring.

MyEnglishLab: Teacher's Resources: extra activities; Useful language bank

7.5 〉 Writing

Report on workplace conflict

GSE descriptors

- Can write a detailed report of work-related events.
- Can write a detailed structured report on work-related topics.
- Can develop a written case to persuade others about the advantages or disadvantages of a course of action.
- Can write linguistically complex and logically structured reports and articles.
- Can write about complex subjects, underlining the key issues and in a style appropriate to the intended reader.
- Can correct structural errors in someone else's written report.

Warm-up

Draw students' attention to the lesson title and ask: *Have you ever written or read a report on workplace conflict? Who might write one? Who might they write it for? What might such a report include?* Elicit ideas around the class but do not confirm answers yet, as students will check their ideas in the exercises that follow.

Lead-in

Students read and discuss an HR report on a conflict in a sales department.

1 You could do this as a whole-class activity or let students discuss in pairs or small groups first and then elicit ideas around the class.

Possible answer

The unsuccessful candidates would feel demoralised, may be quick to criticise their newly promoted colleague (on the basis that they could have done better) and if they are ambitious, they may start looking for a new job in another company. The promoted manager may also have problems moving from a client-facing sales rep role to a managerial team leader role; being good at the first does not automatically mean you are good at the second.

2 Refer students to page 143 and give them time to read the report and answer the questions. Get them to compare answers in pairs before checking with the class. If you did the Warm-up activity, ask students if the report mentions any of their ideas.

1 One left. Two are unhappy.
2 No (not directly: focusing on the poor sales figures rather than personal conflicts).
3 Training in soft skills, working with a mentor.
4 Possible answer: John may have to be reassigned to a different team or a different role; firing him is not really an option as it would send a very bad signal to other ambitious, high-performing sales reps inside the company.

Useful language

Students look at useful language for, and the main structure of, a report on workplace conflict.

3 This is best done as a whole-class activity, checking and discussing answers as you go. After discussing the answers for each item, go through the relevant section of the Useful language box with the class. For question 5, students who have studied academic writing for university entrance can be told that business reports are less formal than academic writing, with shorter sentences and fewer embedded clauses.

Possible answers

1 This is a good structure. The Introduction answers the question 'Why this report?' and the Background gives a factual summary of events to help explain the current situation. 'Action taken so far' is a separate section because action will be the main focus of the SMT: what has been done, leading to what needs to be done now. In general (as here), a Conclusion interprets the facts in the main body of the report and makes a reasoned judgement. It leaves the reader with 'an end to the story' that explains everything. It aims to be remembered long after the details and facts in the report are forgotten. It is not action-oriented. Recommendations (as here) follow the conclusion and are action-oriented. Somebody has to do something. Recommendations should be concise, specific and realistic.

2 Coaching is shorter term, focused on specific learnable skills and can be done by someone inside or outside the organisation. Mentoring is longer term, focused on personal development that may not be easily evaluated and is done by an experienced person inside the organisation. Note also that there is a very specific meaning of 'coaching', often used within the profession, where the coach uses special questioning techniques to get the person being coached to draw their own conclusions about the right course of action (rather than being directly advised by the coach).

3 There is no single, specific term; ideas include having no interest in the work of team members, not knowing what is going on, delegating too much, not taking responsibility and being too hands-off.

4 Soft skills are related to your personality and behaviour. They are non-measurable, subjective skills that are transferrable to any job. They are often called 'people skills' or 'life skills'. Examples include communication skills, a positive attitude, being a team player, flexibility/adaptability, problem-solving, creativity, having a work ethic, interpersonal skills, intercultural awareness, time management, leadership, attention to detail, the ability to learn from criticism and working well under pressure.

5 Students will have their own ideas about formal language, but typical areas to mention are longer, carefully formed sentences contrasting with the short fragments typical of speech, vocabulary with Latin/Old French origins rather than Anglo-Saxon origins (e.g. *receive* vs. *get*, *request* vs. *ask*, *considerable* vs. *a lot of*, *remain* vs. *stay*, *profession* vs. *job*, *opt* vs. *choose*, *finish* vs. *end*, *rapid* vs. *fast*, *provide* vs. *give*, *estimate* vs. *guess*, *assist* vs. *help*, *type* vs. *kind*), specific grammatical constructions like the lack of contractions and a style that is impersonal and measured.

4 Get students to do this individually and then to compare answers in pairs before class feedback. You could list the words/phrases students identify on the board for them to refer to during the writing task.

Extra activities 7.5

A This activity focuses on register and contrasts the formal language in a business report with the more informal language used in speech. Explain the activity and ask students to do it individually. If necessary, do the first item as an example with the class. If there is time, get students to compare and discuss their answers in pairs before class feedback.

1 a F, b I 2 a I, b F 3 a I, b F 4 a F, b I 5 a F, b I
6 a I, b F 7 a F, b I 8 a F, b I

Optional grammar work

The report in Exercise 2 contains examples of prepositions and prepositional phrases, so you could use it for some optional grammar work. Refer students to the Grammar reference on page 122 and use the exercises in MyEnglishLab for extra grammar practice.

Task

Students write a report to tactfully explain a workplace decision.

5A Put students in pairs, explain the activity and allow plenty of time for them to read the notes. If you think your students will find this activity hard, you could allow them to make brief notes of their own and/or refer to the notes in the Coursebook while constructing their summary, but point out that they should not *read* from them. As feedback, invite a few pairs to share their summaries with the class.

5B Explain the writing task and point out that the SMT know nothing of the issues, so the report should include a clear summary of the situation; remind students of their summaries from Exercise 5A. Before they begin, point out the word limit, set a time limit and remind them to refer to the Useful language box and the useful language on the board. If there is no time to do the writing task in class, it can be set for homework.

Model answer
Report: Conflict in the Accounts Department
Introduction
The purpose of this report is to bring to the attention of the SMT an ongoing issue in the Accounts Department. There is a strong clash of personalities between the team leader, Rose and one member of her team, Peter. This is a situation that is getting worse and is beginning to affect the work of that department.

Background
The origin of the conflict was last March, when Peter was appointed to his position as an external candidate. Rose wanted another candidate (Olivia) to get the job. Since then there has been tension and open arguments between Rose and Peter, specifically over the issue of Rose's monthly reports. In order to prepare these reports, Rose relies on figures provided by Peter. Peter is often slow to provide the figures and this sometimes makes Rose late in submitting the reports. Rose is very unhappy about this.

The negative feelings between them became public a few weeks ago in an unfortunate incident. In front of other colleagues, Rose reminded Peter that she needed the figures on time. Peter argued back and said that his deadlines were unrealistic. Strong words were exchanged and there is still tension in the office following the incident.

Action taken so far
Last week, Rose came to see me. She explained the situation as she sees it and requested that we take some action. She does not want to be blamed by senior management for the late reports. I have not spoken to Peter yet.

Conclusion
This is a situation where we have a strong personality-based conflict. In my opinion, there is no solution using the usual discussions and procedures. Having given this matter considerable thought, I cannot see how they can continue to work together.

Recommendations
I believe the only course of action is to move Peter to another position in the Accounts Department, under a new team leader. I would like to give him a fresh start. I will need to have a frank and honest conversation with him in which I explain what is expected of him and the importance of deadlines.

I also need to speak to Rose confidentially. I will assure her that the late reports are not seen as her fault and that we are very satisfied with her overall performance. Do we have a budget to offer her some soft skills training in leadership and assertiveness? It is unfortunate that she spoke to Peter in a negative way in front of colleagues.

Maria Gonzalez
HR Director

5c If students write their reports for homework, this exercise can be done in the next lesson. Put them in pairs and ask them to read their partner's report and discuss the questions. You could then ask them to write a final, improved version of their report, based on their partner's feedback; they could do this in class or as homework.

MyEnglishLab: Teacher's Resources: extra activities; Writing bank; Interactive grammar practice

Grammar reference: Prepositions and prepositional phrases p.122

Workbook: p.38

Business workshop ❯7
International team conflict

GSE descriptors

- Can extract key details from quantitative data in complex business documents.
- Can infer meaning, opinion, attitude, etc. in fast-paced conversations between fluent speakers.
- Can understand complex technical work-related documents in detail.
- Can present detailed, evidence-based arguments during work-related meetings.
- Can confidently argue a case in writing, specifying needs and objectives precisely and justifying them as necessary.

Background

Students read about conflict between international team members of an NGO.

1 Tell students that they are going to read about an NGO and check they understand the term (Non-Governmental Organisation: an organisation which helps people, the environment, etc. and which is not run by the government). At this point, you may wish to teach this vocabulary from the Background: *marginalise, sabbatical, tangible, humanitarian, underpin, cite*. Put students in pairs and give them time to read the text and discuss the questions. Check answers with the class. For question 5, invite different students to share their experiences and views with the class.

1 Running a range of projects in countries with developing economies to support the most marginalised by increasing educational and employment opportunities.
2 Each project uses 12 working professionals from 12 different countries around the world on sabbatical.
3 Internal project team conflicts, with different approaches to communication and collaboration among the project team members, are frequently cited as the major problem.
4 Many projects fall behind schedule or fail to deliver on their planned objectives.

Dealing with international project conflict

Students read survey results on team collaboration, listen to interviews with international team members and read a blog post about culture and conflict.

2 Explain the activity and teach/check understanding of this vocabulary from the questions and survey: *inconsistency, punctuality, orientation*. Give students time to look at the survey results and answer the questions and, if there is time, get them to compare answers in pairs before class feedback.

Possible answers

1 The results indicate poor levels of collaboration. In the specific question on this topic, 49% scored collaboration as poor. Also, the 42% 'Not very' rating of likelihood to recommend indicates a strong dissatisfaction with the project experience. The impact of poor collaboration is also noted as significant, with 47% of those assessed indicating that poor collaboration had a negative impact on the project.
2 Different approaches to work around punctuality and schedules seems to be significant. The survey also implies that a lack of flexibility among team members results in differences becoming an issue (see the 54% reporting that an increase in flexibility would help the project).
3 The high lack of individual learning reported (56% 'Hardly at all') and the appeal for more flexibility in 'other team members' indicates a potential failure of many to learn themselves to be more flexible to manage differences. This also suggests that while differences may be triggering the conflict, the root causes may be a tendency to judge others negatively and a lack of willingness among people in general, irrespective of background, to respond to differences positively and effectively.
4 Extremely seriously. The organisation's projects are being affected in terms of timeliness and quality, reducing impact in-country. Additionally, the negative experience and feedback of those participating may end up damaging the brand and the flow of volunteers into future projects.

3 ◀ BW7.01 Tell students that they are going to hear part of a discussion between Ester Callas, head of HR at Pro-Transfer 12 and two project leaders, Patricia and Jean-Paul. Go through the instructions with them, give them time to read the questions before they listen and check understanding of (*negative*) *stereotyping* in question 2. Ask students to make notes in answer to the questions while listening, play the recording and check answers with the class. In weaker classes, you may need to play the recording twice and/or pause at short intervals for students to complete their answers.

1. Differences connected to time and planning: structured vs. relaxed. People are complaining about each other, leading to frustration and demotivation. This also leads to delays and affects quality of the projects; 80% of projects are behind schedule.
2. Negative stereotyping is when people complain or make negative comments about other individuals using personality or cultural stereotypes to explain the negatives they see in behaviour.
3. Talked about communication as a topic openly in meetings. It was not very successful because people were reluctant to give honest feedback.
4. They did social events such as cooking together. It helped a little.
5. It's a very deep and human, individual issue in the end. Everyone wants everyone else to change and finds flexibility difficult. They find it easy to blame others rather than see the problem in their own behaviour.

Extra activities Business workshop 7

A This activity looks at useful vocabulary from the listening. Go through the words in the box with students before they begin and get them to complete the sentences individually. Alternatively, you could let them complete the sentences using their dictionaries, then clarify meanings during feedback.

1. personality clash 2. moaning
3. negative stereotyping 4. intimidated
5. bad feeling 6. reluctant to speak up
7. buck-passing 8. intransigent

4A Go through the instructions with students and give them time to read the questions. Depending on the level of your class, you may wish to teach some key vocabulary from the text before students read or go through any unknown words during class feedback, letting students attempt the reading task without any vocabulary work pre-reading. Some vocabulary from the text which may be new to students is: *bury your head in the sand, diversity/diverse, homogenous, underperform/outperform, counterpart, hands-on, proactive, insight, emerge, bridge* (v.), *integrate*. If there is time, let students discuss their answers to questions 4 and 5 in pairs before class feedback.

1. She thinks that cultural stereotyping is wrong, but it is not correct to ignore cultural differences as a way to avoid stereotyping.
2. Either a positive or negative one. It depends.
3. Hands-on and proactive management of diversity; this unlocks the different thinking styles and insights in an international team to create positive energy and innovation.

4B Put students in groups of three and tell them that they are each going to read about one aspect of the MBI model mentioned in the text. Refer them to their respective information on pages 127, 132 and 134 and give them time to read while you monitor and help them with any vocabulary questions they may have.

4C In their groups, students now take turns to tell their group about the aspect of the MBI model they read about. Point out that they should use their own words and not read from their texts. When they have finished, they should discuss the MBI model, saying what they like or do not like about it and add any ideas of their own which they think would enhance the model. This part of the activity can also be done with the whole class if time is short: invite different students around the class to share their views on the model and then to suggest ideas.

Possible answers
Mapping
- Take time to build relationships.
- Profile the different cultures of team members using established dimensions of national culture.
- Profile individual psychology to help team understanding. Even exchanging CVs is better than nothing.
- Use a half-day workshop with team members sharing information about their own culture or psychology.
- Regular social events can be used to bond teams.

Bridging
- Find ways to communicate effectively with others.
- Listen more carefully.
- Postpone negative judgements.
- Use a psychotherapist to train project leaders in the art of listening.
- Practise Buddhist meditation to develop a more sensitive thinking style.
- Use 360-degree feedback tools to help individuals give feedback.
- Use of clarifying questions like 'What do you mean exactly?' or 'Why do you say that?' to build better understanding and uncover common ground.

Integrating
- Manage group interaction with formal protocols to ensure everyone's ideas are expressed and heard in an atmosphere of safety.
- Avoid destructive behaviours, e.g. 'Yes, but ... ', arising too quickly.
- Use external coaches and trainers to facilitate diverse teams during the early phases of a project to reduce conflict.
- Train skills to optimise group dynamics, e.g. summarising, interrupting dominant speakers, including quieter individuals to ensure a balanced participation model, using feedback.
- The team should define the rules of engagement and reset these over time.

Extra activities Business workshop 7

B–C Remind students that the sentences in Activity A described different collaboration problems in international teams and if necessary give them time to read the sentences again. Explain that the completed sentences in Activity B propose solutions to those problems. Tell them that they will first match the sentence halves and then match each problem (in Activity A) with a solution (in Activity B). Get them to complete both activities individually and then to compare answers in pairs before class feedback.

B 1 g 2 a 3 d 4 b 5 f 6 e 7 h 8 c
C **Activity A** 1 2 3 4 5 6 7 8
   Activity B 6 3 8 1 5 7 2 4

Task: Introduce a training programme

Students create and present a training programme for new international teams and write a blog post summarising their presentations.

5A Put students in small groups and explain the scenario: they all work for a training company and have been hired by the head of HR at Pro-Transfer 12 to give a training programme to all new international teams. They now need to prepare an opening presentation for their course. Go through the steps students need to follow with them. Encourage them to a) review all the information they have on Pro-Transfer 12, b) think about the ideas and suggestions in the blog post in Exercise 4A and c) refer to any additional ideas they came up with in Exercise 4C. Allow plenty of time for students to prepare their presentations. Set a time limit before they begin and monitor and offer help as necessary during the activity.

5B Join groups together into larger groups and ask them to take turns to give their presentations. In smaller classes, groups could give their presentations to the whole class. As an optional follow-up, the class/groups can then vote on the best training programme.

6 This writing task can be done in class or assigned as homework. Depending on the time available, students could also plan their post in class – individually, in pairs or in their groups from Exercise 5A – and then write their blog posts at home.

Model answer

Developing international project skills

In a recent assignment with Pro-Transfer 12, we supported its international project leadership community by recommending the development of a series of innovative and creative learning solutions which would develop essential cross-border skills. We established a dedicated site on Pro-Transfer 12's intranet entitled 'International Project Competence Place' with engaging, bite-sized digital learning materials and tools which project leaders could use to increase their own awareness of the challenges of working across cultures. Following the Mapping, Bridging and Integrating framework, we created a range of audio, text and video resources, including links to existing web and personal development resources and lists of the latest international management titles for recommended reading.

Some features of the portal include:

- the ability to map and compare different country cultures using online tools.

- suggestions on how to run workshops with project teams to build awareness of cultural differences and synergies in the project.

- tools which individuals can use to test and reduce their cognitive bias.

- guidelines on how to conduct feedback to ensure a learning culture in the project.

- communication tips for meetings and emails to ensure effective decision-making.

- podcasts and videos from internal project leaders sharing insights and best practice.

We believe this dynamic and practical series of bite-sized learning materials, underpinned by research and the real-life experience of Pro-Transfer 12's own project leadership community, is a benchmark in corporate learning and should be of interest to many working internationally or supporting those in a Learning and Development role.

MyEnglishLab: Teacher's Resources: extra activities

Review ◀7

1 1 f 2 c 3 d 4 b 5 e 6 a 7 k 8 j 9 g 10 h
   11 l 12 i
2 1 escalate, resolution 2 criticism, morale
   3 empathy, confrontational 4 blame, irrational
   5 struggle, mediator
3 Possible answers
   2 There is a problem with Zoe's management style.
   3 Luigi was too strong in his reaction.
   4 Can I have a word with you after the meeting?
   5 I will speak to Jan about this issue.
   6 Some employees are not happy with the new arrangements.
   7 You are right, but there is another way to approach this.
   8 There is a misunderstanding.
4 1 c 2 b 3 f 4 a 5 g 6 e 7 h 8 d
5 1 staff morale 2 internal candidate 3 soft skills
   4 mentoring 5 trust 6 confidential 7 defensive
   8 senior management team 9 meet your targets
   10 high-performing

Mindset

Unit overview

	CLASSWORK	FURTHER WORK
8.1 › **The entrepreneurial mindset**	**Lead-in** Students talk about the characteristics of the entrepreneurial mindset. **Video** Students watch a video about the mindset of a successful entrepreneur. **Vocabulary** Students look at vocabulary related to mindsets and being an entrepreneur. **Project** Students discuss personal goals and how they can be achieved.	**MyEnglishLab:** Teacher's resources: extra activities; Reading bank **Teacher's book:** Resource bank Photocopiable 8.1 p.156 **Spoken English:** p.115 **Workbook:** p.39
8.2 › **Mindsets**	**Lead-in** Students discuss the characteristics of a growth mindset and a fixed mindset. **Listening** Students listen to an interview about the characteristics of a growth mindset and a fixed mindset. **Grammar** Students study and practise different verb patterns. **Speaking and writing** Students practise verb patterns by talking and writing about their own mindset.	**MyEnglishLab:** Teacher's resources: extra activities **Grammar reference:** p.122 Verb patterns **Teacher's book:** Resource bank Photocopiable 8.2 p.157 **Workbook:** pp.40–42
8.3 › **Communication skills:** Handling a performance review	**Lead-in** Students read and talk about performance reviews. **Roleplay** Students roleplay a performance review meeting. **Video** Students watch a video of a performance review meeting. **Reflection** Students reflect on the conclusions from the video and their own approach to handling performance reviews.	**MyEnglishLab:** Teacher's resources: extra activities; Interactive video activities
8.4 › **Business skills:** Action learning	**Lead-in** Students read and talk about action learning as a professional development strategy. **Listening** Students listen to an action learning meeting. **Useful language** Students look at guidelines and useful language for effective action learning meetings. **Task** Students hold an action learning meeting.	**MyEnglishLab:** Teacher's resources: extra activities; Useful language bank
8.5 › **Writing:** Self-assessment	**Lead-in** Students discuss the content of a self-assessment for a performance review. **Useful language** Students look at strategies and useful language for writing a self-assessment as part of a performance review. **Task** Students write a self-assessment.	**MyEnglishLab:** Teacher's resources: extra activities; Writing bank; Interactive grammar practice **Grammar reference:** p.122 Verb patterns **Workbook:** p.43
Business workshop 8 › Encouraging personal growth	**Listening** Students listen to a discussion about employee retention. **Reading and speaking** Students read and talk about survey results on job satisfaction. **Task** Students choose a training course to aid young employee retention and write a proposal.	**MyEnglishLab:** Teacher's resources: extra activities

Business brief

The main aim of this unit is to understand the sort of **mindset** that makes a successful **entrepreneur** and examine some of the qualities that help some businesspeople succeed where others fail. It also examines the value of **failures** and **setbacks** in providing opportunities to learn and improve.

Although differences in talents or abilities between individuals undeniably exist, the concept of the **growth mindset** challenges the idea that talent alone is what distinguishes a successful business person from a less successful one. In fact, research shows that there is often little difference in actual ability between one entrepreneur and another; instead, what enables someone to succeed is having the right mindset. This is not just about a person's ideas, but has to do with their attitudes, beliefs, **passion**, flexibility and **persistence**. Those with a growth mindset have been shown to be more open to developing their abilities through risk-taking and to learning through experience, whether good or bad. They are also highly motivated and **in tune with** their intuition, trusting their instincts and not doubting themselves unduly. This helps them maintain their focus and their determination to achieve their aims, even in challenging circumstances.

Educational psychologists distinguish between a **fixed mindset**, which basically says that talent and ability are innate and can't be changed and a **growth mindset**, which says that any person's abilities can be developed through learning, risk-taking and hard work. Studies have shown that college students who exhibit a growth mindset regularly **outperform** students with a fixed mindset, regardless of their economic and social backgrounds and it is their attitude to **setbacks** that seems to be key. In general, people with a fixed mindset are afraid of making mistakes or failing and tend to avoid taking on challenges. Those with a growth mindset on the other hand, are not afraid and instead seem to learn from these experiences and improve.

It has also been shown that in business organisations with a fixed mindset, where people perceived as talented are valued much more highly than others, a culture of secrecy and competition often develops. Employees hide information from each other to win the advantage and rather than collaborate, staff are competitive and will try to outperform their colleagues. Researchers have also found that in such companies, employees who remain in low-level jobs often show a pattern of choosing to pass up opportunities for promotion because they believe they can't change their abilities and therefore their career prospects.

However, some psychologists do believe that people can learn to develop a growth mindset, even if their individual personality tends towards a more fixed one. The first step in this process is being able to imagine oneself as successful: changing attitudes to **failures** and setbacks, both in a business context and in personal life, is key to this process. If an individual can come to see these experiences not as **obstacles** but as an opportunity to learn and improve, this attitude will encourage resilience and flexibility, qualities fundamental to the growth mindset. People who manage to do this also become extremely resourceful, looking at problems in different ways and from different perspectives and becoming more open to new ideas when they present themselves. They start to view challenges as exciting rather than frightening and, as they take more chances and are less afraid of failing, find they appear more confident to others. This in turn encourages people to have faith in them and give them more responsibility. By recognising and practising the attitudes of the growth mindset, people can actually change their profiles in business and even the trajectory of their careers.

Mindset and your students

Both pre- and in-work students will probably have taken psychological tests or quizzes to establish their personality type and attitudes at some point, so the concept of different mindsets shouldn't be foreign to them. Pre-work students may be able to think of examples from their educational careers of people who exhibited a fixed or growth mindset in their attitude to their studies, while in-work students may have seen examples of these differing mindsets among their colleagues and managers and seen how this affects the wider team and ways of working.

Unit lead-in

Draw students' attention to the unit title and teach or elicit the meaning of *mindset*. Then look at the quote with the class and check they understand *declare*. Ask them if they know what a *growth mindset* is. Elicit or give a brief explanation (see Business brief). You could also use this as an opportunity to teach *fixed mindset*, which students will discuss later in the unit. Ask them if they agree with the quote. Why might having a growth mindset be hard? Briefly discuss this with the class, but do not go into detail about the growth and fixed mindsets yet, as students will look at them later in the unit.

8.1 ❱ The entrepreneurial mindset

GSE descriptors

- Can infer opinions in a linguistically complex presentation or lecture.
- Can recognise a speaker's feelings or attitude in linguistically complex speech.
- Can understand nuances of meaning in a linguistically complex presentation or lecture.
- Can answer questions about abstract topics clearly and in detail.
- Can carry out an effective, fluent interview, spontaneously following up on interesting replies.

Warm-up

Discuss these questions with the class: *What traits do you think make someone a good entrepreneur? Can people be taught to be good entrepreneurs? Is there an entrepreneur that you admire? Why do you admire this person?*

Lead-in

Students talk about the characteristics of the entrepreneurial mindset.

1 Put students in pairs and give them a few minutes to discuss the statements. Encourage them to give reasons and examples to support their ideas. As feedback, invite different students to share their views with the class.

❱ **Spoken English**
p.115: Ah, some of us are just worker bees, you know

1 ◀) SE8 Explain to students that they are going to hear people discussing statement 1 which they discussed in Exercise 1. Explain the activity and give them time to read the questions first, so they know what to listen for. Encourage them to make notes in answer to the questions while listening, then play the recording, twice if necessary, and check answers with the class. If you think your students will find this activity hard, you could get them to compare answers in pairs before class feedback, referring to audioscript SE8 on page 164.

Possible answers

1 Fearlessness and (good) ideas.
2 Someone with ideas; someone who turns the ideas into an actual company.
3 Because someone got it wrong, they didn't patent it early enough, or the idea turned out not be so good over the longer term.
4 The 'rare' entrepreneur is someone who is there from start to finish. This is unlike most ideas that are brought to fruition by a team of people.
5 They represent people who have single-handedly made great ideas happen.

Discourse markers

2 ◀) SE8 Refer students to the heading *Discourse markers* and elicit or give a brief explanation: discourse markers are short phrases like *I mean, right, OK* and *anyway* which we use to connect, organise and manage what we say. Explain that the gapped words are all discourse markers and give students time to quickly read through the extracts before they listen. Play the recording, twice if necessary, then check answers with the class.

1 Well   2 you know   3 right   4 you know
5 you see   6 I mean   7 you know

3 Go through the questions with students before they begin and get them to complete the exercise individually and then to compare answers in pairs before class feedback. Alternatively, do this as a whole-class activity, checking answers as you go.

a you know (2, 4, 7), you see (5)   **b** right (3)
c well (1)   **d** I mean (6)

Video

Students watch a video about the mindset of a successful entrepreneur.

2 ▶ 8.1.1 Ask students if they have heard of the Cambridge Satchel Company. Elicit what students know or explain that it is a British company which produces bags and other leather goods. Tell them that they are going to watch a video of Julie Deane, co-founder and CEO of the company, talking about how she became a successful entrepreneur. Explain the activity, give them a minute to read the statements and check that they understand *intuition* and *embrace* (*challenges*). At this point, you may also wish to teach this vocabulary from the video: *drive* (n.), *backfire* (v.), *gut instinct, determination, persistence*. Play the video, then check the answer with the class.

c

3 ▶ 8.1.1 Explain the activity and give students time to read the questions before they watch again. Point out that they should make notes in answer to the questions while watching and not try to write down everything the speaker says each time. Play the video, then check answers with the class. In weaker classes, you may need to play the video a second time and/or pause after answers are given, to give students time to make notes.

Possible answers

1 She studied sciences and feels her love of numbers and logic were very helpful.
2 The growth mindset includes being able to develop one's abilities through risk-taking, as well as learning from any type of experience, which are characteristics typical of many entrepreneurs. Entrepreneurs are often willing to take chances and to learn from both positive and negative experiences.
3 She learnt about people from different cultures, what customers want, how to treat people and how to make them want to come back.
4 She feels that it is necessary to have something that drives a person and helps them to have a 'superpower', to keep them working hard, to maintain their focus and to be able to deal with hurdles when they arise.
5 She thought at first that she wanted to make school bags for children and then realised that she was manufacturing fashion items which meant that the designs, the size of the products and the colours needed changing.
6 The handbags were very popular, meaning that sales skyrocketed and she didn't have enough people manufacturing them, so she had to start a completely new factory.
7 Alongside increased investment, Julie brought in people with more experience to grow the business, but she realised that this was a mistake and admitted that it didn't work for her or her business.
8 She loves what she does because she doesn't like routine; she likes constant challenge as well as having a purpose and a vision of the future.

4 Put students in pairs or small groups and give them 3–4 minutes to discuss the questions. Then invite different students to share their answers with the class.

Extra activities 8.1

A ▶ 8.1.1 Explain the activity and give students time to read the statements. Tell them that they should first decide whether the statements are true or false and then find evidence in the video to support their answers. They could do this by referring to videoscript 8.1.1 on page 150 and underlining the relevant parts of the script and/or watching the video again and asking you to pause each time the evidence for one of the answers is heard. If necessary to clarify answers, play any relevant parts of the video again during class feedback.

1 F (... *the Cambridge Satchel Company manufactures and sells handmade leather bags, which now sell all across the globe.*)
2 T (*But not everyone has the passion or the drive to run their own business.*)
3 F (... *it was a terrific learning experience. ... You can literally pick up something valuable from virtually everything.*)
4 T (*It's this growth mindset which helps successful entrepreneurs to realise their business dreams.*)
5 T (*And then there's the importance of motivation.*)
6 F (*With 24,000 bags on back order, Julie showed amazing courage and entrepreneurial vision. She started a new factory ...*)
7 F (*But that really backfired and that was totally my mistake, that was my mistake in not looking and just thinking ...*)
8 T (*I am so in tune with my intuition, my gut instinct ... If something comes up and it looks wrong, then it is wrong and we're not doing it.*)

Vocabulary: The growth mindset

Students look at vocabulary related to mindsets and being an entrepreneur.

5 Get students to complete the exercise individually, using their dictionaries if necessary and clarify meanings during feedback. Alternatively, go through the words in the box with students before they complete the sentences, then check answers with the class.

1 entrepreneurial 2 skyrocketed 3 vision
4 persistence, setback(s) 5 passion

6 With stronger classes, you could let students use their dictionaries to complete the exercise, then clarify meanings during feedback. With weaker classes, you may prefer to do this as a whole-class activity, checking answers and clarifying meanings as you go.

1 d 2 g 3 a 4 f 5 e 6 b 7 h 8 c

7 Get students to do this individually. Explain that they should use words and phrases from Exercises 5 and 6 and point out that they may need to change the form of some words. Check answers with the class.

1 entrepreneurial 2 passion 3 in tune
4 maintain focus 5 from scratch 6 pick up
7 admit defeat 8 setbacks 9 doubt yourself

8 Put students in pairs or small groups, explain the activity and look at the examples with them. Encourage them to use vocabulary from Exercises 5 and 6 and also point out that the text in Exercise 7 can help them with ideas if necessary. Allow 4–5 minutes for students to discuss in their pairs/groups, then broaden this into a class discussion. You could list students' ideas on the board, for them to refer to when they do Exercise 9B.

Extra activities 8.1

B This activity gives further practice of the key vocabulary from the lesson. Explain to students that the sentences are definitions of the words/phrases in bold and they need to choose the correct option in order to complete the definitions. Check answers with the class, clarifying any errors as necessary.

1 a **2** a **3** a **4** b **5** b **6** b **7** a

Project: Setting a goal

Students discuss personal goals and how they can be achieved.

9A Put students in pairs and go through the instructions and questions 1–6 with them. Give them some time to think about their answers individually first, then ask them to take turns to interview their partner, making notes on their answers.

9B Explain to students that they are going to do this activity in two stages: first, they are going to write a plan for their partner to take forward, using their notes from Exercise 9A and they are going to do this individually. Then they are going to share and discuss their plan with their partner. For the writing task, encourage them to think about the characteristics of the entrepreneurial mindset they looked at in this lesson and to try to use vocabulary from Exercises 5 and 6. If you put their ideas from Exercise 8 on the board, remind them that they can refer to those as well. Depending on the level of your class and, if you think this would help your students, you could also share the model answer below with them, to give them an idea of what they can include in their plan. While students are writing, monitor and offer help as necessary. When they are ready, ask them to share their plan with their partner. They should discuss what they have written, any questions they may still have and give each other feedback on the ideas they both had. You could round off the activity by inviting a few students to tell the class about their goal and the plan their partner devised for them.

Model answer

Vivi would like to start a small business with an app for a new university social programme to help new students when they first begin their studies. She imagines that the app would be on a smartphone and would help students to find social events, apartment rentals, student-friendly restaurants and a range of exclusive student discounts.

She is studying electrical engineering and has the skills she needs to create an app. She would need help in advertising it and will talk to other colleagues or students about this.

The main risk is the amount of time it will take. My partner needs a great deal of time for her studies and for family members, so she will make sure that she also has some free time.

Vivi said she worked with others several years ago on a new computer program and this goal is similar. When they finished, they felt satisfied but realised that there was still work to be done.

Vivi would like to begin this once her major exams are finished, so in the next six months.

She feels that the future lies in apps for phones, but also that many students need a supportive social network and help when they begin their studies, so this could be very useful for them.

She will take the first step by speaking to students about what they need next week.

MyEnglishLab: Teacher's Resources: extra activities; Reading bank
Teacher's book: Resource bank Photocopiable 8.1 p.156
Spoken English: p.115
Workbook: p.39

8.2 ❯ Mindsets

GSE descriptors

- Can infer opinions in a linguistically complex presentation or lecture.
- Can follow presentations on abstract and complex topics outside their field of interest.
- Can comment in detail on the content of a linguistically complex radio programme or podcast in which people describe reactions or opinions.

Warm-up

Write this statement on the board: *There are lots of talented football players out there. What makes the difference is the mindset of your team.* Discuss the statement with the class. Do they agree? Is the same true for the workplace?

Lead-in

Students discuss the characteristics of a growth mindset and a fixed mindset.

1 Put students in pairs, give them a minute to read the questions and teach or elicit the meanings of *glorified* and *worshipped* in question 3. Let them discuss the questions in their pairs for 2–3 minutes, then get feedback from the class.

2 Students could do this in the same pairs as for Exercise 1 or individually. For students who are unclear on the two types of mindsets, explain that having a fixed mindset means feeling we are born with certain talents which cannot be developed or added to; having a growth mindset means feeling that we can develop and that challenges and failure help us to learn. Do not confirm answers at this point, but if you think it will be helpful for students in the listening which follows, collate their ideas on the board (without commenting on whether they are right or wrong).

Listening

Students listen to an interview about the characteristics of a growth mindset and a fixed mindset.

3 ◀) 8.01 Explain the activity and go through the phrases in the box with students before they listen; check that they understand the meanings of *thrive* and *wilt*. You may also wish to teach this vocabulary from the recording: *obstacle, play it safe, binary, spectrum, endorse, outperform, affluent, mentality, innate, cut corners, perseverance, radical, alter.* After checking answers to the listening task, refer students to their notes in Exercise 2. Does the recording mention any of their ideas?

> **Growth mindset:** thrive in face of difficulty, be able to develop abilities, learn from mistakes, try hard when facing challenges.
> **Fixed mindset:** wilt in the face of failures, believe talents and abilities are unchanging, do less well at school, hide information from others.

4 ◀) 8.01 Give students time to read the questions before they listen and play the recording, twice if necessary, for them to check/complete their answers. Check answers with the class.

> **Possible answers**
> 1 Carol would like to know why some people love challenges and others run from them.
> 2 Her research is changing the way people think.
> 3 A growth mindset.
> 4 They can be developed and you never know just how far you can develop them.
> 5 They are frightened they will make one.
> 6 Organisations look for the people they consider intelligent and make them superstars. People also hide information to make themselves look better than others.
> 7 They are afraid they might do something that isn't smart and their parents will be upset or not love them.
> 8 It is more motivating to say they did well because they tried hard.
> 9 Those who have traditionally not done so well at school.

5 Put students in pairs and give them a few minutes to discuss the questions. Encourage them to think about the ideas in the recording as well as their own from Exercise 2. If there is time, invite a few students to share their answers with the class.

> **Extra activities 8.2**
>
> **A** ◀) 8.01 This activity provides students with extra listening practice. Give them time to read the sentences first, then play the recording, twice if necessary and check answers with the class.
>
> > 1 do well 2 do not 3 do not all 4 do not use
> > 5 holds 6 conveys 7 helped

Grammar: Verb patterns

Students study and practise different verb patterns.

6 You could do this as a whole-class activity, checking answers as you go. Then refer students to the Grammar reference on page 122, go through it with them and clarify any points as necessary.

> 1 c 2 e 3 d 4 a 5 b 6 f

7–8 Ask students to do both exercises individually and, if there is time, get them to compare answers in pairs before class feedback. Remind them that they can refer to the Grammar reference on page 122 if they need help.

> 7 1 to be 2 them from finding out 3 him telling
>   4 (that) they know/knew 5 to tell 6 launching
> 8 1 recommended to give giving
>   2 gave to the department an award / gave an award for excellence to the department
>   3 managed changing to change
>   4 how we losing lost
>   5 choose them run to run
>   6 appreciated them to tell telling

> **Extra activities 8.2**
>
> **B** This activity gives further practice of verb patterns. It is a consolidation exercise, so it might be better for students to do it individually. Encourage them to read the whole text quickly before completing the gaps and remind them to refer to the Grammar reference on page 122 if they need help. After checking answers, go over any points that need clarification.
>
> > 1 taking 2 remain 3 trying / to try 4 being
> > 5 find 6 appear 7 ask 8 decides
> > (Not needed: get and mention)

Speaking and writing

Students practise verb patterns by talking and writing about their own mindset.

9A Explain the activity and look at the verbs with students, checking that they understand each one. Remind them that if they do not remember the pattern for some verbs, they can refer to the Grammar reference on page 122. During the activity, monitor and offer help as necessary.

9B Put students in pairs and give them 3–4 minutes to discuss and compare their sentences from Exercise 9A. Then ask them to mark where they think their mindset is on the spectrum. As feedback, invite a few students to tell the class what they found out about their partner's mindset.

9C Explain the activity and tell students that they can use ideas from Exercise 9A. Point out the word limit before they begin and set a time limit for the writing task. If time is short, students can write their texts as homework.

Model answer

I usually **appear** fairly confident and **consider** myself to be a hard worker. I sometimes don't **show** my best side, though and often **postpone** finishing work that is difficult. I try not to **avoid** taking on challenges, but don't always succeed. However, I **manage** to get my work done and **expect** to do well in my chosen career, although I admit that I sometimes question myself.

In order to move towards a growth mindset, I need to accept that talents are not fixed and that I can learn what I need to. I also know that I can face difficulties if I have to and it is not a problem to ask for help. Life-long learning is important and I hope that I never stop doing that. I think my biggest challenge is to learn from my mistakes and to do things differently in future.

MyEnglishLab: Teacher's Resources: extra activities
Grammar reference: Verb patterns p.122
Teacher's book: Resource bank Photocopiable 8.2 p.157
Workbook: pp.40–42

8.3 ⟩ Communication skills
Handling a performance review

GSE descriptors

- Can participate in extended, detailed professional discussions and meetings with confidence.
- Can participate in linguistically complex discussions about attitudes and opinions.
- Can adjust tone or language to build rapport in situations where there may be an unequal power dynamic.
- Can present detailed, evidence-based arguments during work-related meetings.
- Can encourage employees using motivational language.
- Can participate in a fast-paced conversation with fluent speakers.
- Can follow an animated conversation between two fluent speakers.
- Can compare and evaluate different ideas using a range of linguistic devices.

Warm-up

Discuss these questions with the class: *How often do you receive feedback on your work at your place of work or study? Do you find it helpful? Why / Why not? How do you prefer to receive feedback? Do you ever assess your own work?*

Lead-in

Students read and talk about performance reviews.

1A Ask students to complete the exercise individually, then check answers with the class. As a follow-up, if appropriate for your students, you could ask them if performance reviews are common at their place of work and if so, how often they are held.

b The text mentions setting goals for the next year. Although an employee's performance is often assessed against the goals, goals are not normally changed at reviews.

1B If time is short, discuss the question with the whole class. Otherwise, let students discuss in pairs first, then elicit answers around the class. During feedback, you may wish to share some of the information from the Note below with students.

Note

Many people feel that holding a performance review once a year is not often enough and prefer more frequent conversations with their manager.

People dislike annual reviews for many reasons. For instance, managers do not like the time they take up and often have not been trained to give constructive feedback and coach their team. Employees often fear criticism or being compared with their peers and may feel that managers only recall the most recent performance rather than being able to reflect on their performance over a whole year.

There has been a lot of criticism of annual reviews over the years. For example, what is the real purpose of an annual review? Should it be motivated from a corporate-centric or employee-centric perspective? Some research also suggests that performance reviews can actually hinder progress in the workplace by demoralising and discouraging employees.

As a consequence, formal annual performance reviews have been scrapped in many large organisations where they are considered outdated (e.g. Microsoft, Dell). The process has been replaced by more frequent mini-reviews using a different style of review (e.g. what's on track and what's not).

The aim nowadays in many organisations is to make the review process less stressful and more collaborative. It is argued that employees are more motivated and organisations are more successful when managers regularly communicate with their team and help staff to learn, encouraging greater motivation, personal development and improved performance. In order to do this well, though, managers need training in coaching techniques to help them with performance reviews.

Preparation: Holding a performance review

Students read and think about the scenario for a roleplay.

2A If this is the first Communication skills lesson for your class, briefly tell students about *Lifestyle* magazine and the profile of its readership. Otherwise, elicit this information from students before you start. Go through the instructions with students and, if desired, teach this vocabulary from the report: *quarterly, insightful, work to brief, contribute (ideas), reluctant.* Ask students to complete the exercise individually, then check answers with the class.

Positives
- demonstrates an excellent knowledge of the job
- produces well-researched, thorough and insightful reports
- demonstrates 'team player' behaviour
- has an excellent working relationships with whole team
- contributes lots of good ideas

Negatives
- failed to work to brief on one occasion
- has a tendency to miss deadlines; asked for extensions on three assignments in recent months
- tends to rely heavily on email communication
- seems reluctant to have face-to-face and phone conversations
- sometimes lacking in confidence when sent out of the office on assignments

2B Depending on the time available, you could discuss the questions with the whole class, nominating different students to answer each time, or let students discuss in pairs or small groups first, then get feedback from the class.

Roleplay

Students roleplay a performance review meeting.

3A Explain to students that they are going to roleplay a performance review meeting. Put them in pairs, assign roles (or let students choose) and refer them to their relevant information. Give them time to read it, while you monitor and answer any questions they may have. Set a time limit for the preparation stage and point out that students will also need to set goals at the end of the performance review. Let them make notes if they like. During the activity, monitor and provide help as necessary.

3B Students now roleplay their meetings. Set a time limit before they begin and during the roleplays, monitor and note down any points to highlight during feedback after Exercise 3C.

3C Students now reflect on their roleplays; they should do this in the same pairs as for Exercises 3A and 3B. Go through the questions with them and give them 3–4 minutes to discuss in their pairs, then invite different students to share their answers with the class. Finally, highlight any points you noted during the roleplays.

Video

Students watch a video of a performance review meeting.

4 ▶ 8.3.1 Tell students that they are going to watch a video of Donna Johnson and Susan Lam from *Lifestyle* magazine conducting a performance review meeting. Give them a minute to read the questions and ask them to make notes while watching. Play the video, then check answers with the class. During feedback, highlight these points about Donna's approach (if students do not mention them):

- Donna focuses on giving constructive feedback. She does not focus on criticising the negatives; she knows it will demotivate her staff and that their engagement, enthusiasm and desire to contribute to the magazine will suffer. But she does not ignore the negatives either. She tries to frame weaknesses as opportunities to improve, motivate and encourage Susan's learning, personal growth and professional development.

- Donna does not spring any surprises on Susan in the review. This would lead to frustration and, ultimately, distrust in her as a manager.
- The review is a two-way street. Donna does not dominate the meeting. She takes a coaching approach.
- She also allows Susan to challenge feedback and give feedback. Susan is encouraged to open up and feel confident enough to speak.

1 Suggested answer: She uses a coaching approach, encouraging Susan to lead the discussion. She uses a questioning technique to explore issues. She also tries to encourage Susan to have a growth mindset to overcoming her shyness.
2 Because she had not recalled very well why Susan failed to follow the editorial brief and Susan pointed out that actually this had not been the case.
3 She is receptive to Donna's feedback, challenges the feedback at one point and seems willing to put in an effort to improve and sets goals.
4 She is positive and encouraging and offers help and support if Susan wants it.

5A Ask students to complete the exercise individually, using their dictionaries if necessary and get them to compare answers in pairs before class feedback. You could play the video again for students to check their answers or go through them with the class. Do not focus on meanings yet, as students will check these in the next activity.

1 set 2 take 3 procrastinate 4 bitten 5 dreading
6 think 7 take 8 have

5B You could get students to complete the exercise individually and then clarify meanings as necessary during class feedback. Alternatively, do this as a whole-class activity, checking answers as you go. Encourage students to record the words and expressions in their notebooks.

a 7 b 2 c 1 d 3 e 6 f 8 g 5 h 4

Extra activities 8.3

A This activity gives further practice of the vocabulary students looked at in Exercise 5A of the Coursebook. Ask them to complete it individually and remind them that they can refer to the exercise in the Coursebook if they need help. Check answers with the class.

1 take 2 belt 3 set 4 dreading 5 took 6 feet
7 procrastinating 8 bitten

Reflection

Students reflect on the conclusions from the video and their own approach to handling performance reviews.

6A Allow students to work individually first so that they can reflect on their own approach to handling performance reviews. Remind them to think about their answers to Exercises 3C and 4 and the information on performance reviews from Exercise 1A. Encourage them to make notes so they can discuss their ideas with a partner in the next activity.

6B Put students in pairs or small groups to compare their reflections, then round up ideas in a class discussion.

MyEnglishLab: Teacher's Resources: extra activities; Interactive video activities

8.4 ❯ Business skills
Action learning

GSE descriptors

- Can understand in detail discussions on abstract and complex topics among speakers with a variety of accents and dialects.
- Can follow a work-related discussion between fluent speakers.
- Can manage the participants in a fast-moving discussion to keep it on course.

Warm-up
Discuss these questions with the class: *How do you learn best? Do you prefer studying on your own or in a more social environment? Do you work better in a class/team or do you prefer one-to-one lessons/meetings? Which learning experience do you think works better for you: 'learning by doing' or 'learning by reading'? Does it depend on what you are learning?*

Lead-in
Students read and talk about action learning as a professional development strategy.

1A Tell students that they are going to read about a method called 'action learning' which is becoming more and more popular in the business world. Put them in pairs, explain the activity and give them time to read the information and ask you about anything they do not understand. Then ask them to make a list of what they think are the main advantages of action learning. Give them 3–4 minutes to make their lists, then join pairs together into groups of four and get them to compare and discuss their ideas. If there is time, get them to compare ideas with one or two more pairs, then elicit ideas around the class.

Possible answers
- Action learning is a methodology which helps people learn by solving real issues with colleagues.
- It helps to find creative solutions using the experience of colleagues.
- It supports networking across the organisation, improving people's visibility and career opportunities and better understanding across organisational silos.

1B Students now talk about the challenges action learning may pose to an organisation and how they can be overcome. They could do this in the same pairs as for Exercise 1A or, if time is short, as a whole class.

Possible answers
- Time may be an issue. People are busy and may not see it as a priority to give their time to solve other people's problems.
- Managers may be reluctant to let their team participate for the same reason.
- People may feel they lack the expertise to solve others' problems.

Listening
Students listen to an action learning meeting.

2A ◀》8.02   Tell students that they are going to hear the first part of an action learning meeting and go through the instructions and table with them. Play the recording, twice if necessary, then check answers with the class. Before moving on to the next activity, ask students which problem the group choose to discuss (Marco's).

Harry
Type of problem: how to improve customer service
Importance to organisation: need to increase sales with conversion rates in store falling
Desired outcome: new targets for sales staff and a management push to ensure they are met

Marco
Type of problem: how to change store design to reflect growing number of young families with kids as customers; be more customer-friendly
Importance to organisation: to make sure they don't lose sales and to avoid accidents in store
Desired outcome: a family store policy by end of the year

Miriam
Type of problem: need to develop a curriculum of training for customers
Importance to organisation: to reverse dropping of brand position
Desired outcome: offer training in time for Black Friday marketing

2B Discuss the question with the whole class. Encourage students to elaborate.

Possible answers
Lisa facilitates relatively effectively. She gives clear instructions on how people should present their ideas. She gives a clear and brief summary of the problem after each pitch from the meeting participants. And the way of choosing the problem – pointing a finger at the 'most interesting problem' – is a fun method of selection and helps to engage people in the process

3A ◀》8.03   Explain that students are going to hear the next part of the meeting and ask them again which problem the group have decided to discuss (Marco's problem: how to change store design to make it more attractive to and safer for children/families with children). Ask them to note down all the suggestions the speakers make to address the problem and play the recording. Note that students, especially in weaker classes, may need to listen twice or you may need to pause the recording for them to complete their answers.

Need to involve corporate audit and corporate health and safety / have meeting with Head of Health and Safety; meeting with Brand Lead about store concept; check for industry benchmarks on topic of store safety; meeting with store managers; meeting with regional marketing heads to get their ideas; create kids' zones; contact Disney and get information on how they design kid-friendly stores; have Mickey Mouse in stores.

3B Ask students to do this individually. If necessary, play the recording again or let them refer to audioscript 8.03 on page 157.

1, 2, 4, 6, 7

4 ◀)) 8.04 Tell students that they are going to hear the final part of the meeting and need to note down the main learning points for each participant. Play the recording, twice if necessary, then check answers with the class.

Possible answers

Marco: needs to spend more time networking in the organisation; more meetings/discussions outside his own team to change things.

Miriam: needs to discuss potential lack of customer focus with her own team.

Harry: sees the value of action learning; good to meet colleagues; believes more should be done.

Useful language

Students look at guidelines and useful language for effective action learning meetings.

5A Explain the activity and ask students to look at the text. Point out that there are three main stages in action learning meetings and draw their attention to the heading for each step. Give them time to read the information and ask you about anything they do not understand, then get them to complete the matching task; they could do this individually or in pairs. Check answers with the class.

a 6 b 2 c 10 d 9 e 5 f 4 g 3 h 1 i 8 j 7

5B Get students to do this individually and then compare answers in pairs before class feedback. You could list the expressions students identify on the board for them to refer to during Exercises 6A–C. As an optional follow-up, you could ask students what other things they might do as an action learning coach, problem owner or group participant to ensure such a meeting is a success. (Possible answers: clarify and add to the ideas of others to make sure they are explored fully; support by writing ideas on a flip chart to help the meeting leader focus on generating and managing ideas; stay enthusiastic even if the problem is not so motivating for you.) List any ideas students mention on the board for them to refer to when they do the Task that follows.

Possible answers

1 In terms of sequence, can I ask Harry to go first, followed by Marco and then Miriam.
2 What I want from the group today is some ideas on new targets for our store staff …
3 OK, so basically, your challenge is how to improve customer service to improve sales.
4 You're now the Problem Owner and we're discussing kids in stores.
5 Harry, if you were in my shoes, what would you do to handle this?
6 If you mean benchmarking within our industry, I think it's a great idea.
7 Another solution could be to look at benchmarking.
8 Of all the ideas, this was the most useful because it really opened my mind.
9 I'll keep you in the loop and let you know what happens next week.
10 What are the main takeaways for everyone from today?

Extra activities 8.4

A This activity gives further practice of the functional language and techniques students looked at in the lesson. It can be done individually or, if time is short, as a quick whole-class activity, checking answers as you go.

1 h 2 f 3 b 4 g 5 a 6 e 7 d 8 c

B This activity looks at some common multi-word verbs from the listening. Ask students to complete it individually or, in weaker classes, in pairs, using their dictionaries if necessary. Check answers with the class, clarifying meanings as necessary. If there is time, you could also get students to practise the dialogue in pairs.

1 fallen behind   2 end up   3 set up   4 set aside
5 take up   6 come up with   7 wrap up
8 look forward to

Task

Students hold an action learning meeting.

6A Put students in small groups and tell them that they are going to hold an action learning meeting. Allocate roles and give students time to read their information or, if necessary, go through it with them. As far as possible, try to give the role of the action learning coach to a stronger student. Point out that there are specific steps which need to be followed by both the action learning coach and the Problem Owner; refer them to the guidelines in Exercise 5A. Encourage everyone to make notes and remind them to use the useful language from Exercises 5A and 5B. If you did the optional follow-up activity after Exercise 5B, also remind students to refer to the ideas on the board. You could also let them refer to audioscripts 8.02–8.04 on page 157. Allow students plenty of time to prepare for their meetings. Depending on the strength of your class, you might like to group all students with the same role together for the preparation stage, to briefly discuss their roles and brainstorm ideas before returning to their original groups. While students are working, monitor and provide help as necessary.

6B Set a time limit for the meetings and ask students to begin. When they have finished, give them 4–5 minutes to reflect on their meetings. They should talk about what they think went well, what could be improved and what they think of the ideas generated at the meeting. How useful/creative were they? Round off the task by asking students from different groups to share their conclusions with the class.

MyEnglishLab: Teacher's Resources: extra activities; Useful language bank

8.5 ❯ Writing

Self-assessment

GSE descriptors

- Can write about complex subjects, underlining the key issues and in a style appropriate to the intended reader.
- Can employ high-level vocabulary and structures to enhance impact in written correspondence.

Warm-up

Ask students to imagine they have been asked to assess their own performance at their place of work or study. What questions would they ask themselves? Ask them to make a list in pairs or small groups – give them one or two examples if necessary (e.g. *What were my goals and to what extent did I achieve them? Which areas do I want to improve/develop? What support do I need in order to achieve that?*). Give students 2–3 minutes to brainstorm in their pairs/groups, then elicit ideas around the class.

Lead-in

Students discuss the content of a self-assessment for a performance review.

1 Look at the definition of a performance review with the class and explain the activity. If there is time, get students to discuss the questions in pairs or small groups first, then ask different students to share their answers with the class. Otherwise, discuss the questions with the whole class. If appropriate, open the discussion to ask whether students have any reservations about self-assessments. (Concerns might be that self-assessments may remain on file and be accessed by HR or senior management and may contain information that could have long-term career implications. If the manager does not use the information in the self-assessment within clear criteria, then the information it contains, if used wrongly, might have an impact on potential rewards such as bonuses.)

> For the areas usually covered in a self-assessment, see the introductory paragraph in the tips on page 144.

2 This activity is best done in two stages. Start by putting students in pairs and asking them to think about what advice they would give to an employee writing a self-assessment. Give them 2–3 minutes to discuss in their pairs, then elicit ideas around the class and list them on the board. Move on to the second part of the activity: refer students to the tips on page 144 and give them time to read the text and check which

of their ideas it mentions. Help them with any vocabulary questions they may have, then hold a brief class discussion, ticking the tips on the board which were mentioned in the text.

3 Ask students to complete the activity individually and, if there is time, get them to compare their ideas in pairs before class feedback. After checking the answer, you may wish to point out that the style of self-assessment differs from company to company. The model on page 144 provides an idea of the style that some (but not all) companies use.

> Yes, he did.
> **1** *Be prepared:* he gives details of how he achieved the goal by outlining the situation and the tasks he undertook and showing the result of his actions.
> **2** *Don't be modest:* he uses language showing that he was the driving force behind work initiatives (e.g. *I completed, I achieved, I initiated, I amended*). He gives concrete examples of his successes (e.g. *70 percent [of projects] came in ahead of schedule, 15 percent improvement on results, the success of the initiative is verified by feedback*).
> **3** *It's not just about success:* He demonstrates how he solved problems and handled challenges (e.g. *I quickly sourced an alternative, I instructed the communications team to use alternative methods to contacts customers*).
> **4** *Get feedback:* he uses dynamic language (e.g. *consistently demonstrated, proactive attitude, innovative system*). He includes evidence when describing achievements (e.g. *70 percent [of projects] came in ahead of schedule, 15 percent improvement on results*). He demonstrates that he took appropriate action when things didn't go well (e.g. *I learnt from the experience and now ensure … , I quickly sourced an alternative … , I instructed the communications team to …*).

Useful language

Students look at strategies and useful language for writing a self-assessment as part of a performance review.

4 This is best done as a whole-class activity, checking and discussing answers as you go. After discussing the answers for each item, go through the relevant section of the Useful language box with the class. For question 1, check that students understand the meaning of each verb and, if there is time, elicit or give an example sentence for each one. For question 2, list the collocations on the board and encourage students to record them in their notebooks. For question 4, give students time to write their sentences, then elicit a few examples around the class for each one.

> **1** *achieve* + object, *adapt* + *to*, *collaborate* + *with*, *communicate* + *with / communicate* (sth) + *to*, *complete* + object, *exceed* + object, *execute* (+ object), *dedicate* + object + *to/for*, *demonstrate* + object, *design* + object, *gain* + object, *implement* + object, *improve* + object, *increase* + (object) + *by/from/to*, *initiate* + object, *lead* + (object), *mentor* + object, *reduce* + object + *by/from/to*, *supervise* + object, *train* + (object) + *as/in*, *train* + (object) + *to* + infinitive

2 Adverbs and adjectives Mateus used: *innovative, keen, outstanding, proactive, consistently, effectively.* Additional collocations in his self-assessment: *dedicated* (*and proactive*) *attitude, outstanding success, quick and responsive service, increased customer satisfaction, dealing effectively, invaluable lesson, quick action, clear communication, improved significantly.*

3 a cause: *due to*; an effect: *as a consequence, as a result, resulting in*; an example: *such as, examples include*; something else: *verified by.*

5 Get students to do this individually and then to compare answers in pairs before class feedback. You could list the words/phrases students identify on the board for them to refer to during the writing task.

Extra activities 8.5

A–B These activities give further practice of the useful language from the lesson. As they are both consolidation exercises, it would be better for students to complete them individually. You could get them to compare answers in pairs before checking with the class.

A 1 b **2** a **3** c **4** c **5** b **6** a **7** c **8** b
B Possible answers include:
1 During my first month, I implemented/initiated the Green Initiatives training program which was a resounding success.
2 I have consistently demonstrated that I can work collaboratively with international teams.
3 Although / Despite the fact that, the online launch didn't go to plan, the feedback was excellent.
4 I quickly adapted to the new system and completed 100 percent of projects on schedule.
5 I have achieved a lot in my first year and these challenges have helped me to gain (some) valuable insights into my new role.

Optional grammar work

The tips on page 144 contain examples of verb patterns, so you could use them for some optional grammar work. Refer students to the Grammar reference on page 122 and use the exercises in MyEnglishLab for extra grammar practice.

Task

Students write a self-assessment.

6A Go through the instructions with students and set a time limit for this preparation stage. Remind students to refer to the tips and model answer on page 144 and also to use a range of language from Exercise 4, as well as any language they identified in Exercise 5. During the activity, monitor and offer help as necessary.

6B Students now write their self-assessment. Depending on the time available, they could do this in class or as homework. If they write it in class, set a time limit before they begin, highlight the word limit and remind them once again to use language from Exercises 4 and 5.

See Coursebook page 144 for a model answer.

6C If students do the writing task as homework, this exercise can be done in the next lesson. Put students in pairs and ask them to read each other's self-assessment, think about the questions and give their partner feedback. Students could then rewrite their self-assessments, following their partner's suggestions; they could do this in class or as homework.

MyEnglishLab: Teacher's Resources: extra activities; Writing bank; Interactive grammar practice
Grammar reference: Verb patterns p.122
Workbook: p.43

Business workshop ❯8
Encouraging personal growth

GSE descriptors

- Can follow a group discussion on complex, unfamiliar topics.
- Can summarise relevant data or research in support of an argument in a debate or discussion.
- Can present detailed, evidence-based arguments during work-related meetings.
- Can describe the details of problem–solution relationships using a range of linguistic devices.
- Can participate in extended, detailed professional discussions and meetings with confidence.
- Can write about complex subjects, underlining the key issues and in a style appropriate to the intended reader.
- Can write linguistically complex and logically structured reports and articles.
- Can employ high-level vocabulary and structures to enhance impact in written correspondence.

Background

Students read about an advertising agency looking at ways of improving its young employee retention rate.

1 Before students read, you may wish to teach/check understanding of this vocabulary from the background: *pride yourself on sth, retain/retention, CSR* (*Corporate Social Responsibility*), *initiate.* Put students in pairs and ask them to read the text and discuss the questions, then check answers with the class. After feedback, you might also like to share some of the information in the Note below with the class.

1 They are an advertising agency and offer jobs dealing with creating advertisements as well as buying space for them in different media.
2 Their reputation for professional development, CSR and diversity.
3 Many of those classified as millennials only stay a few years before looking for another job, but they comprise 35% of the global workforce.
4 The company carried out a survey to determine staff satisfaction to try and discover which particular problems different age groups of employees are having.
5 They are going to discuss the results of the survey and begin researching ways they can approach the problem.

Note

Attracting and retaining young, talented people in the workforce is a global problem today. The problem with an unbalanced company regarding age is two-fold: older employees begin to retire, taking their knowledge and experience with them and the world of technology is changing so quickly that companies often need younger employees who grew up with technology in order to simply keep up with it. However, those belonging to the so-called millennial generation (those born between 1981 and the mid-nineties), think differently about work than older employees and want to feel more ownership in what they do. They are often interested in finding work they feel is meaningful, they want to have an influence on the company and relationships are important to them, as is personal growth. They also want to work for companies they feel are doing good and business ethics are essential for them. According to research, some 43 percent of millennials do not plan to stay longer than two years in the jobs they are currently in. This means that companies who train young people when they join the company have to start again when they leave, which becomes costly after a time. This is why many companies have been exploring ways to retain their younger talent.

A management meeting

Students listen to a discussion about employee retention, then read and talk about survey results on job satisfaction.

2 ◀) BW 8.01   Go through the instructions with students and make sure they are clear about the scenario and what they need to do. Look at the words in the box with them and check understanding of each one, then put them in pairs and give them time to discuss their ideas. Get brief feedback from the class, then play the recording for them to check their ideas. Discuss the answers with the class.

burnout: There is a popular programme to help people cope with this, seen by Rose as a fix rather than something to solve the problem.
coaching: Louise suggests offering a coaching programme which can help people realise what they need.
communication: Training to work on relationships in the workplace.
CSR: Older employees are happy with it and feel its success reflects on them.
job security: Older employees appreciate this.
mental health: The company stresses the fact that this is important and has developed a multi-faceted programme to deal with it.
personal growth: Young people are looking for this and want to learn new skills both for work and for themselves; they want to feel they can help shape the direction the company is going in and form relationships with others in the workplace.

3 ◀) BW 8.01   Give students time to read the questions before they listen and remind them of the meaning of *burnout* for question 5. Play the recording, twice if necessary, then check answers with the class.

1 The older employees are comfortable with their jobs and will most likely stay, whereas the younger ones are not restricted by their job descriptions and may look for new jobs soon.
2 They like having job security and flexible hours and think that the company's CSR record is positive, as it also reflects on them as part of the company.
3 The younger employees want to follow their passions and dreams, learn new skills, shape the company and experience bonding with their colleagues.
4 They often see themselves as having an entrepreneurial mindset.
5 She feels it is a good idea, but it is a programme dealing with problems after they arise and not in advance, in a preventative way.
6 She promises to gather information about different providers of courses which offer personal growth programmes.

4 Explain the scenario and activity and before students look at the survey results in detail, look at the first column of the table with them and answer any vocabulary questions they may have. Put students in pairs and give them 4–5 minutes to look at the survey results and discuss the questions, then invite students from different pairs to share their answers with the class. Alternatively, if time is short, get students to look at the survey results individually and then discuss the questions with the whole class.

Possible answers

chance to learn new skills, personal growth, creativity, autonomy, working atmosphere, burnout

Extra activities Business workshop 8

A This activity looks at some useful expressions from the listening. You could do it as a whole-class activity, checking answers and then clarifying the meanings of the expressions in bold as you go. Alternatively, get students to complete the activity individually and clarify meanings during class feedback.

1 influences others' opinions of you   2 near
3 introduce a contrast to   4 have an influence on
5 contrasting   6 understand
7 they are not able at the time to judge
8 emphasise their opinion

5 Explain that students are now going to read some statements from the comments section of the survey in Exercise 4 and go through the instructions with them. Put them in pairs and give them time to read the statements and discuss the question. If necessary, also give them time to review the information from the meeting in Exercise 2 by referring to audioscript BW 8.01 on page 161, as well as the survey results in Exercise 4. When they are ready, invite different students to share their answers with the class. Encourage them to give reasons.

Extra activities Business workshop 8

B This activity looks at useful vocabulary from this Business workshop. If time is short, you could do it as a whole-class activity, checking answers and clarifying meanings as you go. Alternatively, get students to complete it individually, using their dictionaries if necessary and clarify meanings during class feedback.

1 f **2** i **3** j **4** e **5** a **6** g **7** c **8** h **9** d **10** b

Task: Choose a course

Students choose a training course to aid young employee retention and write a proposal.

6A Put students in small groups, go through the instructions with them and make sure they are clear about the scenario and what they need to do: before reading the proposals for the three training courses, they should create a set of goals to help retain young talent, based on the issues that came up in the employee satisfaction survey. Point out that they need to think about the survey results as well as the employees' comments from Exercise 5. During the activity, monitor and help students with ideas and/or any vocabulary they may need.

6B Based on the goals students have devised, they now read about three training courses and choose the one they think best matches their set of goals. Start by referring students to the proposals and giving them time to read the texts; monitor as they are reading and help them with any vocabulary or other questions they may have. Then go through steps 1 and 2 with them and make sure they are clear about what they need to do: first, they should discuss the three different courses and think about which issue(s) each one will address. Then they should decide which course would be the best one for Glass & Franks to offer, based on their set of goals from Exercise 6A. Which of the three proposed courses best matches their goals? How? Remind students that they need to give reasons. Set a time limit and ask groups to begin their discussions. As an optional follow-up, groups could then take turns to share their conclusions with the class (or, in larger classes, with another group).

7 Students now write a proposal for the management team at Glass & Franks. Explain the writing task, highlight the word limit and go through the list of points to include with them. In weaker classes, you could let students plan their work in their groups or in pairs, working with a student from their original group. Remind them to divide their proposal into clear sections with headings; if necessary, you could give them an empty 'template' outlining the sections and headings (see model answer below) and ask them to write their proposals based on that. If time is short, the writing task can be assigned as homework.

Model answer

Introduction
Glass & Franks has become increasingly concerned about the employee turnover among their millennial employees. They value young talent, but many of the people in this age group only stay a few years in the company.

Findings
An employee survey showed that older employees were generally happy with their jobs and were glad to be part of the company. The younger workforce, however, do not feel they have the chance to learn new skills and would like to develop both professionally and personally. It is important for them to be able to help create the type of company they want to work for and they would like to build strong relationships with their colleagues.
At the moment, the only programme dealing with the mental health of employees is one on burnout, but management feels that it is necessary to have more preventative training sessions as well.

Conclusion
Proposals from three providers were considered and it was concluded that the training session on improving communication would be the most beneficial. From our point of view, we feel that this course would cover the largest number of areas and help employees to express themselves better and build long-term relationships. In addition, the discovery of one's own resources is a skill which we feel will be welcomed by those who expressed the need to learn and grow.

Recommendation
Our recommendation is to hire Building Bridges to run a training session with up to 30 people, evaluate the course and then decide if it should be run again with a new group.

MyEnglishLab: Teacher's Resources: extra activities

Review ◀ 8

1 **1** a **2** b **3** a **4** b **5** a **6** b
2 **1** setback **2** vision **3** second guess **4** from scratch
3 **1** b **2** b **3** a **4** a **5** b **6** a **7** b **8** a **9** a **10** a
4 **1** challenge **2** solution **3** background **4** sequence
  5 summary **6** benchmarking **7** takeaway
  8 reflection
5 **1** d **2** f **3** e **4** h **5** b **6** c **7** g **8** a

Resource bank

1.1 > Vocabulary

1 Work in pairs. Fill in the correct form of *problem* or *solution*.

1 Our team is working hard to tackle the _____ we are having with the software.

2 The R&D group is used to coming up with innovative _____ and insights which no one else in the industry seems to think of.

3 Being a cutting-edge company, customers expect us to provide them with a high-tech _____ to any difficulty they encounter.

4 The board is discussing the best way of addressing the _____ we had with our last product roll-out.

5 Our client was especially pleased with the well-thought-out _____ to the issue they have been having.

6 It seems to me to be quite an unorthodox _____ but as long as we have no difficulties with our legal department, we can try it out.

7 Their approach to the _____ is different from ours, but it seems to be quite successful.

8 If we don't overcome these _____ , we may find that we cannot continue in business much longer.

9 Our customers have come to rely on us to find digital _____ for their process management.

10 Several clients complained that we had offered them an out-of-date _____ and needed to come back to them with something new.

2 Match the statements (1-10) with the sentences (a-j).

a It is necessary to talk about what happened with the product launch and the subsequent promotion. We need to make sure that something like this does not happen again.

b I am a bit surprised at the ideas they come up with, but as long as they make their customers happy, then I guess it works.

c We are trusted in our field as the people who can help other companies run more efficiently.

d We would like to thank our team for the way they consistently get round any issues we have with our products. We are now far ahead of our competition.

e I will be discussing the matter with our corporate lawyer and if he gives the go-ahead, then it's no problem.

f I keep getting calls from clients about the difficulties they are having installing our products on their computers. We need to come up with a way to fix this.

g The board is meeting to discuss the situation we are in as the outlook at the moment is quite worrying.

h For some reason our customer support tried to help a group of clients with information that was current a few years ago. They were not happy.

i One of our customers called to say they felt we really put time into the way we went about dealing with the difficulties they were having with our product.

j Our reputation means that we need to ensure that whatever we offer our clients is state of the art.

3 Work in pairs. One of you takes statements 1–5, the other statements 6–10. Think of a specific example that could explain the situation. Share your ideas with your partner. Then form new pairs and work with someone who had the same sentences that you did. Compare your ideas and decide which ones are most probable.

1.2 ❯ Grammar

1 Work in pairs or groups and only correct the sentences with errors. Then think about how sure you are about whether the sentence is correct or not and choose a number from 1 (not sure) to 5 (sure). The sentences will then be corrected together and you win or lose the number of points you chose.

Statements	Scoring	Score	
		+	–
1　Many people consider their phones to be the innovation which has most changed their lives.	1 2 3 4 5		
2　The longer you work as a researcher, better you get at discovering new ways of doing things.	1 2 3 4 5		
3　This type of innovative thinking is needed in an industry.	1 2 3 4 5		
4　There is a problem and we really need to let a team know about it before it is too late.	1 2 3 4 5		
5　As far as I know, he works for the Marketing Department.	1 2 3 4 5		
6　I think it is a most out-of-date idea I have heard.	1 2 3 4 5		
7　That was the very high-tech solution I would say.	1 2 3 4 5		
8　What fascinates me most about gadgets is that they can be used for many different things. Being innovative means coming up with lots of the ideas.	1 2 3 4 5		
9　I hope the board approves of our unorthodox solution.	1 2 3 4 5		
10　Can I introduce you to George Smith, my boss?	1 2 3 4 5		
11　Joanna loves her job as programmer for Techwave.	1 2 3 4 5		
	Totals		

2 Work in pairs and write five sentences about activities, goals and plans you both have, using *a*, *an*, *the* or no article in each sentence. Mingle with other students and try to find another pair with similar sentences to yours. For each similar sentence, add 1 point to your final score.

_____
_____
_____
_____
_____
_____
_____
_____
_____

2.1 ❯ Vocabulary

1   **Complete the sentences with the word partnerships.**

1   If companies begin to manufacture products _____ that can be reused, it will greatly cut down on waste.

2   It is time to embrace the idea that the _____ we live in is doing harm to the environment and our planet.

3   Businesses may argue that giving customers the chance to upgrade their devices is an excellent way to _____ .

4   It is clear that industry needs to be profitable, but _____ business people know that we cannot dispose of every product without causing problems for future generations.

5   One of the problems with our landfills today are the _____ feeding into them.

6   When electronic devices are taken apart, it is possible to salvage the _____ inside them rather than just throwing them away.

7   When we produce, use and reuse manufactured goods, this is called a _____ .

8   An innovative material that can repair itself and extend the life of a product is _____ .

9   Environmentally friendly products include those that can _____ the system and be reused in one way or another.

10   Most cities today have _____ where waste is separated and the usable parts are sorted out for reuse.

2   **Discuss these questions in pairs.**

1   Which of these elements of the circular economy are most important to you?

2   Which are the easiest to implement?

3   Which are the most difficult?

forward-	feed	from the	waste	throwaway
circular	recycling	self-healing	drive	precious

thinking	back into	outset	streams	culture
economy	plants	plastic	innovation	metals

2.2 ❯ Grammar

Student A **1** Complete the passive and active sentences on your sheet. Add the agent if you feel it is necessary. Then work in pairs and compare your answers.

Active sentences	Passive sentences
1 The company targeted their product at millennials.	**1** The product _____
2 Some people _____	**2** Some people like having work done for them (by others).
3 Producers are very glad when customers tell them that they like their products.	**3** Producers like _____
4 A company _____	**4** The city is having the streetlights repaired (by a company).
5 Having passed a new recycling law, the government was able to reduce waste.	**5** A new recycling law _____
6 Many people _____	**6** Our firm is regarded as one of the top manufacturers in the field (by many people).
7 We need to streamline the process.	**7** The process _____
8 I hope _____	**8** I hope we will be considered for the award.
9 The R&D Department tested the new product last week.	**9** The new product _____
10 Our HR Department asked _____	**10** Our HR Department had the form signed (by employees).

Student B **1** Complete the passive and active sentences on your sheet. Add the agent if you feel it is necessary. Then work in pairs and compare your answers.

Active sentences	Passive sentences
1 The company _____	**1** The product was targeted at millennials (by the company).
2 Some people like others doing work for them.	**2** Some people _____
3 Producers are very glad _____	**3** Producers like being told / are very glad to be told that customers like their products.
4 A company is repairing the streetlights in the city.	**4** The city _____
5 Having passed _____	**5** A new recycling law having been passed, the government was able to reduce waste.
6 Many people regard our firm as one of the top manufacturers in the field.	**6** Our firm _____
7 We need _____	**7** The process needs streamlining / to be streamlined.
8 I hope they will consider us for the award.	**8** I hope we _____
9 The R&D Department _____	**9** The new product was tested by the R&D Department last week.
10 Our HR Department asked employees to sign the form.	**10** Our HR Department had _____

3.1 ❯ Vocabulary

Find someone who can explain what **make a return** means.	Find someone who can explain what a **high-risk trade** is.	Find someone who can explain what **financial instruments** are.
Find someone who can explain what **interest rates** are.	Find someone who can explain what **foreign exchange** is.	Find someone who can explain what **reap rewards** means.
Find someone who can explain what **deal in currencies** means.	Find someone who can explain what **human resources** are.	Find someone who can explain what **a level playing field** is.
Find someone who can explain what a **backer** is.	Find someone who can explain what **the rate of return** is.	Find someone who can explain what **return on investment (ROI)** means.
Find someone who can explain what **yields** are.	Find someone who can explain what **a ballpark figure** is.	Find someone who can explain what **invest in** means.
Find someone who can explain what **a trade-off** is.	Find someone who can explain what **make an investment** means.	Find someone who can explain what **the trading floor** is.

3.2 > Grammar

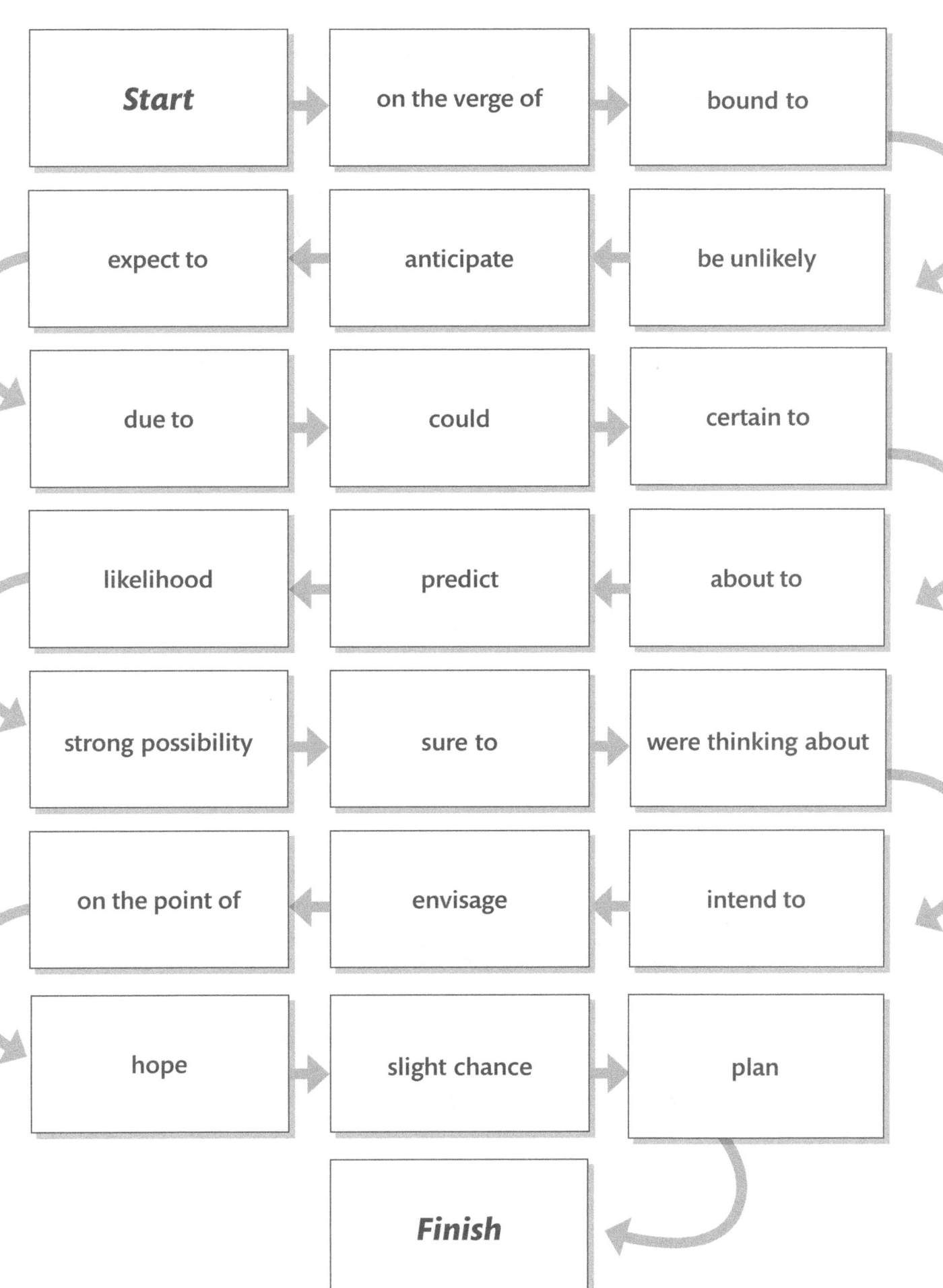

Start	on the verge of	bound to
expect to	anticipate	be unlikely
due to	could	certain to
likelihood	predict	about to
strong possibility	sure to	were thinking about
on the point of	envisage	intend to
hope	slight chance	plan
	Finish	

4.1 ❯ Vocabulary

1 **Read the statements about disruption in the workplace. Make notes in the ovals about which areas of your student/professional life have been or are likely to be affected by these factors.**

1 Companies feel they are more productive when people work in agile cross-departmental teams.

2 The business landscape today is very different from what it was several years ago as companies are abandoning old ideas and taking novel approaches to new ones.

3 Disruptive innovation is moving at a very fast pace today and is a real game changer in many fields.

4 Digital transformation means solving problems in new ways and is injecting new ideas into industry and the way business is done.

5 Established companies are looking for ways to reposition themselves in the market.

6 Traditional methods are being displaced by new ideas, so companies need to dedicate resources towards future-thinking.

2 Work in pairs. Take turns reading the notes aloud from your ovals in a random order. Your partner then guesses which of the statements your notes apply to and says why they think so.

3 Decide together which of the six statements is likely to have the most effect on the workplace and your job in the next decade and for the span of your working life.

4 Discuss with your partner which of these developments you see as positive and which ones you think will be difficult to implement.

4.2 ➤ Grammar

If only	we had more time to develop the product, but we just don't.
I wish	he had told me about the opening at his company, but he never mentioned it.
If only	we had reacted immediately, we could have launched the product before the competition.
Given that	they were the first to use the technology, I would have assumed they are the leaders in the market.
I wonder if	my boss sent the recommendation he promised to write for me.
Suppose they	were looking for someone with your qualifications, would you apply for the job?
Speculating for a moment	that had we begun the new project last year, do you think it would have been a success?
Were they to investigate	the company, a number of irregularities might be found.
Had we realised that	they were working on a similar project, we would have tried to enter the market more quickly.
Let's suppose	that our idea to increase the lifespan of the product works. How will this affect sales?

1 Complete the sentences using the prompts. Read them aloud to your partner and check if your partner agrees that they have been written correctly.

1 If only _____ , we (modal + present perfect) _____

2 Let's suppose _____

3 Suppose _____ (past continuous), _____

4 I wish _____ (past perfect) _____

5 Had I _____ (past participle), _____

6 I wonder if _____

7 Just imagine _____ (past continuous) _____

8 Given that _____ (past simple) _____

9 Let's assume that _____

10 What if _____ (past perfect) _____

5.1 › Vocabulary

Student A **1** Explain your words and phrases to your partner without using the words themselves. Your partner may ask you as many questions as they like and then writes the word or phrase into their grid.

Grid (Student A):
1. t a **c** t i c
2. b **u** y · i n t o
3. p e r **s** u a s i v e
4. t r i c k s · u p · **t** h e i r · s l e e v e s
5. b e · p **o** s i t i o n e d
6. **m** i s l e a d
7. d e c o y · **e** f f e c t
8. w e l l - v e **r** s e d · i n
9. g e t · o n e · o v e **r** · o n · s o m e o n e
a. **n**
b. **g**
c. **a**
d. **g**
e. **e**
f. **m**
g. **e**
h. **n**
i. **t**

2 Write five questions using the words and phrases to ask what retailers do to convince people to buy their products. Ask your partner the questions.

- -

Student B **1** Explain your words and phrases to your partner without using the words themselves. Your partner may ask you as many questions as they like and then writes the word or phrase into their grid.

Grid (Student B):
1. **c**
2. **u**
3. **s**
4. **t**
5. **o**
6. **m**
7. **e**
8. - **r**
9. **e**
a. c a **n** c e l · o u t
b. l e **g** a l · s t a n d i n g
c. p a r **a** d o x
d. w e l l - b e i n **g**
e. r **e** a s o n a b l e
f. m o r e · t h a n · **m** e e t s · t h e · e y e
g. i n c l i n **e** d · t o
h. t a p · i **n** t o
i. i n v i **t** i n g

2 Write five questions using the words and phrases to ask about customers' buying habits in shops. Ask your partner the questions.

5.2 ❯ Grammar

1 **Work in pairs and decide who is 'X' and who is 'O'.**

- Take turns to roll the dice and choose a square which corresponds to the number on the dice (see the bottom of the page).
- Change the sentence using a participle clause, then complete it.
- If you are correct, put your X or O in the square.
- Continue the game until one of you has 'won' four squares in a row. If you roll a 6, you can choose any sentence which is still free.

1 When you eat these foods every day, ...	**2** We called the customer support line and were told ...	**3** If you handle the device carefully, it ...	**4** The product was promoted last week because of low sales figures, and now ...	**5** It hadn't been used properly and it unfortunately ...
6 As we had researched the effects thoroughly, we ...	**7** If you work many hours, you should ...	**8** Because they spoke politely to customers, they ...	**9** The new tariffs took effect yesterday with the result that ...	**10** They introduced their new product last week and began ...
11 Our company is known everywhere and we ...	**12** After they tested the product, they began ...	**13** The man who works at the shop, ...	**14** The tactic which was tried out last week, ...	**15** If you follow the instructions correctly, the product ...
16 I wrote an email but was told I should ...	**17** If you read the information carefully, you ...	**18** The report was rewritten because of the errors, and now ...	**19** It hadn't been updated in years, and it now ...	**20** As they had travelled all day, they ...
21 If you start at a new company, you should ...	**22** Because they told the client what they thought, they ...	**23** Sales decreased with the result that ...	**24** The shop repositioned many items and started to ...	**25** We have an international reputation and ...
26 After they created the market for the product, they ...	**27** When you are a loyal customer, you ...	**28** They created new displays in the shop and then invited ...	**29** As he needed to retrain for a new job, he ...	**30** If you do a good job, you should ...

⚀	= 2 8 13 19 22 29	⚁	= 1 6 9 16 24 28	⚂	= 4 12 15 20 25 30	⚃	= 5 10 14 17 23 27	⚄	= 3 7 11 18 21 26

2 **When you finish the activity, convert any sentences which have not yet been picked into participle clauses and complete them.**

6.1 > Vocabulary

eco-resort	guesthouse	bed and breakfast	destination development
infrastructure	mainland	hotel chain	hotspot
package tour	photobombing	mass tourism	overdevelop
tourist trap	World Heritage Site	record season	tour guide

small hotel where you can sleep and have the first meal of the day	activity of creating successful sustainable places for visitors and developing a brand image for a particular country or area	an environmentally friendly hotel and grounds for holidays which tries to minimise its carbon footprint	a private place where people can pay to stay and have meals
a number of establishments where people can stay which are owned or managed by the same company or person	a place that is popular for entertainment or a particular activity	the basic systems and structures that a country or area needs in order to work properly, for example roads, railways, banks, etc.	the main area of land that forms a country, as compared to islands near it that are also part of that country
a situation when large numbers of people visit the same destination at the same time	to put up too many buildings, shops, roads, etc. in an area	a holiday which includes transportation, accommodation, meals, entertainment, etc.	the activity of deliberately getting into the background of someone else's photograph as a joke
reaching the highest number of people visiting a destination, or tourists staying in a particular hotel or resort	a person who helps visitors and takes them around an area to show them the sights	a negative phrase for a place that many tourists visit, but where drinks, hotels, etc. are more expensive	place selected by UNESCO as having cultural, historical, or scientific significance, legally protected internationally

6.2 > Grammar

1 Write sentences explaining or talking about situations you have personally experienced in the past. Use time expressions or discourse markers when appropriate.

1 A situation that existed before another past action.

_____

2 Two actions or events that were both in progress at the same time in the past.

_____

3 A past event or situation which is still relevant or continuing now (often used with words like *already*, *still*, and *yet*).

_____

4 A life experience that happened before a point of time in the past.

_____

5 Something that is a finished or completed past action or event.

_____

6 Something that happened in the past before another action or event in the past.

_____

7 An action which is still continuing or still true now (often used with *since* and *for*).

_____

8 An event or situation which took place at an indefinite time in the past.

_____

9 An event or situation that was in progress before another finished past action.

_____

10 A situation that existed at the time of another past action.

_____

2 Look at the descriptions of past events. Work through the list with a partner and make a note as to which past tense is needed to express the situation. Do not discuss the sentences you wrote.

3 Work in pairs. Try to guess what your partner wrote about. Ask questions about the situation in order to help you. You have five minutes each.

4 Count up the number of correct guesses per pair. The pair with the highest number is the winner.

7.1 ❯ Vocabulary

Student A

micromanage	likely to cause arguments or make people angry
criticism	performing at the highest possible level
staff morale	the ability to understand other people's feelings and problems
a fine line between	start to discuss a (usually difficult) subject
mediator	not based on clear thought or reason
provocation	say or think that someone or something is responsible for something bad

Student B

irrational	get much worse or more serious (e.g. an argument or fight)
confrontational	behave in a way that is unacceptable socially
resolution	organise and control all the details of another person's work in a way that they think is annoying
on top of one's game	do what you are supposed to do
empathy	an argument between people or groups based on having very different beliefs and opinions
raise the issue with	someone one level higher in rank than you in a company and in charge of your work

Student C

line manager	an action or event that makes someone angry or upset, or is intended to do this
escalate	level of confidence and positive feelings that people working together have
blame	comments that say what you think is bad about someone or something
cross the line	a very small difference
clash	a solution to a problem, argument, or difficult situation
toe the line	someone who tries to end a quarrel between two people or groups by discussion

7.2 ❯ Grammar

1 Work in pairs. Complete the sentences for yourself and for your partner. Then read the sentences that you have written about your partner to him or her. How many times did you guess correctly what your partner thinks?

Me	My partner
1 I may be more/-er _____ than _____ .	**1** _____ may be more/-er _____ than _____ .
2 I think I could _____ in the next few months.	**2** _____ thinks _____ could _____ in the next few months.
3 I would like to be slightly more _____ because _____ .	**3** _____ would like to be slightly more _____ because _____ .
4 I am fairly sure that I _____ at some point in the future.	**4** _____ is fairly sure that _____ _____ at some point in the future.
5 I like _____ to some extent.	**5** _____ likes _____ to some extent.
6 I usually tend to _____ .	**6** _____ usually tends to _____ .
7 I am actually _____ , although this is something that _____ .	**7** _____ is actually _____ , although this is something that _____ .
8 In all probability, I will _____ soon.	**8** In all probability, _____ will _____ soon.
9 It's unlikely that I _____ in the near future.	**9** It's unlikely that _____ _____ in the near future.
10 It is conceivable that I _____ in the next ten years.	**10** It is conceivable that _____ _____ in the next ten years.
11 I was hoping that I _____ by the end of the month.	**11** _____ was hoping that _____ _____ by the end of the month.
12 I thought I might have time to _____ , but _____ .	**12** _____ thought _____ might have time to _____ , but _____ .

2 Write three more sentences about yourself using hedging and tentative language. One of them should be false, the other two should be true. Read them to your partner, who then tries to guess which one is false.

_____

_____

_____

_____

_____

_____

8.1 ❭ Vocabulary

Student A **1** Make some notes about these words or phrases. Then describe them to your partner without using the words or phrases themselves so that your partner can guess the word.

	Notes
entrepreneurial	
skyrocket	
start from scratch	
value (verb)	
pick up	
in tune with	
defeat	

2 Work together to use as many of the words and phrases as you can to discuss mindsets and the concept of being entrepreneurial.

- -

Student B **1** Make some notes about these words or phrases. Then describe them to your partner without using the words or phrases themselves so that your partner can guess the word.

	Notes
passion	
vision	
persistence	
second guess	
mindset	
setback (noun)	
doubt oneself	

2 Work together to use as many of the words and phrases as you can to discuss mindsets and the concept of being entrepreneurial.

8.2 > Grammar

1 Read these questions which all use different verb forms and think about how you would answer them with a few words. Mingle with the class to find people who have similar answers to you. When you have found someone, write their name in the box. When you have five in a row (across, down or diagonally), you can say 'Bingo'.

What's something you would suggest doing this weekend? _____	How long do you usually wait for someone? _____	What sounds like a feasible idea to you? _____	How do you look at the end of the day? _____	What has become unnecessary for you? _____
What do you think is growing more essential every day? _____	What in your life is important for you to keep? _____	What do your friends consider you to be? _____	What is the last item you gave someone? _____	What do you sometimes lend to a friend or colleague? _____
Who was the last person you bought a gift for? _____	What do you bring on holiday? _____	What do you often avoid doing? _____	What is the most unusual thing you have considered doing? _____	What do you appreciate someone doing for you? _____
What can you imagine doing in the next ten years? _____	What was the last activity you agreed to do with someone? _____	What would you like to arrange for your friends to do at the weekend? _____	What is the most recent thing you offered to do for a friend? _____	What was the last thing you asked someone to help you with? _____
What was the last thing that you complained about? _____	What do you expect will be successful for you? _____	When would you wonder where your friends are? _____	What do you spend a lot of time doing every day? _____	What do you love doing? _____

2 Look at the verb patterns in the Exercise 1 and write five more sentences about yourself using these patterns. Discuss them with your partner and explain why you wrote what you did.

_____

_____

_____

_____

_____

_____

_____

_____

_____

1.1 ❯ Vocabulary

- Divide the class into pairs. Hand out a worksheet to each pair.
- Ask students to complete sentences 1–10 in Exercise 1 using *problem* or *solution* in the correct forms. Some sentences can have two equally correct answers.
- Ask students to match the statements in Exercise 2 to the sentences they completed in Exercise 1. Go over the sentences to make sure everyone agrees on the answers.
- For Exercise 3, divide the class into pairs and assign each person in the pair either the first five statements or the second five. Their goal is to come up with specific examples which could explain the statements and the situations. When they have finished, ask them to share their examples with their partners so that each of the ten situations are discussed. When they have finished, tell them to find a new partner who had the same five sentences they had, compare their answers and choose the most probable ones.
- As a follow-up, ask students to look at the statements again and talk about which of these situations they have experienced themselves and which they think might be most common in the field of innovation and technology. They can also discuss which ones they feel could be most positive or negative for a company or business.

> 1 1 problem/problems 2 solutions 3 solution
> 4 problem/problems 5 solution 6 solution
> 7 problem 8 problems 9 solutions 10 solution
> 2 1 f 2 d 3 j 4 a 5 i 6 e 7 b 8 g 9 c 10 h

1.2 ❯ Grammar

- This can be done in groups or pairs. Tell students that they are going to participate in a team challenge and practise articles from Lesson 2.
- For Exercise 1, put students into groups of two or three and give one worksheet to each group. Tell students to look carefully at the statements and correct the ones they feel are wrong. Then ask them to choose between 1 and 5 points depending on how sure they are that they have the correct answer. Go through the answers with the class. For each correct answer, they gain the number of points they chose; for each incorrect one, they lose the number of points they chose. The teams add up their points.
- For Exercise 2, the challenge is to work in pairs and write five sentences about activities, goals or plans using the correct articles in each sentence. Once students have finished, ask them to mingle to find other pairs with similar sentences. For each similar sentence they find, both pairs add one more point to their final scores.
- For a short follow-up spoken activity, students could continue to work in pairs and practise making correct or incorrect sentences, which their partners either correct or agree don't need correcting.
- For a longer follow-up activity, groups could write a short text incorporating the grammar rules and vocabulary from the lesson for another pair to correct.

> 1 1 correct 2 the better 3 an industry 4 a the team
> 5 correct 6 a the most 7 the a very 8 the ideas
> 9 correct 10 correct 11 a programmer

2.1 ❯ Vocabulary

- Tell students that they are going to practise vocabulary from Lesson 2.1.
- Put students in pairs. Cut out the word partnership cards and give students a worksheet each and one set of cards per pair.
- Ask students to put the word partnerships together so that they can be used to complete the gapped sentences. Students check their answers in pairs, then check as a class.
- Give students some time to consider the questions in Exercise 2 and discuss them. Elicit their thoughts.
- Students can follow up by looking up regulations about recycling in their home town or looking for more information about concepts such as the circular economy or self-healing plastic, etc. They could present this information to the class.

> 1 1 from the outset 2 throwaway culture
> 3 drive innovation 4 forward-thinking 5 waste streams
> 6 precious metals 7 circular economy 8 self-healing
> plastic 9 feed back into 10 recycling plants

2.2 ❯ Grammar

- Tell students that they are going to make passive sentences active and active ones passive. They should also consider whether or not they need to add the agent in the passive sentences.
- Hand out a Student A or Student B worksheet to each student. Each student now has five active and five passive sentences which they need to complete. Once they have done this, put students into AB pairs and tell them to compare with their partner and discuss their reasons for changing the sentences as they have done. Check answers as a class.
- As a follow-up, ask students to write four more personal sentences with their partner using these passive voice patterns:
- ▶ phrasal verb with *be* + past participle + preposition
- ▶ *being* + past participle
- ▶ *need* + *to be* + infinitive / *need* + -*ing* form
- ▶ *have* + object + past participle.
- Ask individual students to read their sentences to the class and then discuss them as a class.

3.1 ❯ Vocabulary

- Tell students that they are going to do a class activity to review the vocabulary for finance and investment.
- Copy and cut up the cards. Give each student at least one card (if you have more than eighteen students, make an extra set; if you have fewer, give each student more than one card).
- Ask students to mingle and ask the others if they can explain the meaning of the words on their cards.
- As an extension, put students in pairs or small groups and ask them to divide the cards among themselves. Tell them to each describe the vocabulary terms they have on their cards and get the others to guess what the term is.

backer – someone who supports a plan, especially by providing money

ballpark figure – an estimate, not an exact amount

deal in currencies – buy and sell money used by different countries

financial instruments – investments such as bonds or shares

foreign exchange – the value of two currencies relative to each other

high-risk trade – the act of buying or selling goods, in this case when it is likely that the buyer or seller may lose money on the deal

human resources – the people who work for a company or organisation and the department responsible for managing resources related to employees

interest rates – a charge of payment set to a certain amount that you have to pay on money you have borrowed

invest in – buy shares, property, or goods because you hope that the value will increase and you can make a profit

level playing field – a situation in which different companies, countries, etc. can all compete fairly with each other

make a return – earn money from an investment

make an investment – the use of money to get a profit by buying or doing something because it will be useful or profitable later

reap rewards – to get something such as money from something you have done

return on investment – the profit earned from buying shares, property, etc. in relation to the money spent

trade-off – a balance achieved between two desirable but incompatible features; a compromise

the rate of return – the amount of profit that a particular investment will make, expressed as a percentage

the trading floor – the part of a financial market where shares, commodities, etc. are bought and sold

yields – the amount of profit something produces

3.2 ❯ Grammar

- Tell students that they are going to practise making sentences with different forms of the future.

- Put students into groups of three or four and give a worksheet to each group. Give each group a dice and coins or markers. Tell them to roll the dice and move their counter on the board accordingly. They have to use the word or phrase given in the square they land on to make a sentence expressing attitudes towards the future.

- The group decides if the sentences are correct, calling on the teacher to have the final say if necessary. If a sentence is correct, the student can continue at their next turn; if not, they have to miss a turn. The first person to reach the finish is the winner.

- As a follow-up, they should go through all the words and phrases again once they've finished and make sure they know how to use them correctly. This could be done as a class activity.

Sample answers

about to (for something that is going to happen almost immediately) – *We are about to merge with another company.*

anticipate (to give a future meaning) – *We can't anticipate how the market reacts.*

be unlikely (to express probability) – *He is unlikely to be promoted soon.*

bound to (for something that is definitely going to happen) – *They are bound to lose money on that deal because it is so risky.*

certain to (for something that is definitely going to happen) – *He is certain to have a successful career based on the work he has done so far.*

could (for a future possibility when there is about a 50 percent chance it will happen) – *They could win the innovation award this year if they can get the product to the market on time.*

due to (for something which is going to happen at a fixed or expected time) – *He is due to arrive at 6 p.m. tomorrow.*

envisage (to give future meaning in a sentence) – *We envisage these shares going up by at least 5 percent.*

expect to (to give future meaning in a sentence) – *We expect to see that these commodities will be more valuable in the next few months.*

hope (to give future meaning in a sentence) – *I hope that working with them will turn out to be profitable.*

intend to (to give future meaning in a sentence) – *We intend to expand our business into the Middle East and Africa.*

likelihood (a noun used to express the future) – *There is a strong likelihood that their investment will not turn out well.*

on the point of (for events that seem likely to happen in the very near future) – *They are on the point of investing in a company in Asia.*

on the verge of (for events that seem about to happen) – *They are on the verge of signing a major contract with a new business partner.*

plan (to give future meaning in a sentence) – *We plan to restructure part of the company.*

predict (to give future meaning in a sentence) – *They predict that the market will continue to improve.*

slight chance (a noun used to express the future) – *There is a slight chance I'll be in Singapore for a few days so maybe we can meet.*

strong possibility (a noun used to express the future) – *There is a strong possibility that commodities prices will fall.*

sure to (for something that is definitely going to happen) – *The Central Bank is sure to raise interest rates in the next two years.*

were thinking about (used to express an attitude from a viewpoint in the past) – *They were thinking about buying a company, but decided against it.*

4.1 ❯ Vocabulary

- Tell students they are going to use target vocabulary from Lesson 4.1.

- Give out the worksheet and tell students to read the statements and ask any questions they have about them. This can be done as a general group discussion to make sure all the statements are understood. Then ask them to make notes in the ovals about areas of their student/professional lives which have been impacted or could be impacted in the future by these statements. Get them to think about the way teamwork at university or in companies is usually done, technology they use now which they may not have used earlier, working conditions they have experienced which might change, problem-solving in business or in their courses, the situation today of well-known and established companies, possible future jobs, etc.

- Put students in pairs and ask them to read their notes to each other in random order. Their partners guess which statement the notes are based on.

- Ask them to discuss in their pairs which of these statements they think will have the biggest impact on their working lives. Pre-work students can hypothesise about this using the knowledge they have gained in business and communication skills.

- Students then discuss which of these changes they see positively and which ones they think will have more negative consequences. This can be opened up to a class discussion.

- As a follow-up, students could do some research into a company about how they have dealt with or are dealing with one aspect of these disruptions, e.g. agile teams, repositioning in the market, a member of staff or department tasked with finding new processes for the future, etc.

4.2 ❯ Grammar

- Tell students that they are going to practise hypothetical sentences in a matching game.

- Cut up the cards on the sheet. Distribute one set of cards to each pair of students.

- Tell students to match the sentence halves together. There are two beginnings, *If only/I wish* and *Let's suppose/ Speculating for a moment*, which can each be used for two different sentences. The students need to look carefully at the sentences to see if they refer to past, present or future situations. (*If only* and *I wish* can be present or past, but when talking about a past situation including a modal verb we need *If only*.)

- Hand out the rest of the worksheet to each student. For Exercise 1, ask students write their own sentences based on the prompts and the tenses specified. Ask them to check each other's work in pairs. Any questions the pairs can't answer should be discussed in open class.

- As a follow-up, students can use just the prompts on the cards to practise further with a partner. Their partner can respond to the sentences or simply agree whether or not they are correct.

If only / I wish we had more time to develop the product, but we just don't.

I wish / If only he had told me about the opening at his company, but he never mentioned it.

If only we had reacted immediately, we could have launched the product before the competition.

Given that they were the first to use the technology, I would have assumed they are the leaders in the market.

I wonder if my boss sent the recommendation he promised to write for me.

Suppose they were looking for someone with your qualifications, would you apply for the job?

Speculating for a moment that had we begun the new project last year, do you think it would have been a success? (mainly used with past situations)

Were they to investigate the company, a number of irregularities might be found.

Had we realised that they were working on a similar project, we would have tried to enter the market more quickly.

Let's suppose that our idea to increase the lifespan of the product works. How will this affect sales? (mainly used with future situations)

5.1 ❯ Vocabulary

- Tell students that they are going to practise describing the words and phrases from Lesson 5.1.

- Put students in pairs. Hand out a Student A or B worksheet to each student. Give them some time to think about how they will explain the nine words and phrases they have on their grid to their partner.

- Ask students to explain their words and phrases to their partners, who can ask as many questions as they like, but the person describing the word is not allowed to use the word/phrase itself.

- When everyone has finished, students check their answers in pairs.

- For Exercise 2, tell Student A to write questions about the tactics used by retailers to convince people to buy and Student B to write questions about how the customers feel about the retailers' tactics. When they have finished, tell them to ask each other their questions.

- As a follow-up, students can do some research on up-to-date methods used in the retail business to convince customers that their products are the best. Ask them to share the most interesting ones with the class.

Sample questions
Student A
How do retailers try to get consumers to buy products which are less healthy?
Where are fruits and vegetables positioned in shops?
What is the decoy effect and how do retailers use it?
How do retailers tap into consumers' desire for bargains?

Student B
How do consumers feel about products that are reasonable in price?
What are consumers more inclined to do if they have already bought healthy items?
What products do consumers tend to buy when they consider their own well-being?
Why do consumers think that the healthy effect of one product can cancel out the unhealthy effect of another?
What makes products look inviting to a consumer?

5.2 ❯ Grammar

- Tell students they are going to practise making participle clauses.

- Put students in pairs and hand out a copy of the worksheet and a dice to each pair.

- Explain that the goal of the game is to cross off (with an X or an O) four squares in a row. In order to do this, students roll the dice and if they get a number from 1–5, they have to choose one of the numbers given at the bottom of the chart. They then convert the sentence opening into a participle clause and complete the sentence. If their partner agrees that it is correct, they can put their X or their O in that square. Then the partner rolls the dice and can choose from any of the sentences which have not yet been claimed according to what comes up on the dice. If either student rolls a 6, they can choose any sentence they like.

- When all the pairs have finished their games, go through the sentences with the class to make sure they have done them correctly. In some cases, more than one answer is possible, but the answer key gives the most obvious suggestions.

- Point out that the participle clauses used to show one action taking place after another can be formed in two different ways.

- For Exercise 2, students can continue to practise by converting any prompts which haven't been picked on the board yet and completing the sentences.

1 Eating these foods every day, … / Eaten every day, these foods …
2 Having called the customer support line, we were told …
3 Handling the device carefully, it … / Handled carefully, the device …
4 Having been promoted last week, the product now …
5 Not having been used properly, it unfortunately …
6 Having researched the effects thoroughly, we …
7 Working many hours, you should …
8 Speaking politely to customers, …
9 The new tariffs took effect yesterday, causing … / Taking effect yesterday, the new tariffs …
10 Introducing / Having introduced their new product last week, they began …
11 Being known everywhere, we …
12 Having tested / After testing the product, they began …
13 The man working at the shop, …
14 The tactic tried out last week, …
15 Following the instructions, this product …
16 Having written an email, I was told I should …
17 Reading the information carefully, you …
18 Being rewritten / Having been rewritten because of the errors, the report now …
19 Not having been updated in years, now it …
20 Having travelled all day, they …
21 Starting at a new company, you should …
22 Telling the client what they thought, they …
23 Sales decreased, causing …
24 Repositioning many items, the shop started to …
25 Having an international reputation, we …
26 Having created / After having created the market for the product, they …
27 Being a loyal customer, you …
28 Creating new displays in the shop, they then …
29 Needing to retrain for a new job, he …
30 Doing a good job, you should …

6.1 ❯ Vocabulary

- Tell students that they are going to work with two sets of cards, one with target vocabulary from Lesson 6.1 and the other with the definitions. There are several options for this activity.

▶ Students can simply match the cards to the definitions in pairs.

▶ Students can play a memory game by turning all the cards face down, and turning over two to find matching pairs.

▶ Students can divide the definitions and the words so that each student has a word card or a definition card. Ask them to mingle and read out what is on their card to find the person with the correct matching card.

- When the students have finished their matching activity, tell them to categorise the cards as positive, neutral or negative concepts or things, and give reasons for this. Ask them to discuss in pairs whether these categories are looked at differently by different people, for example by tourists, tour guides or hotel owners.

- As a follow-up, students can look up information about places with eco-resorts, mass tourism, destination development, etc. and give a short talk on these concepts.

bed and breakfast – small hotel where you can sleep and have the first meal of the day
destination development – activity of creating successful sustainable places for visitors and developing a brand image for a particular country or area
eco-resort – an environmentally friendly hotel and grounds which tries to minimise its carbon footprint
guesthouse – a private place where people can pay to stay and have meals
hotel chain – a number of establishments where people can stay which are owned or managed by the same company or person
hotspot – a place that is popular for entertainment or a particular activity
infrastructure – the basic systems and structures that a country or area needs in order to work properly, for example roads, railways, banks, etc.
mainland – the main area of land that forms a country, as compared to islands near it that are also part of that country
mass tourism – a situation when large numbers of people visit the same destination at the same time
overdevelop – to put up too many buildings, shops, roads, etc. in an area
package tour – a holiday which includes transportation, accommodation, meals, entertainment, etc.
photobombing – the activity of deliberately getting into the background of someone else's photograph as a joke
record season – reaching the highest number of people visiting a destination, or tourists staying in a particular hotel or resort
tour guide – a person who helps visitors and takes them around an area to show them the sights
tourist trap – a negative phrase for a place that many tourists visit, but where drinks, hotels, etc. are more expensive
World Heritage Site – place selected by UNESCO as having cultural, historical, or scientific significance, legally protected internationally

6.2 ❯ Grammar

- Tell students that they are going to practise using past tenses.
- For Exercise 1, ask students to work on their own to write sentences which match the situations described. Remind them to use time expressions or discourse markers where appropriate.
- Hold a short brainstorming session with students and write out the names of the different past tenses. Depending on the class, you may want to elicit uses for these or just write the names on the board. Then, for Exercise 2, tell the students to match the descriptions of the past events to the names of the tenses.
- For Exercise 3, in pairs, ask the students to try to guess what their partner wrote about by asking questions. Each student has five minutes to try and guess what their partner wrote.
- Finally for Exercise 4, ask them to count up how many they guessed correctly. The pair with the highest number is the winner.
- As an follow-up activity, students could work in their A/B pairs to write three more statements using different past tenses and referring to a workplace situation.

---

1 past perfect continuous (used with *when*)
2 past continuous (for both verbs)
3 present perfect simple (used with a*lready, still* and *yet*)
4 past perfect simple (often used with the adverbs *ever, never, already, yet* and *just*)
5 past simple (used with … *ago, last* …, dates, etc.)
6 past perfect simple (often used with past simple)
7 present perfect continuous (used with *for* and *since*)
8 present perfect simple (no time is given)
9 past perfect continuous (used together with the past simple)
10 past continuous (used with *when*)

7.1 ❯ Vocabulary

- Tell students that they are going to practise the vocabulary from Lesson 7.1.
- Cut the worksheets into three and hand out a Student A, B or C table to each student. Ask them to form ABC groups.
- Explain that Student A begins by reading the word in the upper left-hand corner (*micromanage*). Tell Students B and C to look for the definition on the right side of the table and whoever has the definition reads it aloud. If it is correct, they tick it off (Student B - *organise and control all the details of another person's work in a way that they think is annoying*). Then the student who has read the definition reads the word directly opposite on the same line (*resolution*) and the other two students search for the definition. They continue in this way until all the words have been read and the definitions have been found. The last definition is in the upper right-hand corner on Student A's table.
- When students have finished, ask them to discuss the two follow-up questions given below. This can be done in open class or in the groups of three with the students reporting back on their discussions. These should engage them again with the vocabulary and make them think about the meanings of the words related to conflict and conflict resolution.

▶ Which of these lead to problems in the workplace?
▶ Which of them are methods of solving problems in the workplace?
- For extra practice, groups could read out definitions and ask the others for the words they have just described.

As a guide for the teacher, these are numbered so that they can be gone through easily with the group if necessary.

A

1	micromanage	18	likely to cause arguments or make people angry
11	criticism	10	performing at the highest possible level
5	staff morale	4	the ability to understand other people's feelings and problems
16	a fine line between	15	start to discuss a (usually difficult) subject
8	mediator	7	not based on clear thought or reason
13	provocation	12	say or think that someone or something is responsible for something bad

B

7	irrational	6	get much worse or more serious (e.g. an argument or fight)
18	confrontational	17	behave in a way that is unacceptable socially
2	resolution	1	organise and control all the details of another person's work in a way that they think is annoying
10	on top of one's game	9	do what you are supposed to do
4	empathy	3	an argument between people or groups based on having very different beliefs and opinions
15	raise the issue with	14	someone one level higher in rank than you in a company and in charge of your work

C

14	line manager	13	an action or event that makes someone angry or upset, or is intended to do this
6	escalate	5	level of confidence and positive feelings that people working together have
12	blame	11	comments that say what you think is bad about someone or something
17	cross the line	16	a very small difference
3	clash	2	a solution to a problem, argument, or difficult situation
9	toe the line	8	someone that tries to end a quarrel between two people or groups by discussion

7.2 ❯ Grammar

- Tell students that they are going to practise different forms of hedging and tentative language.
- Put students in pairs. Give each student a worksheet and ask them to complete the sentences, first about themselves and then about their partner. In the 'My Partner' column, the sentences have spaces to write in the partner's name or 'he/she'. In pairs, tell them to read aloud the sentences they wrote about their partner, who says if they are true or not. Encourage them to discuss their answers using this language.
- If you like, students can win points. One point can be awarded if the sentence a student wrote about their partner is true, two points can be awarded if the sentence is similar to what the partner had written about themselves.
- For Exericse 2, ask them to write three more sentences about themselves using hedging and tentative language. Of the three sentences, two should be true and one should be false. Tell them to read their sentences aloud to their partner, who guesses which one is false.

8.1 ❯ Vocabulary

- Tell students that they are going to practise the target vocabulary from Lesson 8.1.
- Put students in pairs and give each student a worksheet (A or B).
- Ask Student A to explain their words to Student B and give them five minutes to get Student B to guess them. Explainers can give definitions, ask their partner a question that would answered by the word or make a gap sentence, but they can't use any form of the word itself. After five minutes, stop the pairs and tell students to swap roles and ask Student B to explain their words to Student A. They now have five minutes to get Student A to guess their words.
- When the time is up, ask them to show each other their words and discuss any that they were unable to guess.
- For Exercise 2, ask pairs to use the words and phrases to describe different mindsets and the concept of being entrepreneurial.

8.2 ❯ Grammar

- Tell students that they are going to practise verb patterns with a Bingo game.
- Give each student a copy of the worksheet and ask them to quickly write short answers (these can be between one and three words) or to think of what they would write. Have a class mingle and tell students to try to find people who answered the questions in the same way. When they find someone with the same (or a very similar) answer, they write that person's name in the square. The goal is to have five in a row either across, down or diagonally in order to get 'Bingo'.
- When they have finished, discuss the verb patterns as a class and randomly call on students to read out their answers using the patterns.
- For Exercise 2, ask each student to write five sentences about themselves using the different verb patterns and to then discuss these with a partner.

Unit 1 >

1 Read the article about women in business quickly and find the following information.

1 the director of the Cambridge Centre for Neuropsychiatric Research _____

2 a course designed to encourage female scientists to start their own businesses _____

3 a British government agency that has looked into the role of women in science _____

4 the chief executive of the UK's Chartered Management Institute _____

5 a mechanical engineer who attended EnterpriseWISE last year _____

6 a business that is developing technology that should help increase the lifespan of oil pipes _____

2 Read the article again. Complete the gaps (1–6) with the phrases (A–F).

A because of her gender

B why she left to start her own business

C they all want to start a business

D a consulting company specialising in the same field

E women make up 21 percent of the corresponding workforce

F men are still twice as likely as women to be entrepreneurs

3 Find the phrasal verbs (1–5) in the text and match them with the definitions (a–e).

1 to lose out (on sth)   a to belong / try and be a part of something

2 to drop out (of sth)   b to not get something good, when someone else does get it

3 to fit into (sth)   c to depend on, expect someone else to do something for you

4 to rely on (sth)   d to make something possible

5 to allow for (sth)   e to no longer take part in something, give up

4 Look at the words in the box and answer the questions. There are two extra words.

chief executive   stakeholder   counterpart   participant   co-founder   peer   graduate   entrepreneur   director

1 Which two words mean 'a person of the same type'?

2 Which three words mean 'a person who has invested time and/or money in an enterprise'?

3 Which two words mean 'a role in a company'?

5 Choose the correct option to complete the sentences.

1 She _____ on her assistant to keep her diary up-to-date.

   a built   **b** relied   **c** fitted   **d** pursued

2 Once you have invested money in the company, you will become a _____ .

   a stakeholder   **b** co-founder   **c** participant   **d** graduate

3 The company _____ out on a major deal because they didn't get their proposal in on time.

   a dropped   **b** lost   **c** ran   **d** found

4 I would describe Arthur as a/an _____ because he's prepared to take risks and has already started his own business.

   a peers   **b** directors   **c** chief executives   **d** entrepreneurs

5 The number of students who _____ out of university is higher than expected.

   a drop   **b** run   **c** fall   **d** conduct

6 Lidia is my _____ in the Milan office. Our jobs are almost identical.

   a executive   **b** director   **c** counterpart   **d** participant

FT

Science start-ups struggle to bridge the gender gap

Britain's economy loses out on £2bn a year through a dearth of female entrepreneurs

Sabine Bahn insists she is a scientist, not an executive. 'You don't have to be on the frontline. I was never CEO of any of my companies,' says the co-founder of a company that developed the first blood tests to aid the early diagnosis of schizophrenia and bipolar disorder. 'Do what you're good at,' she says.

Professor Bahn, who is director of the Cambridge Centre for Neuropsychiatric Research, is sitting in front of an audience of thirty peers from all over the world. In the room are people working in fluid mechanics, bioengineering and neuroscience. They are a diverse bunch, but share three things: they are all scientists, they are all women and [1]_____.

The course they are attending, EnterpriseWISE, is run by Cambridge Judge Business School and is in its fifth year. The point is to encourage more female scientists working in science, technology, engineering and mathematics (STEM) to start companies, says Shima Barakat, who designed the programme.

Ms Barakat says that about 30 percent of Cambridge's graduates in STEM subjects are women. 'But out of graduates in these fields who go on to start a business, only 7 to 9 percent are female,' she says.

The gap between female and male business owners in the UK has narrowed in the past few years, but [2]_____. The current proportion of self-employed women in STEM is just 14 percent, according to data from the Office for National Statistics.

It is not only the unrealised ambition of potential entrepreneurs that is at stake. Innovate UK, a government agency, estimates that the lack of women in sciences and science entrepreneurship is causing an annual loss of £2bn to the British economy.

But enterprise training directed specifically at women working in STEM is still rare.

That could be a missed opportunity as female scientists find it more difficult to access both promotion in workplaces and funding opportunities as entrepreneurs than their male counterparts, says Ann Francke, chief executive of the UK's Chartered Management Institute.

Ms Barakat says that these kinds of challenges lead many of the relatively few women who pursue STEM research or entrepreneurship to drop out. In the UK, the proportion of female STEM graduates is 25 percent, but [3]_____.

The programme at Judge highlights the achievements of female science entrepreneurs and helps participants, who are used to being a minority within their field, to build a network of other female scientists.

Chiraz Ennaceur attended EnterpriseWISE last year, when she was a programme manager at the Welding Institute, a research and technology organisation for people working in welding and joining. Two months later she left her job and started her first business – [4]_____.

The first company quickly led to a second and, in April, the forty-one-year-old co-founded CorrosionRADAR, a business that is developing a sensor technology to reduce oil pipe corrosion.

'We are conducting field trials in the next twelve months,' Ms Ennaceur says, adding that her team aims to start selling services within eighteen months. The company is targeting oil and gas owners and has started talking to stakeholders such as Shell, BP and the Scottish government.

'Being a mechanical engineer in the oil and gas industry, I have been in male-dominated settings for most of my life, and the higher up I [went] the fewer women I met,' Ms Ennaceur says. She trained first in Tunisia, where she was born, and went on to study for a PhD in mechanical engineering in France.

Ms Ennaceur says she has never been aware of discrimination [5]_____. But spending time in a female-only setting made her view her day-to-day experience in a different light.

'The emotional experience of EnterpriseWISE was shocking,' she says. 'Being with other people in my field who were all women made me realise how, without noticing, I had been adapting to fit into a man's world.'

The language and culture she was used to in business consisted of pre-meeting sports talk, after-work drinks and the seeming absence of childcare responsibilities. 'Most of the men I worked with had children, of course, but it was clear that they relied on their wives,' Ms Ennaceur says.

In the end, the lack of women in her field was one of the reasons [6]_____. She is still managing a team of men, but says the workplace dynamic is more 'down to earth' now that she is chief executive, and allows for more multitasking. 'EnterpriseWISE helped me realise you can do things your way and be successful,' she says.

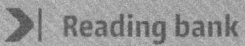

Unit 2 ❯

1 Read the article about the circular economy and decide which two sentences summarise it best.
   a The article is about companies that make cheap disposable goods trying to find ways to produce less waste.
   b The writer is not optimistic about the circular economy because he believes that companies won't be able to recycle or refurbish goods and still make a profit.
   c The writer believes that young consumers are creating pressure to make companies come up with ways to help people shop with an environmental conscience.
   d The writer has little hope that we will be able to solve the problems of packaging and waste.

2 Read the article again and choose the best answer (a, b or c) to the questions.
   1 What is Ikea planning to do to become more environmentally friendly?
      a recycle all their furniture and kitchen units
      b start renting out furniture and kitchen units
      c make better quality and more long-lasting goods
   2 What used to be kept over a decade ago for twice as long as is done today?
      a furniture
      b cars
      c clothes
   3 What could make a company like Ikea go into liquidation?
      a only manufacturing strong, long-lasting products
      b making products from recycled materials
      c reusing and refurbishing goods
   4 According to the writer, what should throwaway containers be replaced by?
      a plastic containers
      b glass containers
      c reusable containers
   5 What is one of the disadvantages of the circular economy for the environment?
      a Some companies will go out of business.
      b Some products will be over-consumed and cause more pollution.
      c The effort that went into producing a product will be wasted.
   6 What is lost through recycling?
      a the work and money that was put into making the product
      b the raw materials used to make the product
      c the environmental damage caused by the manufacturing of the product

3 Find the words in the box in the text and match them with the definitions (a–i).

| **Verbs:** curb   exacerbate   lease   wander   **Nouns:** congestion   heirloom   occupancy   pitfall   velocity |

Verbs
   a to walk slowly past sth _____
   b to allow someone to use something for a period of time in exchange for payment _____
   c to make sth worse, e.g. a problem _____
   d to control or limit sth, especially sth not wanted _____
Nouns
   e valuable object you are going to inherit _____
   f speed _____
   g the act of using or occupying a place (e.g. a house, land or method of transport) _____
   h danger or difficulty, often hidden _____
   i the state of being crowded (e.g. with lots of traffic) _____

4 What is the missing prefix in these words from the text? Which two words are related to throwing away?
   _____ play (noun)   _____ close (verb)   _____ pose of (phrasal verb)   _____ card (verb)

5 Complete the sentences using words from Exercises 3 and 4.
   1 If you want to _____ of unwanted clothes, you can donate them to a charity provided they are in good condition.
   2 The growing availability of cheap goods will _____ the problem of waste worldwide.
   3 There are more advantages than _____ when it comes to buying an electric car.
   4 It can take over an hour to drive across the city because of heavy _____ during rush hour.
   5 Scientists have underestimated the _____ of climate change, which appears to be much faster than initially predicted.
   6 You should try to _____ your impatience with new recruits as it won't help them learn the job any faster.

Ikea furniture does not need to fall apart

The most circular thing about Ikea, the Swedish furniture retailer, has traditionally been the path that it makes customers follow through its superstores to find the goods they have driven there to buy. As they wander along its displays, Ikea wants them to spot other decorations and take them home too.

The 'circular economy' now means something else: the reuse and repurposing of products in different ways. Ikea disclosed this week that it not only wants to recycle more furniture, but plans a trial in Switzerland this year to lease desks, chairs and perhaps kitchens. Instead of acquiring furniture cheaply and later throwing it away, customers might lease it for a while and then upgrade, with the old pieces being refurbished for other users.

No one buys Ikea furniture to pass on to the next generation – it is rarely moved from the spot where it is put together. That is not the point – like fast fashion clothing and other goods made in China, it has been cheap enough to treat as disposable rather than as an heirloom.

But even Ikea shows signs of doubting whether this approach can endure.

The velocity of consumption has steadily risen, partly because companies such as Ikea make buying stuff easy. Sofas and televisions were once hefty household investments but can be bought cheaply now. About a hundred billion garments are made each year – fourteen for each person – and they are kept for only half as long as fifteen years ago, the consultancy McKinsey & Co estimates.

This causes a lot of damage. Each person in the world draws about ten tonnes of raw materials from metals to biomass annually into the economy to support consumption and production, according to the Ellen MacArthur Foundation, which advocates a circular economy. Much of it will end up as waste, given how hard it is to reuse – only 14 percent of plastic packaging is taken for recycling, and far less actually recycled.

Few companies would stay in business if they only made goods that lasted a lifetime. But plenty can do more to limit repetitive consumption. They have selfish motives to try, as Ikea and others are doing: young consumers enjoy buying things but many are environmentally conscious (at least in theory) and are repelled by waste.

Companies can start with packaging, too much of which is made from complex plastics that ends up in landfill or the world's oceans.

More containers should be refillable, like the glass bottles brought to my door by our milkman, and SodaStream®'s carbon dioxide gas canisters for bubbly water.

Packaging is only one of the excesses in the way that products are marketed and consumed. Not only are things bought and disposed of rapidly, but many are used sparingly while their owners have them – in Europe, the average car is parked 92 percent of the time and 31 percent of food is wasted, McKinsey estimates.

Consumers can learn a lesson from the way companies often lease equipment and goods, paying by usage rather than for objects themselves. That applies to photocopiers made by Kyocera and Xerox™, while the flooring company Desso leases office carpets – cleaning as well as fitting.

More things could be rented by individuals, as technology has encouraged. The internet makes it easier to share occupancy of cars and apartments through Uber and Airbnb, and people subscribe to music and other digital services rather than buying discs.

People lease cars for three or four years and there is no reason why more durable goods, including furniture, cannot be rented. Not only does it limit waste but it gives companies an incentive to make things sturdily – better materials would require fewer repairs.

The circular economy has pitfalls, notably the so-called rebound effect: the easier it is to use products, the more intensively this will happen. That is not a problem for furniture, but sharing cars can exacerbate congestion and pollution rather than curbing it.

But the reuse and refurbishment of goods has enormous benefits compared with things being sold once, used for a time and then dumped. At best, recycling involves breaking things into raw materials and, in effect, discarding all the investment and labour that went into their making and marketing.

If Ikea aspires to become circular, there is no reason why others should not follow. It will be difficult to reform the consumption habits of the past few decades but all of us – consumers and companies – can try.

Unit 3 ❯

1 Read the article about saving money and decide which of these statements (a–d) is the best summary of it.

   a The 'Fire' movement reveals the secret of achievable early retirement.

   b The path to future success and happiness can be found in making extreme savings and spending the absolute minimum every month.

   c Although saving is a good idea, trying to live on very little money as proposed by the 'Fire' movement is unrealistic for most people.

   d Trying to save money when you're young is pointless and leads to long-term unhappiness and a life where all you ever do is work.

2 Read the article again and answer the questions.

   1 What does 'Fire' stand for?

   2 What three problems does the writer think make 'Fire' very difficult to achieve?

   3 What habit does the writer suggest you get into that will help you save money?

   4 Why does the writer think it's worth paying for someone to clean her house and deliver her groceries?

   5 What do minimalists hate?

3 Look at the words and phrases from the text and answer the questions.

> parsimony   penny-pinching   frugality   subsisting on next to nothing   thrifty

   1 What do the words have in common?

   2 Find two more phrases in the text that belong to this category.

4 Match (1–6) with (a–f) to make phrases and expressions from the text. Then find them in the text to check.

   1 keep   **a** a thing or two

   2 be music   **b** nature

   3 stay   **c** in the black

   4 be second   **d** a lid on sth

   5 learn   **e** a message to sb

   6 spread   **f** to sb's ears

5 Choose the correct option in italics to complete the sentences.

   1 As students, they were used to surviving on *frugality / next to nothing*.

   2 It's fine to be thrifty, but your *parsimony / penny-pinching* ways are very unattractive.

   3 Your decision to open a new savings account is *magic / music* to my ears.

   4 If you don't *go / stay* in the black, you'll have to pay interest on a loan or overdraft.

   5 We are so used to being careful with money, it's *first / second* nature to us.

   6 The government is trying to *keep / spread* the message about the advantages of opening pension schemes.

The 'Fire' movement and the trouble with penny-pinching

Parsimony is a way of life for those who aim to retire in their thirties and forties – but how practical is it?

If you haven't heard of the Fire movement, you soon will. Originating in the USA, its followers are devoted to extreme forms of money saving to achieve 'financial independence' and 'retire early' (FIRE).

Fire followers budget like crazy and target savings of up to 70 percent of their annual income, which (crucially) they invest for the long term. Once their savings pot has hit the desired level – typically thirty years' worth of living expenses – they keep it invested in low-cost tracker funds, withdrawing no more than 4 percent every year in the hope they will never extinguish their capital.

Having spent years subsisting on next to nothing, their thrifty habits are so ingrained that keeping a lid on day-to-day living costs is second nature.

Young people all over the internet are earnestly raving about how marvellous Fire is, but it's essentially rebranding retirement saving (which, let's face it, sounds terribly dull) as an empowering lifestyle choice.

It is music to my ears to hear of young people getting into the savings habit early – all the more time for their investments to compound – but I feel compelled to throw a little fat on the fire.

The first problem is property. Most young people I know would struggle to save even 20 percent of their annual income, as they spend more than 50 percent of it on rent. Others manage to save – plenty still live with their parents – but intend to use their pot as a property deposit, rather than a retirement fund.

The second issue is having children – good luck staying in the black when that happens.

And the longer-term worry is the maths. Catchy concepts such as the '4 percent rule' look increasingly optimistic as global growth stutters, and bond yields turn negative. I'm a fan of investing in low-cost trackers, but I'm not expecting this performance to be repeated over the next ten years.

For all of these reasons, I think everyone needs to get 'fired up' about spending less and saving more to give themselves more options in later life. Saving 70 percent and retiring in your 40s might not be achievable; saving a bit harder and retiring in your 60s still could.

My final beef is the frugality. If you take money saving to the extremes that some Fire devotees claim to, then the whole thing becomes self-defeating.

There is no way that some of the 'young retired' stars of Fire videos on YouTube could manage this lifestyle if they had a full-time job! The money-saving tactics they employ are often extremely time-consuming – walking everywhere, making their own clothes and boiling up endless pots of low-cost pulses.

They might have 'retired' in their thirties – but it sounds like they've just swapped one form of drudgery for another.

I save as much as I can, but examining my own spending data (a thrifty habit I recommend) shows exactly where I sacrifice money saving to buy greater convenience. This includes paying someone to clean our flat; taking a taxi to cut a journey time in half; having online groceries delivered. The time this saves me makes it worth the money.

But consumer habits are increasingly shaped by the cost to our planet, not just the cost to our pockets.

In the old days, we may have tried to hide our penny-pinching ways. Today, virtue signalling has made being thrifty fashionable.

Another closely related YouTube tribe are the minimalists, to whom clutter is anathema. There are some great money and planet-saving lessons in the '12 things I've stopped buying' video by vlogger Conni Biesalski.

Fast fashion is number one on her 'no shopping list'; physical gifts are number two. She prefers to give the gift of spending time on shared experiences, arguing this is a far better way of sustaining a friendship than a scented candle (I'll second that).

Policymakers and financial providers could learn a thing or two from the Fire movement's appeal. Make saving into pensions and Isas attractive by keeping the rules as simple as possible; speed up initiatives such as the pensions dashboard that will better enable savers to plan for the future, and think of snappier ways to spread the savings message to young people.

Unit 4 >

1 Read the article about inventions and choose the best subtitle for it.

   a A Tokyo museum puts together an international collection of technologies that changed the world
   b A Tokyo museum is providing the key to successful technological advances
   c A Tokyo museum is trying to identify the technology that has changed the way we live
   d A Tokyo museum submits a definitive classification for technological innovation
   e A Tokyo museum is creating an objective and unbiased collection of technological advancement

2 Read the article again and choose the best answer (a, b, c or d) to the questions.

   1 What does the writer think of the museum's choice of technological advances?
      a It is interesting and unusual.   **b** It is weird and controversial.
      c It is balanced and objective.   **d** It is ordinary and dull.

   2 What sort of things can be chosen for inclusion in the museum?
      a gadgets and machines only   **b** inventions and advances in medicine, industry and technology
      c advances in heavy-industry and technology   **d** any global industrial advance that has changed lives

   3 Which of these inventions is famous for being very sturdy and unlikely to break if dropped?
      a Sony's robot dog.   **b** the Yamaha® D-1 electric organ
      c the Sony Walkman™.   **d** the Casio™ G-Shock watch

   4 Which of these is particularly popular with visitors?
      a the VHF antenna   **b** the Motoman® industrial robot
      c the TR-808 rhythm composer   **d** extrusion-moulded joints for 500kV cables

   5 How does the writer see the list improving in the future?
      a It will become more evenly distributed between hardware and software.   **b** It will include fewer crowd-pleasers and be more objective.
      c It will become more subjective and the standard of entries will rise.   **d** Eventually, it should equip us with a means of quantifying the effect of technology on humanity.

3 Find the adjectives (1–5) in the text and match them with the definitions (a–e).

   1 ever-expanding   **a** difficult to find, define, achieve or answer
   2 coveted   **b** essential, important because other things depend on it
   3 pivotal   **c** becoming increasingly bigger
   4 astounding   **d** very surprising, amazing
   5 elusive   **e** popular, wanted, desired

4 Complete the phrases with a verb in the box. Then find them in the text.

assume	claim	play	prize	qualify	quibble	represent	transform

   1 to _____ their spots on the roster   **5** to _____ an era
   2 to _____ the concept of disruption   **6** to _____ to be the inspiration for a song
   3 to _____ a notable role   **7** to _____ hardware over software
   4 to _____ for a place in the history books   **8** to _____ with some entrants

5 Complete the sentences using words from Exercises 3 and 4 in the correct form.

   1 They are the most _____ devices because they are the ones that everybody wants.
   2 I don't think they can _____ that their computers are the most best on the market as their competitors have produced an even more innovative device.
   3 The solution to the problem is so _____ I don't think we'll be able to find it.
   4 Although the prototype is interesting, it isn't user-friendly and we don't think it _____ for further development.
   5 I don't want to _____ with your choices, but they do seem rather odd.
   6 The recently appointed director has now _____ his position on the board.
   7 The new chairman has _____ the company from an old-fashioned firm into contemporary, forward-looking business.

Business Partner C1   © Pearson Education 2020

FT

Walkmans™ and world firsts: Japan's gadget hall of fame

What do a nineteenth-century placard cast from pig iron, Sony's discontinued-then-resurrected robot dog and the Yamaha® D-1 electric organ have in common?

The answer, of course, is that they have all now assumed
5 their spots on the roster of Japan's Essential Historical Materials for Science and Technology – that country's quirkily compiled, ever-expanding hall of fame for the contraptions that really count.

Without ever quite saying it, this list represents the
10 concept of disruption before Silicon Valley co-opted the word. Only the best Japanese gadgets need apply. The coveted places are awarded, at the rate of about twenty per year, by experts at Tokyo's National Museum of Nature and Science.

15 There are several categories of qualification, any one of which can put an era- or industry-defining piece of tech in contention. To make it on the list, a Japanese gadget (or medicine or industrial advance) must have played a notable role in improving people's way of life or
20 creating new ways of living; it must represent a pivotal moment in scientific or technological development; it must represent an important moment in the relationship between society and tech; it must demonstrate a 'uniquely Japanese scientific or technological development from
25 an international perspective' – i.e., to have beaten the Americans, Brits or Germans to some coveted beachhead of postwar progress.

The 285 positions that have been awarded since the hall of fame opened in 2008 are held by a fabulous range
30 of inventions that date from the start of Japan's great modernisation. This year's entrants include the Casio™ G-Shock watch (the Casio pocket calculator was inducted some years ago). It qualifies for a place in the history books for its 'astounding shock resistance'.

35 Some inventions, such as the 1882 cement-grinding mill, transformed an era when 'tech' was something measured in tonnes and involved iron axles the size of a tree. Others, such as the PCM processor that essentially made home video possible, were the invisible progenitors of the
40 digital age.

There are plenty of obvious crowd-pleasers – such as the Motoman® industrial robot. The Sony Walkman™ was the 109th item to be inducted on to the list, and quite obviously belongs as a 'changing the way we live'
45 contender: what other gadget, after all, can claim to be the inspiration for a song (*Wired for Sound*) by Cliff Richard? The list is consistently surprising and endlessly worthy of perusal. It's very easy to enjoy it for what it is. But you can also read it as a tightly defined history of the past 150
50 years of Japanese technology – and a further reminder of how Japan prizes hardware over software.

There are numerous world firsts, including the first inverter air conditioner for home use, the first VHF antenna, the first radio phone and the first mobile phone
55 with built-in camera. Separated by decades, each has a serious claim to have changed the way we live or, at the very least, to have inspired someone to have asked 'what on earth did we do before the …?'

It is possible to quibble with some entrants. The TR-808
60 Rhythm composer from 1980 makes it in for 'allowing freedom to program a rhythm pattern for an entire song and had a great effect on the music scene'. Other hall-of-famers, such as the extrusion-moulded joints for 500kV cables, require some fairly technical knowledge
65 to understand how they changed the world.

Perhaps inadvertently, the list also raises an elusive question: Is it ever possible to precisely measure and compare the impact of two completely different technologies? In its current form, it isn't quite equipped to
70 do so: it appears to be compiled using scientific standards but remains, in reality, a subjective, impressionistic painting of progress. It leaves open, however, the clear prospect of becoming much less so over time as the bar to entry becomes higher and as those selecting the new
75 entrants are forced to apply stricter standards. At some point, this list will provide a means to quantify the human impact of tech, just not yet.

Unit 5 ❯

1 Read the article about food quickly and find what the numbers refer to.

1	500,000	**6**	380 million
2	20	**7**	2011
3	7	**8**	150 million
4	10	**9**	223
5	more than 50 years ago	**10**	459

2 Read the article again and match the sections (A–F) with the statements (1–6).

1 There is a lack of clarity regarding what is permitted in innovative food production.

2 The world will be affected by how China plans to feed its population in the future.

3 A Chinese venture capital firm has invested in a company that uses tiny six-legged creatures to make food.

4 The USA is ahead of Asia in the production of new and revolutionary foods.

5 People are less likely to buy insect-based food if they can taste and see the insects.

6 There hasn't been any significant progress in food production since people began farming on a large scale in the 1960s.

3 Choose the best meaning (a or b) for the phrases in bold according to the text.

1 **Venture capitalists** are involved in …
   a investing money in new businesses.
   b investing money in well-established companies.

2 An **accelerator** is a group or company that …
   a invests in start-ups and helps them to grow.
   b focuses on encouraging disruptive innovation.

3 If you have a **minority stake** in a company, it means …
   a your own most of the shares.
   b you own less than half the shares.

4 **Seed funding** involves …
   a investing in the food and agriculture business in exchange for shares in the business.
   b investing in a company at the initial stages of its development in exchange for shares in the business.

5 A **sovereign wealth fund** is …
   a owned by the state.
   b owned by a group of individuals.

6 If a company makes an **overseas foray**, it means it …
   a has decided to suspend all international business
   b is becoming involved in business in another country.

4 What is the missing word in these phrases? Which item could make you sick if you ate it?

sustainable _____

_____technology sector

_____ safety issues

_____ contaminants

_____ security

_____ related start-ups

lab-grown _____

5 Find the words in the box in the text. Then answer the questions.

> alternative meat   arable land   beverage   chemical fertiliser   condiments   dietary requirements
> insect-based snacks   lab-grown meat   pesticide   plant-based burger patties

1 What can be eaten by vegetarians? _____ and _____

2 What are used to give flavour to food and make it tastier?_____

3 What do famers use to stop their crops being destroyed by certain insects and vermin? _____

4 What is added to the soil to make crops grow stronger? _____

5 What are needs connected to food and drink? _____

6 What do you call places that are suitable for growing crops for food? _____

7 What is another word for a drink? _____

8 What do you call small meals made from tiny creatures with six legs and a body divided in three parts? _____

9 What do you call meat produced from animal cells by scientists in a laboratory? _____

Asian investors begin to bet on bugs as the future of food

Ventures back silkworms and crickets in the race to meet region's dietary needs

A TOKYO – An emerging group of Asian venture capitalists are exploring whether insects and lab-grown meat can help meet Asia's growing appetite for food.

5 For Shanghai-based accelerator Bits x Bites, the search for sustainable food has extended to silkworms. Last year, the company invested in Bugsolutely, a local start-up that is developing insect-based snacks.

10 Silkworms are grown to spin silk, with the worms usually disposed of when they are done producing. Bugsolutely, which made its name selling cricket-based pasta, noted that the waste was rich in protein, vitamins and minerals. This prompted the company 15 to infuse snacks with silkworm powder in popular flavours like salt and vinegar.

B 'After a lot of consumer research, we started to learn that the younger generation doesn't want to see whole insect pieces,' said Matilda Ho, managing 20 director of Bits x Bites, in a recent interview in Tokyo. 'Also, since no one can taste the insect flavour … it's easy for people to adapt to the new concept. We are excited about how this product can bring traction to the market.'

25 Bits x Bites was launched in 2016 as China's first accelerator specialising in the food technology sector. Backed by local condiments company Shinho, it normally provides no more than $500,000 in seed funding to promising ventures in exchange 30 for minority stakes. It also supports their growth through mentorship programs and connecting them to potential investors.

C As Asia's population expands, so will the challenges in meeting its dietary requirements. 35 China has already overtaken the USA in the number of obese people, according to a 2016 study by *The Lancet* medical journal. The country has also been plagued with numerous food safety issues.

'China represents 20 percent of the world 40 population but less than 7 percent of its land is arable,' said Ho. 'How we feed the population for the next decade will have global implications.'

Snacks made from insects are only part of the equation. Bits x Bites has invested in ten start-ups 45 ranging from Alesca Life, which sells mini farms that fit in shipping containers, to Inspecto, a developer of portable devices that detect pesticides and other food contaminants. They are all tackling an industry that observers say has been slow to embrace cutting-edge 50 technology.

D 'More than half a century ago, the "green revolution" solved the food security issue with chemical fertilisers and pesticides that enabled mass production,' said Satoshi Koike, President of Tokyo-55 based farming technology start-up Vegetalia. 'Since then, there has been no innovation in the way we produce food.'

Food-related start-ups have taken off in the USA, where preference for organic food has spread. 60 California-based Impossible Foods, which sells plant-based burger patties that taste like real meat, has raised more than $380 million since its establishment in 2011. Tyson Foods, the largest meat packer in the USA, set up a $150 million venture arm 65 in 2016 to invest in start-ups that develop alternative meat.

E The number of investors worldwide backing food and beverage start-ups doubled from 223 in 2015 to 459 in 2017, according to CB Insights. While some 70 prominent Asian investors like Singapore sovereign wealth fund Temasek Holdings and Li Ka-shing, one of China's richest men, have invested in Impossible Foods, the region lags behind the USA, where large food companies have begun actively pursuing the 75 latest innovations.

It remains uncertain whether the funds can create homegrown champions that can compete against well-funded U.S. rivals, many of which are eyeing Asian markets. Impossible Foods recently entered 80 Hong Kong by tying up with local restaurants, marking its first overseas foray outside the USA.

F Regulation is another obstacle. Insects, for example, are generally not approved for consumption in most Asian markets, and regulations concerning 85 lab-grown food are even less clear. Even the U.S. Food and Drug Administration recently announced that it will hold a meeting in July to share 'initial thinking' about how it will regulate food made from harvesting animal cells.

90 'I don't think we will be able to see everyone starting to eat organic in the next decade,' concedes Ho. 'It's going to be a long-term game.'

Unit 6 >

1 **Read the article about tourism and answer the questions.**
  1 What event caused the Italian government to ban cruise liners from the centre of Venice?
  2 What two things do cruise ships do which cause damage in Venice?
  3 Why are protestors doubtful about the amount of money defenders of cruise ships claim passengers spend in Venice?
  4 What two things do companies like Airbnb do to harm cities like Venice?
  5 What has been done in Paris and Barcelona to help protect these cities from damage caused by tourism?

2 **Read the article again and decide which sentence best describes the views expressed in it.**
  a Tourists should be charged to visit cities, transport should be banned in city centres, the cost of travelling should rise, and there should be a limit to the amount of tourist accommodation available.
  b There should be better planning for mass tourism in heavily visited cities, more stringent regulation with regard to transportation and accommodation, as well as tourist taxation to help with maintenance and to prevent damage.

3 **Find the adjectives (1–6) in the article and match them with their opposite meanings (a–f).**

  1 savvy   **a** ready, well planned or organised
  2 ill-prepared   **b** modest and prim and proper
  3 transient   **c** unheard of, unknown
  4 saucy   **d** ignorant, inexperienced
  5 fabled   **e** permanent

4 **Find the verbs in the box in the text and use the context to choose the best answer (a, b or c) to the questions.**

| to lose out on sth   to blot out sth / to blot sth out   to venture on to sth   to stick around |
| to mitigate sth   to stir sth   to levy sth   to penalise sb for sth |

  1 Which of these can't you lose out on?
    a promotions   **b** profits   **c** redundancies
  2 What wouldn't you normally blot out willingly?
    a good memories   **b** an ugly view   **c** sad thoughts
  3 If you venture on to something, what does it usually involve?
    a no risk   **b** an element of risk   **c** boredom
  4 If you don't stick around somewhere, what have you done?
    a left   **b** stayed where you are   **c** given up
  5 Which of these don't you usually mitigate?
    a disaster   **b** risk   **c** prosperity
  6 What isn't usually stirred?
    a controversy   **b** benefits   **c** emotions
  7 Which of these aren't levied by governments?
    a taxes   **b** fines   **c** salaries
  8 What are people usually penalised for?
    a breaking laws   **b** sticking to the law   **c** following rules

5 **Complete the sentences using words from Exercises 3 and 4.**
  1 Although the government has tried to _____ poverty by creating more jobs, there hasn't been an increase in wealth yet.
  2 They would like me to _____ all memories of the disaster and start a new life.
  3 Many cities have _____ populations of students who come for a few weeks at a time.
  4 The government is planning to _____ a new tax on diesel cars, which will make them less economical for the consumer.
  5 Removing these benefits will _____ the most disadvantaged in society and increase the gap between the rich and poor.
  6 She's a _____ politician who knows what she's doing and has a lot of experience.
  7 The _____ city of El Dorado is supposed to have existed in South America, but it has never been found.
  8 We decided not to _____ on to the ice because we weren't sure how thick it was.

FT

Venice cruise crash is a sign of the risks of tourism

Proper infrastructure and planning will keep cities from losing out on the economic benefits

Venice has a history of successfully managing threats from the sea. Throughout the Renaissance, the Most Serene Republic fought the Ottoman navy for dominance in the Mediterranean. Turkish galleys and galliots have
5 long since given way to a new threat: cruise liners which tower over the City of Masks. After a collision involving the MSC Opera, the Italian government has ordered these floating tubs of fun to steer clear of the city centre.

Mass tourism is not a new challenge for Venice or
10 other cities, though it is a stiffer one. The U.S. trade war may have contributed to a slowdown in Chinese tourism overseas in the short term, but the country's travelling middle class are here to stay. Ever larger packs of camera-toting, trinket-purchasing tourists of all nationalities bring
15 rewards to savvy cities, and risks to the ill-prepared.

Leviathans such as MSC Opera, nearly 300 metres long and with a maximum capacity of more than 2,500 passengers, do more than blot out the skyline. As early as 2006, protesters warned the volumes of water they
20 displace are eroding the sinking city's foundations and threatening the local ecosystem. In 2017, cruise ships in European waters emitted more than 60 kilotonnes of sulphur dioxide, a chemical compound linked to acid rain and lung cancer, according to a report released in June.

25 There are also doubts about the benefits their passengers bring to Venice. While the cruise lobby claims they inject €280 million a year into the local economy, activists argue that many choose to eat and drink aboard their ship, only venturing on to land for a matter of hours.

30 The problems caused by tourism extend beyond floating hotels and their passengers. Upstarts such as Airbnb pose an existential threat to the hotel industry and can cause disruption in the housing market. People renting their properties do gain a financial benefit. The
35 danger is if housebuyers simply flip the purchase: renting out their property rather than sticking around to help build a community.

There are several ways to mitigate the negative impacts of tourism. Size restrictions and tighter emission
40 controls on cruise ships would reduce both the potential environmental hazard and limit the number of tourists entering the city through this route.

In Venice, the entry charge for those on day trips, implemented in May 2019, should also be maintained,
45 despite the controversy it stirred. The city already levies a tourist tax on those who stay in hotels, as do many other cities. It is unclear why more transient visitors should get special treatment. Their time in the city, using public facilities and spaces, generates the need for cleaning and
50 repairs.

Finally, Venice should look at measures taken in other tourist hotspots to protect their unique character. In Barcelona, the government has stopped granting new hotel licences. The deputy mayor of Paris has called for a
55 ban on coaches in the city centre, suggesting tourists take more eco-friendly routes which cause less congestion. These steps are not about penalising visitors. They are about preserving the authentic nature of the city.

The rise of cheap international travel has opened up
60 opportunities to millions. This has not been a universal boon. The collapse of some British coastal resorts speaks to changing travel patterns and changing tastes. A saucy seaside postcard and a plate of fish and chips are not enough these days. People want an experience and that
65 means exploring the unfamiliar, if not the exotic. Digital media has made the world a lot smaller. What has not changed is the cost of keeping the fabled cities running. Therein lies a Venetian lesson.

Unit 7 ❯

1 Read the article about bullying in the workplace quickly and match the people's names (1–6) with the types of company/institution they represent (a–e). Two people work for the same type of company/institution.

1 Amy Edmondson	**a** Parliament
2 Laura Cox	**b** Consulting firm
3 Linda Aiello	**c** Technology company
4 Emilie Colker	**d** Design company
5 Wim Vandekerckhove	**e** Higher education institution
6 Wendy Addison	

2 Read the article again and answer the questions.
1 Why did nurses make fewer mistakes even though they didn't have very positive feelings about their workplace?
2 What problem does Susan Fowler's case illustrate?
3 What aspect of people's achievements should be analysed before they are promoted to managerial positions?
4 Why are 360-degree-style appraisals sometimes unreliable?
5 What kind of training is recommended to help workers deal with bullying?

3 Read the definitions of verbs that can be examples of bullying. Then find a verb in the article for each definition. The first letters have been given to help you.
1 to make someone or something seem small or unimportant – b_____
2 to describe someone or something as a very bad type of person or thing, often unfairly – b_____ (somebody as a troublemaker)
3 to make someone's life unpleasant, for example by frequently saying offensive things to them or threatening them – h_____
4 to make someone feel ashamed or stupid, especially when other people are present – h_____
5 to refuse to accept someone as a member of a group – o_____

4 Find the nouns (1–5) in the article and decide which of them is not a person. Then match them with the definitions (a–e).

1 whistleblower	**a** the quality of being honest and telling the truth, even when the truth may be unpleasant or embarrassing
2 candour	
3 underling	**b** an insulting word for someone who has a low rank – often used humorously
4 high-flier	**c** someone working for an organisation who tells the authorities that people there are doing something illegal, dishonest, or wrong
5 bystander	
	d someone who watches what is happening without taking part
	e someone who is extremely successful in their job

5A Complete the phrasal verbs with a word or phrase in the box. Then find the verbs in the text.

out   to   up   up to

1 pander _____   2 root _____   3 dry _____   4 kiss _____

5B Match the phrasal verbs in Exercise 5A with the definitions (a–d).
a to come to an end (e.g. of a supply)
b to try to please someone in order to get them to do something for you
c to find out where a particular kind of problem exists in order to get rid of it
d to give someone anything they want in order to please them, even if it seems unreasonable or unnecessary

6 Choose the correct option in italics to complete the sentences.
1 As a manager, don't be afraid to speak openly to your team about their performance – always discuss it with them with *underling / candour*.
2 Our competitors have paid out significant sums to silence would-be *whistleblowers / high-fliers*.
3 Instead of boosting her employees' morale, she tends to speak to them in a way that *harasses / belittles* their importance.
4 If the prime minister doesn't intervene, foreign investment may *dry up / root out*.
5 We can't create an ideal customer for our products; all we can do is *pander / bystander* to the customers we have.

FT

Nurture a workforce that values ideas and contributions

Banish workplace bullying and strengthen your organisation

While investigating the relationship between nursing performance and team working, Harvard academic Amy Edmondson made a curious discovery. Her data showed that the lower a team's morale, the fewer errors its nurses
5   made.

That was surprising – until the penny dropped. Nurses who were constantly criticised and belittled by managers had merely learnt to hide mistakes.

That discovery in the 1990s, which Professor
10 Edmondson describes in a new book *The Fearless Organization,* prompted her research into 'psychological safety'. When workplaces are ruled by fear, she says, workers stop learning, innovation dries up and pandering to power replaces candour and useful debate.

15 Organisations may be aware of the damage caused by bullying, but it is hard to root out. Whistleblowers regularly report difficulties in highlighting problems, while a new study by City & Guilds Group, which promotes skills in the UK workforce, reveals that bullying is widespread. The
20 research found that 52 per cent of UK staff employed by large global organisations have encountered bullying.

Many organisations publish dignity and respect policies. Yet, as an independent inquiry into bullying and harassment of House of Commons staff highlighted last year, policies
25 are worth little without institutional commitment. The UK report, led by Dame Laura Cox, said that people remain silent because they fear being 'disbelieved', 'ostracised' and 'branded a 'troublemaker'.

They may also lack faith in the reporting process and
30 doubt the organisation's willingness to act, especially if the bully or abuser is considered a high-flier. That was Uber engineer Susan Fowler's experience after she reported her manager to HR for alleged sexual harassment.

Taking action over managers who humiliate or harass
35 others sends a message to the whole organisation. It is also a reminder to selection panels to pay heed to the character of those they promote into management.

For some organisations that may mean looking at how star performers achieve their success, says Linda
40 Aiello, who heads international HR at Salesforce, the U.S. technology company. 'As [people] go through the organisation, "the how" and the behaviours that surround [achievement] become more and more important.'

So-called 360-degree-style appraisals that gather
45 feedback from team-mates and juniors should in theory stop the rise of colleagues who kiss up to superiors and kick down underlings. Yet allegations by former Facebook employees that staff forge opportunistic friendships around appraisal time, highlight that such practices can be gamed.

50 Emilie Colker, executive director at international design company Ideo, recommends collecting feedback continuously, not just at bonus time, and noting who co-workers want to work with on their projects. Over time it becomes apparent that 'certain people are asked for a lot'.

55 At the same time, Wim Vandekerckhove, reader in business ethics at the University of Greenwich, recommends systematically recording complaints and concerns. 'If a pattern becomes visible it may be that the manager needs an additional skill, or it may be that being
60 a manager isn't for them.'

To build a high-achieving culture based on teamwork, Ideo emphasises via statements and promotion policies that employees' success will also depend on helping others. 'Even if you're a rock star you still won't progress unless
65 you're making other people successful,' Ms Colker says.

Employers can help teach workers how to challenge mistreatment, says Wendy Addison, founder of consultancy SpeakOut SpeakUp. 'Bystander training', for instance, teaches how to be an ally when others need help, and how
70 to ask co-workers for back-up.

Unit 8 >

1 Read the article about female entrepreneurs in Asia quickly and match the numbers (a–e) with the things (1–5) they refer to.

a	100 million tugrik	**1**	the percentage by which Lhamour's earnings went up in 2016
b	2014	**2**	the percentage of value added tax and income tax charged in Mongolia
c	83 percent	**3**	the amount of money raised by Lhamour to finance new products in September
d	sixty million	**4**	the number of animals kept by nomadic herders in Mongolia
e	10 percent	**5**	the year Lhamour was launched

2 Read the article again. Complete the gaps (1–6) with the phrases (A–F).

A and aims to build a laboratory to study raw materials

B referring to small and medium-sized enterprises

C they are a good example for this industry

D that is rapidly attracting attention from foreign buyers

E employing only part-time workers

F which employs 38 full-time workers at a factory

3 Complete the expressions with the words in the box. Then find them in the text.

> asset   distribution   duties   exempt   overheads   raise   source

1 to utilise an _____ (in order to create a product)

2 to _____ revenues by … percent

3 to _____ materials

4 to eliminate customs _____ on sth

5 to keep _____ low

6 to reduce _____ costs

7 to _____ a company from sth (e.g. a tax)

4 Find the adjectives (1–6) in the article and match them with the definitions (a–f).

1	nomadic	**a**	following what is acceptable to society; conservative, traditional
2	value-added	**b**	travelling from place to place
3	nascent	**c**	beginning to exist, new
4	comparable	**d**	beginning to become successful (often used to describe an artist or writer)
5	budding	**e**	similar, equivalent
6	conventional	**f**	describes sth that is worth more because it has been improved in some way

5 Choose the correct option in italics to complete the sentences.

1 One way for the company to *keep / reduce* its overheads low is to use home workers rather than bringing people into an office or factory.

2 There are many *nascent / budding* entrepreneurs who would like to start their own business and make their fortunes, but only a few will succeed.

3 The business owners should consider how best to *raise / utilise* the assets that they have.

4 We have been unable to *source / raise* enough raw material to produce our new range of products.

5 Could you show me *comparable / conventional* figures for the same period of time last year?

6 In this country, charitable foundations are *eliminated / exempted* from paying income tax.

Mongolian women drive rise in organic cosmetics

Entrepreneurs seek to expand overseas sales of animal-based oils and creams

A group of female entrepreneurs is developing a new use for the tens of millions of yaks, goats and sheep on Mongolia's vast steppes – a range of organic cosmetics and skin creams, often based on traditional recipes,[1]_____.

Lhamour, the biggest of several companies launched to exploit the potential of locally made organic cosmetics, raised 100 million tugrik ($53,000) in September from a bond issuance launched to finance new products, and is already exporting to Taiwan, South Korea, Hong Kong, Singapore and Belgium.

The company, established in 2014, makes soaps from yaks' and goats' milk. Its animal-based products are widely available in retail outlets in Ulaanbaatar, but Chief Executive Khulan Davaadorj said significant investment was needed in animal husbandry to allow the industry to grow.

'We need investment, innovation and technology to utilise this asset [livestock] to create value-added products that can compete internationally,' said Khulan, thirty, who founded Lhamour after resigning from a job in Mongolia's wind power industry.

Fast-growing Lhamour, [2]_____ and office building in Ulaanbaatar, remains small by the standards of the international cosmetics industry. But the company raised revenues by 83 percent in 2016 to $220,000, and is targeting a further increase in the near future.

Half of Mongolia's three million people are nomadic herders, and their sixty million livestock offer many potential opportunities for the development of organic products. However, sourcing materials remains a significant problem for the nascent cosmetics industry.

'It is difficult for us to source quality raw materials on a constant basis as many of our suppliers have never supplied in [this way],' said Khulan. 'Mongolian animal husbandry is still very traditional [and] suppliers have a traditional mindset.'

Lhamour's success has prompted the emergence of a number of other companies making organic cosmetics.

Battsetseg Chagdgaag, the thirty-five-year-old co-founder of Gilgerem, another organic soap maker, began selling soaps made from natural materials such as sea buckthorn (an oily berry) in 2016. Battsetseg said she is also about to launch a soap made from camels' milk, having been inspired by Lhamour and Goo, another organic skin care company that started operations at about the same time as Lhamour.

'I am proud of [Lhamour and Goo]; I respect them,' she said. 'We call them the older sisters, and [3]_____. They spent their energy and finances to make everyone understand these organic handmade soaps, which paved the way for my business.'

Battsetseg, who sells her soaps for $2–$3 each – half the price of comparable Lhamour products – keeps overheads low by operating from a two-room basement, [4]_____, and selling her products in local supermarkets to reduce distribution costs.

Battsetseg urged the Mongolian government to help promote entrepreneurial companies such as hers by exempting them from the country's 10 percent value added tax and from income tax, also levied at 10 percent. She said the government could also help by reducing or eliminating customs duties on some imported ingredients required to make her products.

'Our soaps could have been sold even cheaper and could compete against imported soaps if the government exempts some of the taxes and really supports SMEs,' she said, [5]_____.

All the organic cosmetics entrepreneurs said that much more research was needed to develop the industry. Battsetseg, who was a previously a journalist, marketing manager and fashion stylist, said she is 'addicted' to researching chemicals [6]_____ so she can produce more unique products.

Commentators say there is no shortage of budding entrepreneurs in Mongolia, suggesting that the initial success of the organic cosmetics industry may draw in more competitors, especially if export sales continue to grow.

'Young and educated Mongolians are increasingly going into start-ups, and are disinterested in becoming conventional salary men and women [because of low wages] in both private and public sector,' said Gerel Orgil, CEO of East Maven, a public relations agency based in Ulaanbaatar which was one of Mongolia's most successful business start-ups in 2012.

Unit 1

1 **1** Professor Sabine Bahn **2** EnterpriseWISE **3** Innovate UK **4** Ann Francke **5** Chiraz Ennaceur **6** CorrosionRADAR
2 **1** C **2** F **3** E **4** D **5** A **6** B
3 **1** b **2** e **3** a **4** c **5** d
4 **1** counterpart and peer **2** stakeholder, co-founder, entrepreneur **3** chief executive and director
5 **1** b **2** a **3** b **4** d **5** a **6** c

Unit 2

1 a and c
2 **1** b **2** c **3** a **4** c **5** b **6** a
3 **a** wander **b** lease **c** exacerbate **d** curb **e** heirloom **f** velocity **g** occupancy **h** pitfall **i** congestion
4 *dis* is the missing prefix, dispose of = throw away, get rid of; discard = get rid of (both related to throwing away); display = putting something on show; disclose = make information (usually secret or new) known
5 **1** dispose **2** exacerbate **3** pitfalls **4** congestion **5** velocity **6** curb

Unit 3

1 c
2 **1** 'financial independence' and 'retire early'
2 trying to save when rents are high, starting a family/having children, the maths: low-cost trackers may not perform as well in the future
3 examining your own spending data
4 because it saves time
5 clutter
3 **1** They are all connected to saving money.
2 Two more phrases: money-saving tactics, savings pot
4 **1** d **2** f **3** c **4** b **5** a **6** e
5 **1** next to nothing **2** penny-pinching **3** music **4** stay **5** second **6** spread

Unit 4

1 c
2 **1** a **2** b **3** d **4** b **5** d
3 **1** c **2** e **3** b **4** d **5** a
4 **1** assume **2** represent **3** play **4** qualify **5** transform **6** claim **7** prize **8** quibble
5 **1** coveted **2** claim **3** elusive **4** qualifies **5** quibble **6** assumed **7** transformed

Unit 5

1 **1** the amount in dollars provided in seed funding to promising ventures in exchange for minority stakes
2 the percentage of the world population that is Chinese
3 the percentage of land in China where you can grow crops
4 the number of startups Bits x Bites has invested in
5 when the 'green revolution' solved the food security issue
6 the amount in dollars Impossible Foods has raised since it started
7 the year Impossible Foods was founded
8 the amount in dollars Tyson Foods contributed to a venture arm in 2016 to invest in start-ups that develop alternative meat
9 the number of investors worldwide backing food and beverage start-ups in 2015
10 the number of investors worldwide backing food and beverage start-ups in 2017

2 **1** F **2** C **3** A **4** E **5** B **6** D
3 **1** a **2** a **3** b **4** b **5** a **6** b
4 food; *food contaminants* could make you sick
5 **1** alternative meat and plant-based burger patties **2** condiments **3** pesticide **4** chemical fertiliser **5** dietary requirements **6** arable land **7** beverage **8** insect-based snacks **9** lab-grown meat

Unit 6

1 **1** A collision involving the cruise ship MSC Opera.
2 They move a lot of water about (which erodes the foundations of the city and threatens the local eco-system) and they emit sulphur dioxide.
3 Because cruise ship passengers tend to eat and drink on board and only spend a few hours in the city.
4 They damage the hotel industry and they disrupt the housing market.
5 Barcelona has stopped granting new hotel licences and Paris has called for a ban on tourist coaches in the city centre.
2 b
3 **1** d **2** a **3** e **4** b **5** c
4 **1** c **2** a **3** b **4** a **5** c **6** b **7** c **8** a
5 **1** mitigate **2** blot out **3** transient **4** levy **5** penalise **6** savvy **7** fabled **8** venture

Unit 7

1 **1** e **2** a **3** c **4** d **5** e **6** b
2 **1** In fact, they didn't make fewer mistakes, they simply didn't reveal them because they were bullied by their superiors.
2 Her case shows why some victims of bullying are unwilling to report the problem, especially when the bully is a high-flier. When she said she was being harassed by her manager, the company's reaction made her lose faith in it.
3 The way in which they achieved their success.
4 Because sometimes people manufacture results to score well – they befriend others without expressing the true motivation behind it. The other person might think it is genuine.
5 Training in helping others to protect themselves against bullying, e.g. 'bystander training'.
3 **1** belittle **2** brand **3** harass **4** humiliate **5** ostracise
4 **1** c **2** a **3** b **4** e **5** d
5A **1** pander to **2** root out **3** dry up **4** kiss up to
5B **a** dry up **b** kiss up to **c** root out **d** pander to
6 **1** candour **2** whistleblowers **3** belittles **4** dry up **5** pander

Unit 8

1 **a** 3 **b** 5 **c** 1 **d** 4 **e** 2
2 **1** D **2** F **3** C **4** E **5** B **6** A
3 **1** asset **2** raise **3** source **4** duties **5** overheads **6** distribution **7** exempt
4 **1** b **2** f **3** c **4** e **5** d **6** a
5 **1** keep **2** budding **3** utilise **4** source **5** comparable **6** exempt

1 › Reports

Lead in

The content and style of reports may differ from company to company. A business report is written in a formal style. It may contain information about financial performance and the successes and challenges that an organisation has experienced, or present research findings and give recommendations. A proposal is a type of report that formally proposes different recommendations as solutions to a problem or situation.

Reports can be long documents, so the main information is usually paraphrased at the beginning in the executive summary. In addition, sub-headings such as Introduction, Findings, Recommendations, Conclusion or Future outlook, etc. are also used to make the report easier to read.

Reports may be read by decision-makers, shareholders, or stakeholders outside of the company. It is therefore essential that key ideas are expressed clearly. It is also common in reports to use formal language, such as noun phrases, impersonal structures, language for making recommendations, as well as appropriate linking words/phrases.

Useful language

Organisation

Section	Function	Examples
Executive summary	The summary is found at the start of a report and it tells the reader what type of information the report will contain. It is usually short (less than 150 words). The summary can be written after the rest of the report is complete. In academic research papers, it is called the abstract.	This report looks at the ways that innovation can be improved in the workplace. This report explores global trends in the circular economy. Each year over 3000 small companies close due to problems with financing new ideas.
Introduction	States the purpose of the report and briefly outlines what the report intends to do.	The purpose of this report is to outline the advantages and disadvantages of using green energy and recommend ways to improve environmental strategy in line with company policy. The aim of the report is to look into the cause and effect of conflict in the workplace and suggest solutions for both employers and employees. This report looks into the main reasons for customer service complaints and proposes changes to staff training in order to resolve the issues.
Conclusion	Restates and summarises the main message of the report. Bullet points or numbered lists may be used to highlight key information.	In this report, we have looked at ways to improve productivity and outlined key areas which need to be improved. In conclusion, it is essential to develop the use of new digital technologies in training and development.

Main body of a report

	Examples
Describe problem, reason, result	The main problem is plastic waste, which is destroying sea life. The key issue is lack of housing to attract a skilled workforce. The new logo was unsuitable due to its style and colours. This could have a negative impact on brand loyalty. The company, however, has difficulties with cashflow and needs large amounts of capital to expand. They need more funding to continue. There is also cause for concern as they have no other projects in the pipeline.
Report findings	Our market research confirmed that our brand was immediately recognisable. The findings indicate that any delays should be minimal. None of the participants felt that the changes were positive. Over half those surveyed reported that they would be unlikely to invest in this start-up. Recent research has demonstrated that customers are influenced by opinion leaders in social media. There is a big gap in the market as there is currently no drug to treat dementia.
Make recommendations	To improve delivery times we should employ more drivers and change our main delivery routes. It is suggested that health and safety procedures are reviewed annually in order to reduce these risks. As a result, a number of recommendations have been made to use green energy in our factories. We strongly recommend adopting the proposed measures in the next six months. For these reasons, I would recommend investing in this start-up at the earliest opportunity. My recommendation is that we do make an investment in this technological start-up.

Research report

The content and style of reports may differ depending on the situation. For example, a research report would usually be written using a standard template designed for a particular field or R&D department. All reports should be written using a formal style and clear language. Longer reports may contain a summary, introduction/background, main body with findings, recommendations and a conclusion. It is important to use topic sentences to give each paragraph structure, and to use suitable linking words or phrases for cohesion. It is best to avoid repetition by using synonyms, pronouns or determiners. However, it can be useful to use repetition at times to emphasise and clarify key points or findings.

Model report

Biotech company with drug to treat dementia in phase two trials

Background

This is a company that was spun out of a university biochemistry department. Its founders are three PhDs. They are working on a drug that slows the progression of Alzheimer's (dementia) by slowing the production of beta-amyloid, a protein associated with the disease. Further research may lead to even better treatment and this lab is certainly a world leader in the field.

Our exit strategy would be the normal one for a biotech start-up – selling the business to a large pharmaceutical company. There will be no shortage of potential buyers if the drug gets to phase three trials.

SWOT

Strengths	Weaknesses
CEO/CFO are external appointments Phase 1 trials passed, phase 2 looking good	Running out of cash Lack of other drugs in pipeline
Opportunities	**Threats**
Gap in the market Ageing population	Other labs doing similar research

Report

This company is strong in human resources. Not only are there excellent scientists working in the lab, but the management team is also very good. They used their first-round venture capital funding to appoint a CEO and CFO from outside the company. Both managers impressed us when we went to visit the company. And the drug itself looks very promising. It has passed phase one trials and is now in the middle of phase two. The initial results show very positive clinical outcomes.

On the negative side, the company does have difficulties with cashflow. Their early funding has supported them until now but they only have three months' working capital to pay salaries and overheads. They need more funding to continue. There is also cause for concern that they have no other drugs in the pipeline. We would be betting on just one drug.

There is a lot of money to be made by whoever can find a successful treatment for dementia. There is a big gap in the market as no drug currently exists to treat dementia, and the world's ageing population gives a huge potential market. There are many other labs working in this area and so competition is strong. However, there will be room in the market for more than one drug.

Recommendation

My recommendation is that we do make an investment. The risk is no more than usual for a biotech start-up, but the potential profits if the drug proves successful are huge.

Structure and cohesion

Useful language

	Examples
Topic sentences	This is a company that was spun out of a university biochemistry department. There are two main issues. Firstly, … This company is strong in human resources. On the negative side, the company does have difficulties with cashflow. My recommendation is that we do make an investment.

Linking words to express cause, consequence, contrast, a supporting point, order, or a result, etc.	**Furthermore**, two of the founders are skilled in computer programming. (*supporting point*) The company has experienced some financial difficulties lately **due to** increased overheads. (*cause*) **However**, this start-up does have problems with staff retention. (*contrast*) They should widen their portfolio **so that** they do not rely on only one kind of product. (*result*) My recommendation **therefore** is that we do not invest in the company at this stage. (*consequence*) I recommend examining their business plan **before** making a final decision. (*order*)
Avoiding repetition using synonyms, pronouns, or changing verbs to nouns	**Worldwide** demand is expected to be significant, so our **global** sales are likely to increase. (*The entrepreneurs'…*) **Their** business model has been successful so far. **It** (*The business model*) should continue to be so as long as **it** remains subscription-based. We made several recommendations in our last report. None of **these** measures have been implemented to date. (*these refers to several recommendations*) They **launched** an innovative product. (*verb*) The product **launch** was successful. (*noun*)
Using repetition	The world's ageing population gives a **huge potential** market … the **potential** profits if the drug proves successful are **huge**. (*repetition of key words*) **Creativity** is key. **Creativity** leads to innovation. (*repetition for clarity and/or emphasis*) It took them longer than planned because they developed a pilot website to showcase their concept. **It** was highly original. (*Here, repetition of 'it' is unclear – the website or the concept?*)

Financial report

The Executive summary comes at the start of a report. In a financial report, the Executive summary reviews the table of figures line by line, comparing them and giving reasons for them. It is common to use vague language to give approximate figures, instead of repeating exact figures from graphs or tables. In the final paragraph, the report analyses the impact on future investment plans and makes predictions, providing a provisional budget forecast.

Model report

Quarterly budget report for the period ending 30th June (in $000s)

	Budgeted	Actual	Variance Favourable (Unfavourable)
REVENUES			
Sales income	150,000	156,000	6,000
Cost of sales	82,000	80,000	2,000
Gross profit	68,000	76,000	8,000
OPERATING EXPENSES			
Salaries	30,000	32,000	(2,000)
Marketing costs	18,000	18,000	–
General expenses	9,000	11,000	(2,000)
Total operating expenses	57,000	61,000	(4,000)
OPERATING PROFIT	**11,000**	**15,000**	**4,000**

Executive summary

In this second quarter ending 30 June gross profit was up by $8,000,000 on the budgeted figure, an increase of around 12 percent from the forecast $68,000,000. This was due to two factors. First, sales increased significantly, the most likely reason being a general improvement in consumer confidence as a result of the better economic environment. Second, cost of sales went down slightly as we are now seeing the benefits of the investment we made in automation last year and fewer workers are needed in the plant.

This increase in gross profit was partly offset by a small rise in operating expenses. There were two reasons for this, both related to the decision to open a new regional office in the north of the country. First, we recruited new employees to staff the office and this pushed our salary costs higher. Second, there were additional general expenses associated with the opening of the new office, such as rent, utilities and other overheads. These two factors caused our total operating expenses to rise by $4,000,000 on the budgeted figure. Marketing costs were in line with budget and so had no impact on costs. In general, this increase in total operating expenses is modest, given the opening of the new office.

Overall, operating profit rose from a forecast $11,000,000 to an actual $15,000,000. This is a considerable increase and puts us in a very good position with our investment plans. We have CapEx planned for a new factory in Slovakia, and this will help us become a major player in the European market. We also have significant OpEx investments that we want to make in the area of R&D – our competitors all have bigger R&D programmes than us and we need to catch up.

After discussions with the senior management team, I am able to give some provisional forecasts for the rest of the year. Revenue is expected to grow strongly as the economy improves and consumer spending rises. We are also in the lucky situation that costs are likely to be unchanged – we have already budgeted for both the factory in Slovakia and the larger R&D department. Therefore, it is probable that operating profit will continue to go up. I am optimistic about the future, partly because of our expected growth in Europe, and partly because the economy looks like it will continue to improve.

Useful language

Comparing figures	Gross profit was up by $8,000,000 on the budgeted figure. There was a substantial increase/decrease in sales. Last year saw a rise/fall in profits of 15 percent.
Vague or approximate language	Sales increased significantly. Cost of sales went down slightly. This led to a slight/small drop in cost of sales. This increase in total operating expenses is modest. Marketing costs were in line with budget. We also have significant OpEx investments. It was an increase of around/approximately 12 percent. Operating profit rose by almost a third.
Giving reasons	This was because revenues dropped slightly. This was because of a slight drop in revenues. This was (partly) due to a rise in OpEx. As a result of considerable investment in new equipment, we expect a significant fall in profits. Investment in new equipment resulted in a fall in operating profits. Rising profits were offset by rising costs. (*Rising profits and rising costs had an opposite effect and so the final result was more or less the same.*)
Making predictions	Net profits might/may/could rise. (*less formal*) Net profits are likely to/expected to/forecast to rise. (*more formal*) It is (highly) likely/probable (that) net profits will rise. (*more formal*) I expect operating expenditure will fall next quarter. I am optimistic (that) our net profits will be slightly higher.

Reporting on sensitive issues

Some reports are more factual, while others are more difficult to write because they involve describing a negative outcome or a challenging topic. When writing about sensitive issues, describe the facts or findings in the Background section. Background gives a factual summary of events to help explain the current situation.

'Action taken so far' or what has been done, has a separate section because action will be the main focus of the Recommendations section, which will outline what needs to be done next. In general, a conclusion interprets the facts in the main body of the report and makes a reasoned judgement. It leaves the reader with 'an end to the story' that explains everything. The conclusion should be remembered long after the details in the report are forgotten. It is *not* action-oriented. Recommendations follow the conclusion and *are* action-oriented. Recommendations should be concise, specific and realistic.

When writing business reports, it is common to use more formal language to describe something carefully without offending people, e.g. changing the verb from the active to the passive form and using the impersonal *it*.

Model report

Introduction

The purpose of this report is to bring to the attention of the SMT an ongoing issue in the Accounts Department. There is a strong clash of personalities between the team leader, Rose, and one member of her team, Peter. This is a situation that is getting worse and is beginning to affect the work of that department.

Background

The origin of the conflict was last March, when Peter was appointed to his position. Rose wanted another candidate to get the job. Since then, there has been tension and open arguments between Rose and Peter, specifically over the issue of Rose's monthly reports. In order to prepare these reports, Rose relies on figures provided by Peter. Peter is often slow to provide the figures, and this sometimes makes Rose late in submitting the reports. Rose is very unhappy about this.

The bad feelings between them became public a few weeks ago in an unfortunate incident. In front of other colleagues, Rose reminded Peter again that she needed the figures on time. Peter argued back and said that his deadlines were unrealistic. Personal words were exchanged, and there is still tension in the office following the incident.

Action taken so far

Last week Rose came to see me in my role as HR Director. She explained the situation as she sees it and requested that we take some action. She does not want to be blamed by senior management for the late reports. I have not spoken to Peter yet.

Conclusion

This is a situation where we have a strong personality-based conflict. In my opinion there is no solution using the usual discussions and procedures. Having given this matter considerable thought, I cannot see how they can continue to work together.

Recommendations

I believe the only course of action is to move Peter to another position in the Accounts Department, under a new team leader. I would like to give him a fresh start. I will need to have a frank and honest conversation with him where I explain what is expected of him and the importance of deadlines.

I also need to speak to Rose confidentially. I will assure her that the late reports are not seen as her fault, and that we are very satisfied with her overall performance. Do we have a budget to offer her some soft skills training in leadership and assertiveness? It is unfortunate that she spoke to Peter in a negative way in front of colleagues.

Maria González

HR Director

Useful language   Careful, formal language

	Less formal/Informal	Formal/More formal
Passive forms, formal expressions and impersonal structures	The SMT requested this report. The SMT know about the problems. There's bad feeling in the office. She left because she didn't like the long hours. John was making the job difficult. I think/believe … … considering all the reasons I've just given. You haven't finished the report. If you can … We don't think that's a good idea.	This report was requested by the SMT. The SMT are aware of the issues. There has been a drop in staff morale. There are increasing tensions in the workplace. She gave the long hours as the main reason for leaving. The job was being made difficult by John. From an HR perspective/the organisation's point of view … … for the reasons mentioned previously/above. It appears/seems that the report hasn't been completed. If it were possible to … We think it would be better to …
Example vocabulary	ask for a lot of Bad news! because of … change get give guess help let you know important thing job problem stay tell	request a great deal of/considerable Unfortunately, … I'm afraid … As a result of … transition receive achieve obtain provide estimate assist confirm urgent matter position/profession challenge, issue remain inform

Proposals

The format and style of proposals may vary, but they are a form of formal correspondence. An internal proposal can be brief and may be part of an email. Proposals sent outside the company are generally more detailed, as they may be used to apply for grants, loans or funding for a project.

In business proposals, it isn't necessary to wait until the end to make recommendations. The main recommendation is outlined at the beginning for emphasis so that the reader has a context for the proposal. Proposals are aimed at an official person or group, such as decision-makers or shareholders who need to consider the implications carefully before making a decision. It is useful to use some persuasive language in proposals. However, writers of business reports should show that they have a balanced viewpoint that is based on careful consideration of the facts.

All proposals contain an outline of the purpose, some background followed by detailed information, the recommendations, as well as a summary of the recommendations. Some proposals may have a Business Case section including more detailed reasons, benefits, costs, risks, staffing needs and the timescale of a project.

Model answer

Review of suppliers for our South East Asia assembly plant

Background

We have just one supplier, QualTec, for our South East Asia assembly plant. Firstly, the business relationship has been generally satisfactory up to now, with both positives (their quality levels are good) and negatives (their delivery times are not always reliable). But clearly, the fact that they are the sole supplier adds significant business risk for us. Secondly, it should be noted that all our main competitors have several suppliers.

I believe that we need to review the whole situation in SE Asia in relation to our supply chain. Firstly, we should give a strong warning to QualTec that they need to be much more consistent with their delivery times. Secondly, we need to look for a second supplier anyway, regardless of the situation with QualTec.

Business Case for reviewing SE Asia supply chain

QualTec have been our business partner for many years. There have been both positives and negatives in this relationship. Their quality levels are reasonable without being excellent. Other areas are very positive, in particular their input of accurate data into our IT system which helps us with our supply-chain management. However, on the minus side there are issues with their late deliveries to our assembly plant. We operate a very tight just-in-time assembly line with just two days of stock at the factory gate. As a result, any late delivery typically causes serious problems.

In fact, these points are secondary. The main issue is that QualTec have a monopoly position as our sole supplier. It is true that there are certain advantages in only having one supplier to deal with, but on the other hand it gives them too much bargaining power over price and other terms.

In summary, I believe that the situation in SE Asia needs to be the subject of a management review. Of course, we could simply talk to QualTec about the delivery times and continue as we are; however, the situation with having just one supplier definitely leaves us in a considerably weaker position.

Recommendations

I therefore recommend that we give QualTec a formal warning that unless their delivery times become more reliable we will end the business relationship. In addition, we should look for a second supplier so that we have greater control over the supply chain, for example, in areas such as responding to fluctuations in demand at our assembly plant and being able to negotiate on the price of supplied parts. In relation to immediate action, my recommendation is that a senior SCM team member needs time allocated to deal with all this, and they can report back at the March meeting.

Useful language

Using linking words helps the reader to follow your ideas and enables you to build a case for your arguments.

Measured language is used to prevent the writing from seeming too certain. We can use: modal verbs; adverbs, e.g. *mostly*, *relatively*; modifiers, e.g. *a little*, *somewhat*, *quite*; and adverbial expressions, e.g. *to some extent.*

Linking words	As a result/Therefore, any late delivery causes serious problems. I therefore recommend that we give QualTec a formal warning. In addition/Furthermore, we should look for a second supplier. In fact/reality, these points are secondary. For example/For instance, in areas such as responding to fluctuations in demand. In relation to/As regards immediate action, my recommendation is that a senior SCM team member needs time allocated to deal with all this. In summary/In conclusion, I believe that the situation in SE Asia needs to be the subject of a management review.
Building an argument	Firstly, the business relationship has been generally satisfactory. Secondly, it should be noted that all our competitors have several suppliers. It is certain/true (that) there are certain advantages in only having one supplier. On the other hand/On the minus side, it gives them too much bargaining power over price and other terms. However/Nevertheless, there are issues with their late deliveries. But clearly, the fact that they are the sole supplier adds significant business risk for us.
Measured language	Any late delivery typically/generally/usually/sometimes causes serious problems. Having just one supplier leaves us in a considerably/significantly/ somewhat/slightly weaker position. We should look for a second supplier so that we have greater/a little more control over the supply chain (to some extent/degree).

2 ❯ Minutes of a meeting

Lead-in

When writing minutes of meetings, sentences tend to be short and written in a semi-formal style as these documents will be read quickly. However, they are not always just for internal purposes and may be forwarded to clients, partners and suppliers who attended the meeting.

A formal meeting consists of several parts and the minutes will reflect the main sections. The order for meeting minutes will generally have this structure:

1 Names/initials of attendees and apologies from those not attending
2 Matters arising from previous minutes
3 Agenda items with decisions taken
4 Action points – what people need to do and by when
5 Any other business (AOB)
6 Confirmation of date, time and place of next meeting

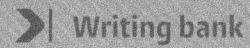
Model answer

Meeting: **Project review – construction of new plant**
Client: **RELAY**
Date: **9 November 2020**
Time: **14:30**
Present: **HK, YD, SW, ML**
In chair: **PG**
Apologies: **DS, PA**

1 Matters arising

The shortage of workers is now solved because we are working with a new agency. However, we will need to pay higher wages to attract skilled workers from abroad. YD asked how much more money this will be. There was some discussion but no clear figures.

2 Scope

The client now wants the warehouse to be 20 percent larger. There are some planning issues still to be resolved with City Hall (see next item). Obviously the client expects to pay an increased cost, but the meeting questioned whether this should be 20 percent directly in proportion to the extra warehouse area. HK to investigate this and prepare a new budget for the larger warehouse.

3 Schedule

Mr Stephens at City Hall says that he expects a response to the planning application in the next few weeks. SW stated that there are no immediate issues regarding/in relation to the schedule as work on the warehouse has not started yet.

4 Cost

The issues of higher wage costs and an expanded warehouse area were discussed previously. In relation to wages, the meeting felt that there will be some hard negotiations with the employment agency as they will want to keep a big percentage fee for themselves.

5 Overall progress

The project is on track. It was pointed out that the client's business is obviously going well, given their request for extra warehouse space. Therefore we may be able to charge them more than 20 percent for the extra construction costs involved in building the larger warehouse.

6 AOB

Maria in the Accounts department is leaving the company after twenty years. There will be a small leaving party this Friday, 12 November, at 13.00 in meeting room 3.

7 Next meeting

The next meeting will be on 12 December at 14:30 in meeting room 1.

Useful language

When writing notes for meetings, abbreviations, such as *CEO* and *AOB*, and short forms are often used. Some abbreviations are also used in the full minutes, e.g. for months (Mar = March), initials for people's names or positions (MD = Managing Director), names of departments, and frequently used terms in the business, e.g. w/house = warehouse.

It is also usual to omit unnecessary words in notes and, to a lesser extent, in the minutes. This is known as ellipsis. Words frequently omitted are those that do not carry meaning and are 'grammatical' words, e.g. articles, auxiliary verbs, pronouns and certain prepositions.

Writing notes during a meeting

Abbreviations and short forms	cc, CEO, doc, info, Jan, HQ, MD, mtg, rm, tech
	hv = have, incl = including, obvs = obviously, nt = not, re = regarding/about, shd = should, w/ = with, wd = would, v = very
	& = and, % = percentage, > = greater than
Ellipsis	Time of next mtg 9am = The time of the next meeting is 9.00 a.m.
	Agreed BM shd … = It was agreed that Brad Miller should …

Writing minutes

When writing full minutes, reporting verbs are often used to report what was said. The passive form makes the style more impersonal and formal and it can also be used in order not to repeat the subject, e.g. *All participants agreed.* vs. *This was agreed.* But it is best not to overuse passive forms in business writing.

Style features, names, action points, impersonal 'it', and passive form	LD to investigate this and write a short report by 11 Mar. (*name + action point + date*)
	It was agreed *BM* should meet with the client asap. (*It + passive form + abbreviations*)
	It was suggested *that* LD investigates this and writes a short report. (*It + passive form*)
Expanding notes into sentences	Architect wants to … → We have had a request from the architect to …
	Client v worried re … → The client is very concerned about …
	Main reason extra lab costs → The main reason for this is the extra labour costs.
	But hv to monitor closely → However, the situation has to be monitored closely.
Reporting verbs (active/passive forms)	PE reported (that)/It was reported that the actual costs to the end of Feb are 2.6 percent over budget.
	The chair suggested (that)/It was suggested (that) the small cost overrun can be absorbed in the profit margin projected for this project.
	Mr Nakamura argued (that)/It was argued that we should look for another supplier, given the quality issues.
	Participants agreed (that)/It was agreed that NW should speak to Mr Ivanov directly and report back to the next meeting.

3 ❯ Emails

Lead-in The style and tone of emails may differ from company to company and country to country. Some points to consider when keeping a positive tone that develops the relationship with a business partner are:

1 Focus on your audience. For first contact, or when there hasn't been contact for some time, it is best to use semi-formal language and the tone should be friendly and polite, especially in the opening paragraph, in order to create rapport. Language that is very formal may seem polite but too distant.

2 If you are answering questions, make sure you answer them appropriately in terms of language and amount of detail. Bullet points or numbering are useful for highlighting key information.

3 If you need to address negative points, apologise and give good reasons for any problems, delays or misunderstandings.

4 In the closing paragraph, as in the opening, it helps to end on a positive note with some social English, reference to future contact and continuing the business relationship.

5 Always reread the email before sending. Check all the important details have been included, and that language, spelling and punctuation are correct.

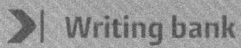

Model answers   Initial email: customer to supplier

✉ ‹

To: jia.yumei@cityevents
From: anna.meier@greenleaves
Subject: Sales conference in Singapore next year June 4–7

Hi Yumei,

This is Anna Meier from Greenleaves. Do you remember me? City Events helped us to organise a very successful sales conference in Singapore last year. We were in contact at the time and it was a pleasure working with you.

The good news is that we want to organise another similar conference next year, with the delegates staying four nights, June 4–7. However, we don't want the same hotel. We had issues with the size of the conference hall and the food supplied for the buffet lunches was bland and unimaginative. Next year, we would like City Events to find a more suitable hotel for this event, using your local knowledge.

I have a few questions:

1 How much extra will you charge to find a hotel? We will make the booking from our end. We simply need you to research the alternatives and check on the facilities and suitability. To give an idea of the high service level we expect from the hotel, last year we had an issue with the flower display on the main stage. We asked for a large display of fresh flowers next to the speaker's podium, but the flowers were never changed and looked tired by the end of the conference.

2 Last year the gala dinner on the final night was held at the hotel. Next year we would like to go to an outside restaurant, so can you also research this? The restaurant needs to seat 200 people in a separate private area. There must be a varied menu suitable for all tastes (including special dietary requirements), and good food.

I would appreciate an early reply. If you need any more details, please don't hesitate to get back to me. It will be great to work with you again, and I am sure that together we can organise another fantastic event.

Best wishes,

Anna Meier

Executive Assistant to the VP (Sales)

Reply: supplier to customer

✉ ‹

To: anna.meier@greenleaves
From: jia.yumei@cityevents
Subject: Re: Sales conference in Singapore next year June 4-7

Hi Anna,

Yes, of course I remember you! I was at the conference in my role as Event Coordinator and met many of your colleagues. I also remember the impressive opening talk given by the Vice-President of Sales.

We would be delighted to work with you again as your local event management partner. Thank you very much for choosing us a second time. Of course we can do all the research you require – we have very good local knowledge of hotels and restaurants based on over twenty years of organising conferences and promotions here in Singapore. I understand completely your point about the poor quality buffet lunches last year. If you wish, we could look into the possibility of external caterers providing lunch, regardless of which hotel you choose.

With regard to your specific questions:

1 Unfortunately, for next year we will have to increase our price to you by 15 percent. There will be extra work doing the research you have asked for (on-site visits, etc.). In addition, our staff costs are now higher than before. This is in line with what is currently happening in the labour market in Singapore.

2 Yes, we have in mind a restaurant that will be an excellent setting for your last night gala dinner. It is a venue that specialises in corporate events and can hold 200 people. After dinner, delegates can move upstairs, where you will have exclusive use of their rooftop bar. The venue is quite expensive, but we have used it before and the food is excellent.

Finally, let me apologise about the flowers on the stage of the conference hall. As you know, this was covered by the contract you had with the hotel and was not under our immediate control. But as the event organiser, we should have talked to the hotel as soon as the issue was flagged up by you. I will make sure that next year we monitor all aspects of the event and respond quickly if you raise any concerns.

I am sure that your sales conference will be a great success. We look forward to welcoming you here in Singapore next June.

Regards

Jia Yumei,

City Events

Useful language   Initial email: customer to supplier

Positive opening	I would like to thank you for your help in finding/for all your hard work in making … We were very happy with the event you organised for us last year and … We were in contact (last year) and it was a pleasure working with you.
Reason for writing	We have opened bookings for another group to visit Edinburgh. I'm writing because I have a few questions about …. The good news is that we want to organise a similar conference next year.
Requirements	We would like you to organise everything else, including … We simply need you to research the alternatives. We would like to … , so can you also research this?
Questions	Can/Could you confirm your prices will remain the same? How much extra will you charge to find a hotel? Can/Could you let me know how many people you are expecting?
Negative points	Apparently, there was some confusion over … We had issues with the size of the conference hall … Last year we had an issue with the flower display … There is just one issue from last time that I need to mention.
Next steps	Feel free to contact me if you have any questions. I would appreciate an early reply. If you need any more details, please don't hesitate to get back to me. Do let me know if you have any queries.
Positive closing	We look forward to working with you again. It will be great to work with you again. I hope that we will have a successful, long-term business relationship.

Reply: supplier to customer

Friendly opening	Great/Good to hear from you again. It was a pleasure to work with you on … Yes, of course I remember you!
Positive start	We would be delighted to work with you (again) as your local event management partner. Thank you very much for choosing us a second time. Those dates work very well for us.
Help and alternatives	Let me know if there is anything I can do this end re (regarding) … If you wish, we could look into the possibility of external caterers … If you wish, we can discuss these issues (of …) when we speak on the phone.
Responding to questions	(Yes,) I'd like to confirm our prices will remain the same for your next visit. Unfortunately, for next year we will have to increase our price by 15 percent. Yes, we have in mind the exact venue for you.
Dealing with negative points	Thank you for the feedback on … . Unfortunately, … I'm afraid there was a double-booking and they assure me it will not happen again. I apologise again for the confusion last time. I understand completely your point about the poor quality buffet lunches. Finally, let me apologise about the flowers. I will make sure that next year we monitor … and respond quickly if you …
Friendly closing	I am sure that your sales conference will be a great success. I'm looking forward to working with you (again). We look forward to welcoming you here in Singapore next June.

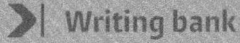

4 › Advertising copy

Lead-in Examples of persuasive texts include advertising copy, cover letters, proposals, websites or blogs. Advertising copy includes information for the product, describing not only key features, but also the benefits of these features to the customer. It's a good idea to use a variety of persuasive techniques: rhetorical questions, repetition of sounds, key words or structures for emphasis, contrasting ideas, three memorable points, emotive language, positive adjectives, a testimonial from an expert or opinion leader, a catchy slogan, if appropriate, and ending with a 'call to action'.

Look at the two models of persuasive advertising copy. Which do you prefer? Why?

Model answers ## Model 1: more conventional, with a mix of facts and feelings

Our microwavable *Cosy Slippers* are the ultimate in comfort while you watch TV, work at your computer or get ready for bed. Your nice, warm feet will feel so good.

- Choice of colours: cream, navy blue or black.
- Available in three sizes (S, M, L).
- Non-slip soles.
- Easy to use – just place in the microwave for 90 seconds and the heat lasts for an hour.
- Natural seeds inside the slippers retain heat.
- Not machine washable. Wipe with a damp sponge.

Cosy Slippers are the treat you deserve. The warmth of the luxurious, soft fabric will aid relaxation and relieve tired and aching feet. Also make an ideal gift – your daughter would love these, just as much as your granny.

Click here to order

Model 2: more openly persuasive and more emotive

You are at home. Maybe watching TV, or chatting to loved ones, or using your tablet. There is just one problem – your feet are cold! Wouldn't it be nice to have the comfort of some new slippers? But not just any slippers. You need the warmth and luxury of our super-soft microwavable *Cosy Slippers*.

Yes, that's right – microwavable slippers! Our *Cosy Slippers* contain natural flax seeds – just pop them in the microwave for 90 seconds and the slippers retain their heat for a whole hour. The luxurious, soft fabric and non-slip soles will add to your comfort. Get ready to relax. Bedtimes and weekends are going to feel soooooo good.

And why not give *Cosy Slippers* as a gift as well? Imagine their smiling faces as they see what you bought them. Best. Present. Ever.

Get ready to relax

► View Cosy Slippers here ◄

Useful language

Persuasive language in advertising copy

1 Write an opening 'hook' to get the reader's attention. This could be using a rhetorical question that addresses the customer's needs or desires. Rhetorical questions can create interest and then offer a solution.

2 Use a variety of persuasive techniques, such as repetition, contrast, tripling and emotive language.

3 Think from the customer's point of view. Use informal and personalised language to connect with the target audience using, for example: personal pronouns (*you, your, we,* etc.), ellipsis (omitting grammatical words) to sound more familiar and imperative forms to sound more direct.

4 End with a 'call to action'. This may be contacting someone for information, clicking on a link, signing up for an event, or ordering/buying the product or service.

Rhetorical questions	Where would you like to go today? We fly to over 200 destinations ... Does it sound like Paradise? Wouldn't it be great if there was a solution? Why buy only one? And why not give Cosy Slippers as a gift as well?
Repetition	Deluxe Vegan Nut Roast ... with perfect pecans, crunchy cashews and wonderful walnuts. With our Business Class seats you'll arrive relaxed and refreshed. Give them a present they will use again and again. Nice and large, nice and warm, nice and lightweight
Contrast	Our Slanket Deluxe comes in four colours – bright and bold or calm and classic. Offering both elegant, sophisticated watches and contemporary, minimalistic styles.
Three points/ Tripling	No added salt, no added sugar, no added nasties of any kind! Best. Present. Ever. We are organised, flexible and creative.
Emotive language	Your nice, warm feet will feel so good. Your daughter would love these. Imagine their smiling faces when they see what you bought them. It will keep you super-cosy all evening long as you watch a box set or chat with your bff.
Customer's viewpoint	Imagine yourself lying on a white sand beach, a cocktail in your hand as you listen to the gentle sound of the waves. As soon as you feel the softness of this luxury fleece, you'll know what we mean. Why not buy one for yourself and another for a loved one?
Call to action	Get ready to relax. Click here to order. View Cosy Slippers here. For information, contact us on 902 451 397. Book now!

5 > Self-assessment

Lead-in The style and content of self-assessment as a part of performance reviews may differ from company to company. A performance review is a formal assessment section which will usually be carried out once a year. It may include a self-assessment which provides employees with an opportunity to highlight their accomplishments and contribution to the organisation as well as to outline challenges and describe how they were dealt with.

How the self-assessment is completed may vary, for example online or using a standard hard-copy template. The style should generally be formal or semi-formal, using clear, descriptive language. When describing achievements, it is good to include specific examples or facts to support statements wherever possible. The aim is to demonstrate a positive impact in a wider context (to the department or organisation) so it is useful to link individual attainments to performance goals or company values.

Most self-assessments also require some evaluation of actions taken when things did not go to plan. Negative language should generally be avoided. Instead, one approach is to include a brief outline of the challenge followed by a focus on how the situation or problem was dealt with and, where appropriate, a description of a positive outcome.

Model answer

In my first year in the role of communications supervisor for Kombi Telecoms, I have consistently demonstrated a dedicated and pro-active attitude with a keen focus on meeting goals. I completed a project management course and immediately used some of the key strategies I learnt in my new role. As a result, 99 percent of projects were completed by the agreed deadlines and over 70 percent came in ahead of schedule. This was a 15 percent improvement on results from the previous year. The new project management strategies which I instigated were an outstanding success. Despite the fact that one project required an extension when the supplier was unable to fulfil an order, the delay was minimal as I quickly sourced an alternative and also negotiated a lower price. As a consequence, the cost of the project came in at 12 percent below anticipated costs.

Another key goal has been to improve customer communication, both in terms of quality and speed of response. I achieved this by implementing an innovative system to deal with customer enquiries. It uses an online chat mode on the company website, resulting in a quick and responsive service and increased customer satisfaction. Client surveys recorded that nine out of ten respondents felt that their queries had been dealt with quickly and efficiently and they were therefore more likely to recommend our company as a result. There was a minor technical glitch when the feature was initially launched. The process of dealing effectively with this challenge provided an invaluable lesson for future online communications improvements. While the technical department resolved the issue, I instructed the communications team to use alternative methods to contact customers and issued an apology on the website. As a result of the quick action and clear communication, no complaints were received. I learnt from the experience and now ensure that all future upgrades have a beta launch before going live to our wider customer base.

Finally, in line with both company values and my own performance goal, I initiated regular conference calls in order to improve communication between teams operating in global branches of the company. Due to the lively discussion and enthusiasm of participants, the early meetings were large and sometimes overran. In order to make the meetings more effective, I amended both their size and length so that they were smaller and shorter. The success of the initiative is verified by feedback from the branches reporting that the teams now feel more connected and communication has improved significantly.

Useful language

Using dynamic language helps to give energy to your writing and enables you to add emphasis to the accomplishments that you want the reader to focus on. The statements are strengthened when backed up by examples or data that illustrates the positive aspects of your contribution.

When things don't go to plan it is useful to describe the situation, briefly say what didn't go well and then include a positive statement outlining how you reacted or the solution you arrived at. It can also be useful to indicate any lessons that have been learnt from dealing with challenges.

Action verbs and powerful adverbs and adjectives.	I have consistently demonstrated a dedicated/flexible and pro-active attitude with a keen focus on meeting goals.
	I completed a project management course and immediately/effectively used key strategies in my new role.
	The new project management strategies which I instigated were an outstanding/resounding success.
	I achieved this by implementing an innovative system to deal with customer enquiries.
	I initiated/organised regular conference calls in order to improve/streamline communication between teams operating in global branches of the company.
Giving examples and describing results	As a result, 99 percent of projects were completed by the agreed deadlines and over 70 percent came in ahead of schedule.
	As a consequence/Due to this, the cost of the project came in at 12 percent below anticipated costs.
	It uses an online chat mode on the company website, resulting in a quick and responsive service and increased customer satisfaction.
	Client surveys recorded that nine out of ten respondents felt that their queries had been dealt with quickly and efficiently and they were therefore more likely to recommend our company as a result.
	The success of the initiative is verified by feedback from the branches reporting that the teams now feel more connected and communication has improved significantly.
Describing challenges, action taken and lessons learnt.	Although/Despite the fact that one project required an extension when the supplier was unable to fulfil an order, the delay was minimal as I quickly sourced an alternative and also negotiated a lower price.
	In order to make the meetings more effective, I amended both their size and length so that they were smaller and shorter.
	(On the other hand,) the process of dealing effectively with this challenge provided an invaluable lesson for future online communications improvements.
	I learnt from the experience/gained valuable insights and now ensure that all future upgrades have a beta launch before going live to our wider customer base.

1 ❯ Presentation skills

Lead-in

Presentations may need to be adapted to fit different contexts. Here are some questions to research before you prepare presentations for different audiences.

1 What strategies will be most effective in engaging the interest of your audience? Are they likely to respond better to information that is presented visually or to verbal strategies such as the use of strong vocabulary, repetition or rhetorical questions?

2 What tone will work best? Tone can describe the overall mood or feeling set by the presenter. Some audiences may prefer a colloquial approach with the use of anecdotes or stories to highlight key points. Others may expect a more serious, fact-based delivery. Tone can also describe the style in which the presentation is delivered; the inclusion of pauses, changes of speed and varying falling and rising delivery can create impact and interest, but a more neutral delivery might be better in some contexts.

3 How can questions be used to allow the audience to participate in the presentation? Does the structure of the presentation encourage questions during the talk or at the end? What questions can you ask to awaken curiosity about the topic under discussion or to find out how much the audience knows about it? How can you respond to questions in a way that creates a closer relationship with the listener?

Starting a presentation

	Formal/Semi-formal	Less formal
Opening and welcoming the audience	It's a pleasure to be here today. Can I start by extending a warm welcome to everyone?	It's great to be here today. Hi, everyone. Thanks for taking the time to be here.
Providing a brief outline of the talk	Let me start by giving you a brief outline of the topics we'll be looking at today. The presentation consists of (three main parts/sections).	We'll begin with (an overview of the main challenges facing our sector) and then go on to look at/ discuss (the strategies that can provide some solutions). The talk will be divided into (three main parts).
Informing the audience when they can ask questions	I will be happy to answer any questions during the presentation. There will be an opportunity after the presentation to ask questions.	Do feel free to ask questions as we go along. We've set aside time after the presentation to answer any questions.

Engaging an audience

Departing from convention	Today, I want to do something a little bit different (and start the presentation with a brief demonstration). Let me close by asking an unusual question.
Challenging assumptions	The issue is not about (working harder), but about (working smarter). The problem is not (a lack of choice). The problem is (quality).
Creating a sense of urgency	We really have to (communicate our message more effectively). If we don't act quickly, we won't (bridge the gap between us and our competitors).
Building rapport	We all want to (achieve the best possible outcome). I believe we share (a common outlook).
Getting the audience to talk to one another	Have a quick chat with the person next to you. I'll give you (two minutes). Can you discuss these (three questions) with your neighbour for (five minutes)?
Communicating optimism and confidence	I'm convinced that (we) can succeed in (overcoming these challenges and we'll be a leaner, stronger company). There's little doubt in my mind that (you) will succeed.

Quoting interesting data	Let me read out these (customer suggestions) for you; they're certainly (illuminating).
	Have a look at these statistics published last month in a leading (science journal).
Using powerful vocabulary	In fact, that's totally wrong.
	It's really amazing feedback.
	It's an incredible achievement.

Using questions

Awakening curiosity	Before I show you, what do you think it will be?
	What's the most common customer complaint in business today?
Using rhetorical questions	So how do we find out what they need?
	So how will we build a new culture?
	How could this happen?
	What was going on?
Checking for questions during the presentation	(Are there) Any questions before we move on?
	Please stop me if you have any questions.
	If there are no further questions, I'd like to move on.
Inviting questions after the presentation	Does anyone have any questions?
	Any questions or comments?
	I'm happy to answer any questions now.
Commenting on questions	That's a great question. And I think there's a simple answer.
	That's a very interesting question.
	For me, the answer is clear. No one knows about us.
	I'll tell you: you slow down your learning and you make more mistakes.
	I can tell you in very simple language …

Responding to different types of questions

A question you don't know the answer to	That's a good question, but I'm afraid I don't have (that information to hand) at the moment. I'll find out and (send you the data by email).
	I'll need to check and get back to you on that.
A question which requires clarification	I'm sorry, I didn't quite get that. Could you run that by me again?
	So let me summarise your question to be sure I understood it correctly. You want to know about (the long-term impact of this strategy)?
A question better answered by another person or department	Let me put you in touch with (our finance team) who can answer that query in detail.
	(Veronica), perhaps you'd like to answer that question, as you've done a lot of research in this area.
An angry question	That must be very frustrating. One option would be to (arrange a video call so we can discuss these issues with the team).
	I can see that you feel very strongly about this. What outcome would improve the situation?
A question you want to address later	That's a good question and I'll be addressing this point in detail later.
	Thanks for highlighting that. We'll come on to (the technical specifications) in a moment.

Storytelling in presentations

Making clear an intention to tell a story	To begin, I'd like to tell you about (something that happened recently). So this story begins with (two penniless students and one crazy idea). That reminds me of (an event that occurred when I was starting out in this industry). I'd like to share a story about (why this organisation was founded).
Giving information about what happened	It was the hottest part of the day. There were three of us taking part in a race across the desert. We were already dangerously low on water and provisions when, without warning, the car broke down.
Adding detail to the story	I just wanted to (get out of the rain, get home and get into some dry clothes). I was (halfway up the mountain) when (the wind increased, the skies darkened and snow began to fall).
Saying when, how and why something happened	The problem with all of this, (Ursula informed me after the meeting), was that (I was so focused on the information I had prepared that I had stopped listening to what the client really wanted). No sooner had I (arrived) than there was (a transport strike and there was no way to get to the conference where I was supposed to give my talk). That very morning they had (made an incredible discovery on this very topic).
Describing emotions	I have to tell you, I was really shocked and disappointed. To my relief, the visa arrived in time for my business trip. I couldn't have been more delighted to hear the news. I'll be honest with you, it was absolutely terrifying.
Introducing a turning point	And then all of a sudden, we got the news that our idea had been put forward for a prestigious award. By an amazing coincidence, the CEO of the company also had his flight delayed and we were both stuck at the airport. So we got talking. In the meantime, the marketing team had come up with a game-changing idea.
Reflecting on what happened	And you know what the funny thing is, (because of all the disasters that had already happened that day, I was no longer nervous when I sat down in the interview). It just shows that (there is more than one road that can lead us to our goal).
Highlighting the moral of the story	So, this has taught me an important lesson about (what makes a good manager – you have to lead by example and really care about your team). At the end of the day, (there were no winners or losers in this story – we had all experienced what it was like to see things from the other person's perspective). So, I guess the moral of my story is that (we shouldn't be afraid of failure, because without it we'll never learn resilience).
Concluding the story	To cut a long story short, (that single branch turned into a nationwide chain and then a global franchise). In the end, (the people who believed in the company were the ones that really mattered).
Linking from story to current context	And I think we can all learn from this because (we all forget from time to time how important it is to evaluate risk). I think everyone here can identify with what it feels like to (be nervous on your first day in a brand new job).
Introducing the main subject of the presentation at the end of the story	So that's what I want to talk about today: (how we can understand the variety of risks facing our industry and, equally, the importance of taking chances). And that's why we want to (introduce this mentoring programme).

Closing a presentation

Summing up	So what have we covered? Let me summarise the main strategies outlined in this talk.
	To recap, the key points can be summed up as (reduce, reuse, and recycle).
Closing	In conclusion, our sector will benefit from all the technological innovations we've looked at today.
	I think we've covered all the key points. So, let's wrap things up there.
Thanking the audience	Thank you for taking the time to attend this presentation.
	Many thanks for your attention. It's been a pleasure talking to you today.

2 ❯ Meeting skills

Lead-in Some business practices may differ from country to country. Here are some questions to research if you are attending meetings in a different country.

1 Is the process likely to be similar or different to your previous experience of meetings? Will the context be structured and formal or more relaxed and informal?

2 What strategies will you use to engage positively with other participants? How will you deal with opinions that are different to your own?

3 If you are leading the meeting, are the other participants likely to be confident and assertive when putting forward ideas, or will they require encouragement? What techniques will you use to ensure participation?

4 Is it anticipated that participants will already have in-depth knowledge of the topics under discussion? What might need to be explained and highlighted in the context of the meeting? What will need to be prepared, researched or read in advance as background?

Introductions and opening

Welcoming participants	It's great to see everyone here this morning.
	Thank you very much for joining the meeting today.
Introducing participants	Does everyone know each other?
	Before we begin, could we go round and say a little about our roles?
	Most of you know each other but you may not have met (Carlotta) who is joining us from (Head Office) today.
	(Khalid) sends his apologies. Unfortunately, he can't be here today because (he's attending a conference in Brazil).
Stating the meeting objectives	The main focus of this meeting is (to agree on the best way forward for our shareholders and stakeholders).
	You'll see from the agenda that we are here to (decide on the best location for the new warehouse).
	The aim of this meeting is to (outline the technological challenges that our industry will face in the future and discuss long-term strategies).
	We're here today to (clarify the implementation stages related to the upcoming merger and identify any action required by each of your departments).
Opening the meeting	Everyone is here now, so shall we begin?
	OK, let's start by looking at the first point on the agenda.
	As usual, we'll start by (referring to the minutes of the last meeting).
	In terms of sequence, can I ask (Mario) to go first, followed by (Luisa) and then (Elana).

Communicating effectively

Engaging positively with other participants	Thank you very much for joining the meeting, I really appreciate it. It's great to have everyone on the line today. I clearly need to slow down and talk this through carefully. I know this is complex so (stop me if you'd like me to clarify anything). I understand and respect your concerns (about the situation). I want to give us time to discuss (this) in detail as there are different perspectives on it.
Establishing process and context	Can we just go through (the policy together to identify what needs to change)? Can we start with a quick look at (the findings and discuss the implications)? In fact, we discussed this in detail (with our clients and have previously addressed all these concerns). Just by way of background, (I've prepared some visuals to compare output for this year compared to last year). If you look at (page 21), you'll see that (the changes to procedure are outlined clearly). I think the key information is (in the final paragraph).
Focusing on others' concerns and understanding	This is where it is very relevant for you because (of your concerns that currency exchange rates might add costs). This is a situation which affects us all. Does this make sense? Let me throw the mic/floor open to any comments people might have. I don't want to dominate here, so can I ask others to comment first?
Ensuring clarity	I'm not saying that this will be easy. Let me be clear here, this doesn't mean (that we're going to lose our competitive advantage). We simply want to (provide clarity for our customers). In a nutshell, (we have limited options).

Discussing change

Specifying the negative impact of change	These changes have resulted in (a sharp drop in productivity). We've triggered (a lot of animosity within our international teams). This generated a lot of anxiety and negativity. It's been very disruptive. There's a lot of uncertainty and confusion about what's happening.
Exploring current and past possibilities	I think if we'd done that, (communicated the potential positive outcomes), we wouldn't be having (these issues today). If we hadn't (opened our first plant in Brasilia), we wouldn't be sitting here now talking about (global expansion).
Analysing past actions critically	I just think we could have handled things better. We should have (carried out more research). I think (our recent expansion into property management) has been part of the problem here.
Looking ahead to future developments	We're hoping to (have the project completed by end July). By the end of the month, we'll have (moved over to flexible working hours for the majority or the workforce). We'll have (introduced our new fleet of electronic delivery trucks) by the start of next year. I think we're on track with (the planned relocation schedule).
Expressing learning points from changes	My main takeaway from all of this is to (trust your instincts, but back that up with hard facts). The biggest insight for me is that (statistics are only part of the story).

Contrasting and exploring views on change	Strategically, it was the most logical option. Environmentally, it was the worst decision possible.
	Did we do enough to communicate to staff the positive outcome of these changes, or did we simply assume that they would share our viewpoint?
	Contrary to (the fears of some experts), the latest research demonstrates that (these crops will grow well in a cooler climate).
	The market research data indicates that customers (prefer the colours of the new logo). On the other hand, they feel that (the design is less attractive).

Discussing and solving problems

Summarising the problem	The problem I want to discuss today is (the danger that social media can pose for your future career).
	So, basically, your challenge is (how to reduce costs without reducing quality).
Asking questions to provide more detail	Could you give us a little more background about (how this finance will be raised)?
	That's an interesting point about (transport options). Could you go into more detail about (how that might affect delivery times)?
Asking participants to suggest ideas to solve a problem	From your perspective, what would be (a fair way to resolve this)?
	If you were in my shoes, what would you do to (improve morale)?
Responding positively to ideas	It's an interesting idea to (have an inter-departmental lunch once a month).
	If you mean (greater diversity within our industry), then I think it's a great idea.
Suggesting possible solutions	In my view, you need to (contact the supplier and renegotiate the contract).
	Another solution could be to look at (alternative energy sources).
Summarising main ideas	As we're running out of time, let me summarise very briefly. So, in terms of a plan (we're going to delay the launch until September).
	Of all the ideas, this was really the most useful because (it introduced a new perspective on the issue).
Thanking for participation	Many thanks for all your great ideas, particularly around (the training courses).
	Your input is valuable and much appreciated.
	Thank you all for your contributions and for taking the time to participate.
Talking about updates	I'll let you know how things go.
	I'll keep you in the loop and let you know (the outcome of the call to the supplier).
	At the next meeting, we'll feedback on the action points decided today.
Encouraging reflection	What do you see as the main learning points from the discussion today?
	What are the main takeaways for everyone (from this meeting)?
	So, what will you do differently after today?
	How can you implement these ideas into your working day?

Ensuring collaboration	**Including quieter individuals**	(Nidra), you haven't said anything yet. Any ideas? (Cosmo), let me come to you to make sure everyone has the opportunity to speak. What are your thoughts? (Aisha), you haven't commented so far. What's your view on this?
	Quietening louder individuals	(Ramon), can I stop you there? I want to hear from some of the others. That's a good point and we'll return to it later. First, let's open up the discussion (to get some other opinions).
	Making people engage with others' ideas	(Liliya), what do you think of (Imari's) idea? (Wei), what's your response to (Kesha's findings)? Can we hear some thoughts on (Hakim's concerns)? Would anyone like to add to that suggestion?
	Paraphrasing to ensure understanding	So, you're saying (the best approach would be to postpone the launch)? If I've understood correctly, your main point is (that this data might not be reliable)?
	Asking for clarification	What do you mean, exactly? Can you think of any examples to support that view? I'm not sure I follow. Could we go over that again? Could you just clarify that last point? Do the figures refer to (this quarter or the whole year)?
	Reminding people of the process	Remember, no arguing. We're just exploring ideas at this stage. First, we'll gather all the suggestions and then we can talk about them in more depth. We need to prioritise the tasks before we decide who will action them.
	Offering support	Shall I (take notes on the flipchart as you speak)? Does anyone need any (paper or pens to make notes)? Would you like me to help (set up the equipment)?
	Managing different opinions positively	I think both your perspectives are valid because (they will result in a similar outcome). Let's not dismiss (the idea of hiring a consultant) too quickly. Why don't we talk it through a little more?

Closing a meeting	**Summing up**	That's all we have time for today. Let's sum up the main points. To recap, we've (agreed on a plan to move forward).
	Closing	Thank you for coming, everyone. It's been a very productive meeting. We appreciate your contributions and ideas.

3 ❯ Negotiation skills

Lead-in   Some business practices may differ from country to country. Here are some questions to research before you prepare for a negotiation.

1 What differences might there be between negotiation strategies? Is the other person likely to come to the table with a win–win or a win–lose attitude? Are they likely to communicate their aims clearly or will you need to discover these? Will mutual requirements be discussed openly or are there areas where there could be ambiguity and uncertainty?

2 Is the negotiation more likely to start with small talk and relationship-building or will you be expected to get down to business quickly? Will it be considered a strength or a weakness to show empathy for each other's position and make concessions?

3 How can you establish commonality to build trust and rapport? What are the likely needs and motivations of the person you are negotiating with? What can you do to find common ground?

4 What questions can you ask to learn more about the other participants and discover what is in their minds? What questions are you likely to be asked? Which questions could help gain information, enhance participation, reduce tension by finding out about the other person's viewpoint, keep the negotiation on track and reach agreement?

5 How might views, attitudes and approach differ in the negotiation? How are the other participants likely to react to views that are different from their own? What can you do to avoid seeming to contradict or undermine the other participants and causing possible offence? Is there any behaviour which could make it seem as though you are prioritising your views or values over those of the other participants? How might attitudes to negotiation differ between your culture and that of the person you are negotiating with? How might misunderstandings be avoided?

6 How might the approaches to disagreement and conflict differ? What mediation skills might be useful to discuss differences in a way that leads to a positive outcome? What strategies can help make people feel listened to and that their concerns are being taken seriously? Are other participants likely to show emotion or remain neutral? What cross-cultural information would help you read your fellow participants more clearly and avoid misinterpretation?

Opening

	Formal/Semi-formal	Less formal
Welcoming someone	Thank you for coming today. It's a pleasure to meet you. Good to meet you/see you again. Can I offer you something to drink?	Thanks for agreeing to meet today. Can I get you a tea or coffee before we start?
Outlining objectives	(Today), we need to agree on (the product range). I'd like to have a brief discussion about (the proposed launch date). Shall we begin by discussing (the new trade regulations)? We need to find a way to move forward with the (changes to the terms and conditions).	Let's try to agree on (some training dates for your team). Can I have a quick word about (the project deadline)? I'd like to start by (hearing what you think of the new design). We have to find (a compromise we can both work with).

Exploring options

Building relationships	Are you the (Justin who's just returned from working on the volunteering project)? How did it go? Have you been working with (Aiko) long? And do you work a lot on these kinds of (projects)?
Clarifying understanding	Was the situation with my company and what we're looking for, clear? You want to (amend the schedule), yes? How do you mean, exactly? And you want to (increase the quantity), did you say?
Eliciting opinions	How does that sound? Could this work for you? What are your thoughts on this? I'm keen to hear what you think about (the relocation package).
Challenging and providing alternative views	Is that (actually) true? I heard the opposite. Did you know that (the latest research presents an alternative view)? Perhaps there's another way to look at this? I'm not sure that's factually correct. Do you know the source of that information? Actually, (they no longer use fossil fuels). I think (the data) might have been misinterpreted.

Focusing attention on the most important topic	Can we come back to the main issue? Can I just focus on that key point? I'd just like to return to (your main point). So, would you say that this is (your key requirement)?
Understanding others' needs	Is your problem here (the impact this will have on your team)? Can you outline your concerns? We want to make sure that we understand your main requirements.
Highlighting and celebrating agreement	Yes. It's a win–win, isn't it? Sounds good? It looks like we have a deal.
Identifying weakness in a position	Wouldn't it be difficult to (deliver that in the current time frame)? Isn't it problematic to try to (change vendor at this stage)?
Suggesting simplification	Can't we just (cut out the middle stage)? Wouldn't a quick solution to this be to (transfer stock from another branch)?
Exploring decision-making authority	Do we need to involve (the board of directors) if we want to (move ahead with this)? Who else may need to be consulted on this? Who has authority to sign off on this?
Suggesting a summary	Should we recap? Can we go over this one more time?

Bargaining

Proposing solutions	Just as an idea, how about (hiring a larger venue)? Supposing we (look at the data again to see if there's anything we've missed)? Just thinking a little out of the box, (we could do away with any packaging. It cuts costs and would be a greener option).
Presenting contrasting views	To play devil's advocate, why (do we need to change a strategy that seems to be working)? On the other hand, it would (significantly cut our expenditure). Looking at this from another perspective, it could (increase employee motivation).
Exploring ideas to reach agreement	Would you be willing to consider (reducing the shipping costs)? What if you (reconsidered the price of the components) and in return we (extended the delivery date)? So if we (agree to this change in the materials), you'll commit to (the new terms and conditions)? Would you consider making concessions on (the payment terms)?
Stating limits to a concession	With the proviso that (you can commit to those dates) … Given that (the requirements have now changed) … If we agreed, it would be on the understanding that (we would review again in three months). Say we were to agree, we would need (an assurance that there would be no further amendments).

Building trust-based relationships		
	Demonstrating trust	I'm happy for you to take decisions without consulting me for the sake of speed. I'm happy to just go with your idea on this; let's try it and see. We knew you had everything under control.
	Showing empathy	Great to see you again. Hope you're well and (have recovered from your accident). I think this solution meets all your needs and expectations. I imagine it's quite demanding to (travel for work so much)?
	Being reliable	As I promised when we last spoke, (I've compiled a list of the potential locations). I would be able to commit to (delivery within six weeks). Well, I'm very happy to stick to the promise I made when we met about (completing the project by the end of the month).
	Demonstrating integrity	I want to be fair about this, so let's agree to meet halfway. It's important that the solution is fair. Our main objective in business is to help you reach your goals,
	Proving competence	As you have most expertise in this, it's best if you decide. My experience of similar projects tells me to (do more research before making a final decision). Obviously you know your business better than we do, but I'm sure we can help.
	Establishing similarity	I think we both share the view that (this is a win–win situation). I think we've both seen this situation before. It seems we have a lot in common when it comes to (outlook). So, you prefer to look at the data in detail? Same here. You play (squash)? Me too. You've lived in (Japan)? So have I! You don't want this to drag on and neither do I. What a coincidence! I originally trained in (accounting) too.
	Sharing information	To be perfectly honest, I don't have an answer to that question. I'm very happy to provide more documentation. I can send you (the reports for background information). We have more data on this, which I'd be happy to share.
	Being open	To be honest with you, I'm feeling concerned that (more staff would be required to complete this effectively). I feel a little nervous, but also excited. I have to say, I'm feeling a little under time pressure.

Mediating conflict		
	Clarifying the mediator role	My role today is simply to help you understand each other and reset your collaboration in the sales team. I've been asked to step in, so I'm having a chat with you both separately and then we can meet … What I want to do today is support you both in understanding each other and agree a way forward.
	Stressing common objectives	I think everyone is motivated to (implement these improvements). Everybody wants to resolve this. (Listening to you both), I think we can find common ground and a way to move forward. I honestly think there's a lot of common ground. I think the underlying motivations are actually very similar.

Identifying and sticking to ground rules	It's important that our meeting remains constructive and focused on a solution. Let's share views openly, stay positive and find some solutions to move forward. Please allow others to speak without interrupting.
Discovering individual views	What's your perspective on (the changes to working hours)? Can you tell me from your point of view what's been happening? What's your take on the situation?
Forcing people to listen to each other	Can you respond to what (Tia) has just said about (her concerns)? But can you see where he is coming from? What would be your answer to that point?
Challenging perceptions	I'm not sure this communication is intended to (criticise your working methods). Have you actually discussed this with (anyone)? Be careful of assumptions. I think this communication style may be creating a misunderstanding. We need to be careful not to jump to conclusions.
Confirming the perceived point of conflict	Let me summarise what I'm hearing. So, for you, is the issue mainly about (client confidentiality)? So, for you, (the lack of a clear chain of command) is the real issue? So, is the key problem for you about (sharing information)?
Exploring solutions	What do you suggest as a solution to this? Any other suggestions from your side? How can we move forward? How can we work together to move beyond this point? What would need to happen for us to move beyond this? How can we resolve this?
Proposing a way forward	Would it be acceptable for you to (discuss this with your team and then feed back)? So, given this, can we decide on a way forward? Can we discuss some logical next steps? What would you like to happen next?
Summarising the final agreement	So, to recap, as a way forward, we agree that (changing the meetings from monthly to weekly) would be useful. So, we have agreed to (discuss this with our teams) and come up with a solution. So, we'll try that solution for the time being and then we can review (at the end of the month).

Closing a negotiation	**Reaching an agreement**	It's been a productive discussion. I feel/In my opinion, this is a very positive outcome.
	Failing to reach an agreement	I'm sorry that we couldn't reach an agreement on this. Perhaps we'll be able to reach an agreement at a later date.
	Outlining the next steps	Let's keep in touch to discuss how to proceed. I'll send an email outlining what we've agreed today.